Tourism, Climate Change and the Geopolitics of Arctic Development:

The Critical Case of Greenland

Tourism, Climate Change and the Geopolitics of Arctic Development:

The Critical Case of Greenland

Derek Hall

Seabank Associates

CABI

CABI is a trading name of CAB International

CABI
Nosworthy Way
Wallingford
Oxfordshire OX10 8DE
UK

Tel: +44 (0)1491 832111
Fax: +44 (0)1491 833508
E-mail: info@cabi.org
Website: www.cabi.org

CABI
WeWork
One Lincoln Street
24th Floor
Boston, MA 02111
USA

Tel: +1 (617)682-9015
E-mail: cabi-nao@cabi.org

A catalogue record for this book is available from the British Library, London, UK.

References to Internet websites (URLs) were accurate at the time of writing.

ISBN-13: 9781789246728 (hardback)
9781789246735 (ePDF)
9781789246742 (ePub)

DOI: 10.1079/9781789246728.0000

Commissioning Editor: Claire Parfitt
Editorial Assistant: Ali Thompson
Production Editor: James Bishop

Typeset by SPi, Pondicherry, India
Printed and bound in the USA by Integrated Books International, Dulles, Virginia

Contents

Figures

All photographs are by the author except Fig. 4.7, source anonymous

Tables

Boxes

Abbreviations

(US)$	(US) dollar
AAC	Arctic Athabaskan Council
AC	Arctic Council
ACAP	Arctic Contaminants Action Programme
ACE	adventure, cultural, eco- (tourism)
ACIA	Arctic Climate Impact Assessment
ADS	Approved Destination Status
AECO	Association of Arctic Expedition Cruise Operators
AEPS	Arctic Environmental Protection Strategy
AHDR	Arctic Human Development Report
AIA	Aleut International Association
AMAP	Arctic Monitoring Assessment Programme
AMM	Arctic marine mammal
AMO	Atlantic Multi-Decadal Oscillation
APPG	All-Party Parliamentary Group for Greenland
ASM	Arctic Science Ministers meeting
ATTA	Adventure Travel Trade Association
AWG	Arctic Winter Games
AY	Arctic Yearbook
AZRF	Arctic Zone of the Russian Federation
BCEG	Beijing Construction Engineering Group
bn	billion
BRI	Belt and Road Initiative
C	Centigrade
CAFF	Conservation of Arctic Flora and Fauna
Can$	Canadian dollar
CCCC	China Communications Construction Company
CEDAW	Convention on the Elimination of All Forms of Discrimination Against Women
CIA	(US) Central Intelligence Agency
CLIA	Cruise Lines International Association
cli-fi	climate fiction
CO_2	carbon dioxide
Covid-19	Coronavirus pandemic

CSR	corporate social responsibility
cu	cubic
DKK	Danish Kroner: with the equivalence of (as at 1 March 2021): €0.13, £0.12, US$0.16, Can$0.21, Norwegian Krone 1.41, Icelandic Króna 20.59
DMO	destination management/marketing organisation
EEA	European Environment Agency
EEZ	exclusive economic zone
EIA	environmental impact assessment
EIB	European Investment Bank
EPPR	Emergency Prevention, Preparedness and Response (Arctic Council Working Group)
EU	European Union
EXIM	Export-Import Bank
FEMS	Federation of European Microbiological Societies
FPIC	free, prior and informed consent
G-60	1960s settlement centralisation policy in Greenland
GCI	Gwich'in Council International
GCM	global climate model
GDP	gross domestic product
GNP	gross national product
IAATO	International Association of Antarctic Tour Operators
IASC	International Arctic Science Committee
IBA	impact and benefit agreement
ICC	Inuit Circumpolar Council
ICES	International Council for the Exploration of the Sea
ICOMOS	International Council on Monuments and Sites
IMBIE	Ice Sheet Mass Balance Inter-comparison Exercise
IMO	International Maritime Organization
IPCC	Inter-governmental Panel on Climate Change
IPO	Indigenous peoples' organisation
IPTRN	International Polar Tourism Research Network
IPY	International Polar Year
ISME	International Society for Microbial Ecology
IUCH	International Union for Circumpolar Health
IUCN	International Union for the Conservation of Nature
IWGIA	International Working Group for Indigenous Affairs
KGH	*Den Kongelige Grønlandske Handel* (The Royal Greenlandic Trade)
km	kilometre
KNI	*Kalaallit Niuerfiat* – the former KGH
LNG	liquified natural gas
m	metres
MARPOL	International Convention for the Prevention of Pollution from Ships
mm	millimetres
mn	million
MNR	Ministry of Natural Resources (China)
MOSAiC	Multidisciplinary Distributed Observatories for the Study of Arctic Connections
NAO	North Atlantic Oscillation
NATO	North Atlantic Treaty Organisation
NGO	non-governmental organisation
NSR	Northern Sea Route
OCTs	Overseas Countries and Territories (related to the EU)
OECD	Organisation for Economic Cooperation and Development
OSPAR	Convention for the Protection of the Marine Environment of the North-East Atlantic

OW	open-water ship
PAME	Protection of the Arctic Marine Environment Working Group (Arctic Council)
PC6	Polar Class 6 ship (ice-breaking capability)
PRC	People's Republic of China
PSR	Polar Silk Road
RCEP	Regional Comprehensive Economic Partnership
REE	rare earth elements/metals
SDG	sustainable development goal
SDWG	Sustainable Development Working Group (Arctic Council)
SCTF	Scientific Cooperation in the Arctic (Arctic Council Task Force)
SNIJ	sub-national island jurisdiction
SOE	state-owned enterprise
SOLAS	International Convention for the Safety of Life at Sea
sq	square
t	tonne(s)
T-MOSAiC	Terrestrial Multidisciplinary Distributed Observatories for the Study of Arctic Connections
tn	trillion
TUARK	Tromsø-Umeå-Arkhangelsk-Rovaniemi-Kingston
UNCLOS	United Nations Convention on the Law of the Sea
UNDP	United Nations Development Programme
UNDRIP	United Nations Declaration on the Rights of Indigenous Peoples
UNEP	United Nations Environment Programme
UNESCO	United Nations Educational, Scientific and Cultural Organization
UoA	University of the Arctic
US/US(A)	United States (of America)
USGS	United States Geological Survey
VFF	visiting friends and family
VFR	visiting friends and relatives
WCRP	World Climate Research Programme
WHS	(UNESCO) World Heritage Site
WoA	Women of the Arctic
WTTC	World Travel and Tourism Council

Institutions and Place Names

Institutions

Greenlandic	Danish	English
Ilisimatusarfik (Kalaallit Nunaata Universitetia)	*Grønlands Universitet*	Greenland University
Inatsisartut	*Landstinget*	Greenland's Parliament. Meeting at Inatsisartut, an islet in central Nuuk
Kalaallit Nunaat	*Grønland*	Greenland
Naalakkersuisut	*(Grønlands regering)*	Greenland government
(Kalaallit Nunaanni) Naatsorsueqqissaartarfik	*Grønlands Statistik*	Statistics Greenland
Namminersorlutik	*Sevstyre (Hjemmestyre)*	(Self-rule) Legislature (Home Rule)

Place names

Greenlandic	Danish
Aasiaat	Egedesminde
Igaliku	Garðar
Ikerasagssuaq	Prins Christian Sund
Ittoqqortoormitt	Scoresbysund (Kap Tobin)
Kangerlussuaq	Søndre Strømfjord
Kulusuk	Kap Dan
Maniitsoq	Sukkertoppen
Nanortalik	Nanortalik
Narsaq	Nordprøven
Narsarmijit	Friedrichsthal/Frederiksdal
Narsarsuaq	(Narsarssuak/Bluie West-1)
Nunap Isua/Umanarssuaq	Kap Farvel
Nuuk	Godthåb
Nuussuaq	Kraulshavn
Paamiut	Frederikshåb
Pituffik	(Site of Thule air base)
Qaanaaq	Thule
Qaqortoq	Julianehåb
Qasigianguit	Christianshåb
Qassiarsuk	Brattahlíð
Qeqertarssuaq	Godhavn
Qeqertarsuatsiaat	Fiskenæsset
Saqqaq	Solsiden
Sisimiut	Holsteinsborg
Tasiilaq	Ammassalik/Angmagssalik
Upernavik	Upernavik

Contributors

Mette Simonsen Abildgaard is Associate Professor in the Department of Culture and Learning at University of Aalborg, A.C. Meyers Vænge 15, A, 0, 2450 København SV, Denmark, where she is part of the research group CIRCLA (Centre for Innovation and Research in Culture and Living in the Arctic). Her work takes place in the intersection of Techno-anthropology, Media Studies and History of Technology, and centres on the cultural history of infrastructure and telecommunication in Arctic and Nordic countries. masab@hum.aau.dk

Derek Hall is a partner in Seabank Associates, Maidens, Ayrshire, KA26 9NN, Scotland. Formerly head of department and professor of regional development at the Scottish Agricultural College, and visiting professor at HAMK University of Applied Technology in Finland, his interests include the interface of geopolitics and tourism. His most recent previous book is *Brexit and Tourism: Process, Impacts and Non-Policy* (2020, Channel View). Pseudonymously, he also writes fiction. derekhall@seabankscotland.co.uk

Rong Huang is Associate Professor in Tourism Marketing at the University of Plymouth, Cookworthy Building, Drake Circus, Plymouth, PL4 8AA, Devon UK. She has undertaken research into experiential tourism particularly in relation to Chinese international students and Chinese tourists. rong.huang@plymouth.ac.uk

Carina Ren is Associate Professor at Aalborg University, Department of Culture and Learning, A.C. Meyers Vænge 15, A, B3, 2450 København SV, Denmark, and platform co-ordinator for AAU Arctic. Carina researches tourism development in the nexus of social and climate change in a Greenlandic and Arctic context, with a particular focus on cultural innovation. Carina has published in leading tourism and Arctic journals and is the co-editor of several special issues and books, recently *Collaborative Research Methods in the Arctic* (with Anne Merrild Hansen, 2020, Routledge). ren@hum.aau.dk

Kirsten Thisted is Associate Professor, Department of Cross-Cultural and Regional Studies at the University of Copenhagen, Karen Blixen Plads 8, 2300 København S, Søndre Campus, Denmark. She has researched and published widely on Greenlandic culture, narratives and representations of Greenland society and indigeneity. Kirsten has been leader of the international project 'Denmark and the New North Atlantic: Identity Positions, Natural Resources and Cultural Heritage'. Her most recent publications include *Denmark and the New North Atlantic: Narratives and Memories in a Former Empire* (edited with Ann-Sofie Gremaud, 2020, Aarhus University Press/ Oxbow Books). thisted@hum.ku.dk

Preface

Greenland (Kalaallit Nunaat) attracts superlatives. It is the world's largest island (Australia being considered a continent), and is the globe's most sparsely populated country. It is also experiencing climate warming at twice the rate of much of the rest of the world, with dramatic local and global consequences.

While the focus of this book is the context, nature and role of tourism in Greenland, it is set within an overlapping geopolitical frame of: (a) the heightening climate crisis; (b) Greenland's trajectory towards political independence from Denmark; (c) its concept of economic 'self-sustainability' in supporting this trajectory; and (d) growing international interest in, and competition for, Greenland's natural resources and infrastructure projects. The last in its turn partly reflects improving land and sea accessibility afforded by climate change, which paradoxically both challenges and encourages Greenland's concepts of sustainable development, within which tourism plays an ambivalent role: while elements of global and local tourism have been seeking to create a more responsible sector, within Greenland's development trajectory tourism appears to be supporting a sustainability ideology that ignores, or at best camouflages, the climate crisis.

In 2019 two events, occurring within weeks of each other, highlighted the significance of these processes, projecting Greenland into the forefront of the world's attention.

Unprecedented high early summer temperatures triggered accelerated ice sheet and sea ice melting, which sent unusually large volumes of cold freshwater into the North Atlantic, heightening concern over global sea level rise and interference with the north Atlantic circulatory system.

Two months later, attention was unnecessarily diverted when US president Trump offered to buy Greenland from the Danish government. Trump's offer reflected a geostrategic interest in the increasing maritime accessibility of Arctic waters resulting from polar sea ice reduction, combined with the knowledge that increased warming is rendering potentially more accessible the considerable range of minerals – not least rare earth elements – that have hitherto lain unexploited beneath Greenland's snow and ice on land and below its icy Arctic waters. This appeared not a little ironic, given the then US president's sustained climate crisis denial. Although Trump's gesture was summarily rebuffed, it drew further global attention to Kalaalit Nunaat.

As a potent symbol of the Arctic's dynamic image in global consciousness, therefore, the territory of Greenland has become a critically important arena. Yet until relatively recently it remained poorly represented in the English-language literature. One of the core purposes of this book is to articulate these critical processes through the lens of tourism development – one of the three pillars of Greenland's economic future – and the socio-cultural, (geo)political, environmental and economic dimensions it embraces and illuminates.

Cross-cutting these major themes, two events in 2020 further highlighted Greenland's distinctive yet sensitive position – its perceived peripherality, yet its confirmed presence in the global mainstream.

In response to the Covid-19 pandemic, the territory was able to physically isolate itself from the rest of the world in a comprehensive lockdown, thereby foreclosing on any international tourism activity for the short summer season, and beyond. For Greenland, being both insular and relatively remote – characteristics holding some ambivalence for tourism and economic development that are addressed in this volume – as well as being comprised of some 74 self-contained communities, such a lockdown was both relatively easy and highly effective.

The second event of 2020, also relating to a global phenomenon, was the red daubing of the prominent Nuuk statue of Hans Egede, the 'Dano-Norwegian' colonial settlement founder and proselytiser. The spray-painted English language declaration 'DECOLONIZE' adorning the statue's base drew Greenland into the intensified debate concerning the ownership and representation of 'heritage' by statues and other symbols of colonialism and slavery, while it specifically drew attention to Greenland's colonial past, its semi-independent present and its possible politically independent but likely (inter)dependent future.

The central themes of this book therefore employ the role of tourism and travel as a lens through which to examine climatic, societal, economic and geopolitical change in the Arctic as specifically articulated in the experience of Greenland. The 'critical' situations in which Greenland finds itself reflect external perceptions of the global climate crisis and geostrategic manoeuvring over Arctic resources, and domestic considerations of socio-economic development and increased sovereignty. The volume thereby highlights the close and often critical interrelationships between the local, regional and global. A recurring observation is the paradox – one of several – of a region hitherto regarded as peripheral but which is becoming increasingly central to global concerns, with tourism-related dynamics reflecting such centrality.

In this way, the book aims to:

- emphasise the critical role of change in the Arctic in general and in Greenland in particular;
- highlight critical interrelationships between tourism, climate change and the geopolitics of Arctic development, where 'geopolitics' is interpreted as applying at a number of scales from the interpersonal and quotidian to the global geostrategic; and
- provide a critical examination of Greenland's post-colonial tourism development path, as the territory becomes the focus of increasing global interest.

In recent years, the recognition of the critical importance of the changing nature of the Arctic has generated a growing literature. In relation to Greenland, the work of Danish/Greenlandic researchers has long been notable, if not always available in the English language. Collaborative hubs such as CIRCLA – the Centre for Innovation and Research in Culture and Living in the Arctic – at Aalborg University have been important. The results of such long-term research co-operation and intimate knowledge of the region are now more accessible to English-language audiences. Collaborative works such as Hansen and Ren (2020), Thisted and Gremaud (2020), and Johnstone and Hansen (2020b), the *Arctic Yearbook* (e.g. Heininen, Exner-Pirot and Barnes, 2020b) and a special Arctic-themed issue of the *Journal of Tourism Futures* from 2020 are building on regionally-based journals such as the relatively recent *Arctic Review* (2012+), *Journal of the North Atlantic* (2010+), *Polar Journal* (2011+), *Scandinavian Journal of Hospitality and Tourism* (2001+), and the longer-standing *Arctic Anthropology*, *Polar Geography*, *Polar Record* and *Polar Research*.

Greenland's significance for the UK was highlighted in November 2020 when Westminster parliamentarians established an All-Party Parliamentary Group (APPG) for Greenland – the first outside the Kingdom of Denmark. The Group's stated aims were to strengthen UK–Greenland relationships in political, economic, social, cultural and scientific spheres, to raise greater awareness about Greenland within the UK and to promote closer co-operation with the Greenlandic parliament (*Inatsisartut*) (Jonassen, 2020b). This appeared to be genuinely more than just a post-*Brexit* gesture.

Book Structure

The book is organised into three parts with a total of 13 chapters.

Part 1, Arctic Context, acts to provide both the contextual framework and significant themes for the book's subsequent focus on Greenland and is comprised of three chapters. Chapter 1, 'Framing the Arctic', draws upon both positivist constructions and conceptual representations of the Circumpolar North. Most of these 'imaginaries' have been constructed by outsiders. To counterbalance this bias, the chapter also highlights the changing role of indigeneity and the ways in which native Arctic peoples' voices are now being heard and how the position of women is gaining prominence. Chapter 2, 'The Changing Role of the Arctic: Transforming Peripherality', provides a brief overview of Arctic dynamics, highlighting the interrelationships between climate change and the geopolitics of development, and the role of the Arctic Council in mediating between these dynamic forces. Chapter 3, 'Arctic Tourism: Sustainability, Resilience and Identity' interrogates the Arctic context for tourism and the tourism context for the Arctic, focusing on the interplay of sustainability and identity, and the research themes that have emerged.

Part 2, 'Dynamic Greenland', opens with Chapter 4, 'Evolving Greenland'. This addresses the nature and evolution of Greenland as an inhabited Arctic territory. It evaluates the island's colonial, neo-colonial and dependency roles, and contemporary tensions in its pursuit of political independence. Chapter 5, 'Greenland's "Self-Sustainability"', focuses on the hesitant evolution of Greenland's recent economic development, its interrelationship with aspirations for political independence and the philosophy of economic 'self-sustainability' that has emerged in the face of growing global climate crisis pressures.

'Greenlandic Identity and Culture within Development Processes' is the title of Chapter 6. It addresses the nature of Greenlandic traditional and contemporary culture, and tackles debates over identity, language and indigeneity. In so doing, the chapter features a short invited contribution from Kirsten Thisted: 'The Renegotiation of Greenlandic Identity'.

Chapter 7 is 'Tourism Supporting Greenland's Aspirations'. As one of the three central economic planks for Greenland's future, tourism is interrogated through the evolving interrelationships between the country's political trajectory and its tourism administration and development. The chapter examines the pre-Covid numerical basis upon which Greenland's tourism future is being projected.

Chapter 8, 'Peripherality, Tourism and Geopolitical Dimensions of Accessibility in Greenland' examines in conceptual and logistical terms the international and domestic dimensions of Greenland's peripherality, and addresses geopolitical consequences arising from attempts to ameliorate these.

In Chapter 9, 'Fulcrum of Climate Change?' the nature and implications of climate and climate change for tourism development in Greenland are highlighted. This is exemplified in relation to two dimensions of tourism's role: as climate education and as climate experience.

Chapter 10, 'Tourism and Imagery: Soft Power, Branding and Cultural Disconnection' examines the nature of tourism's soft power and Greenland's brand and tourism imagery. The chapter addresses the difficult area of cultural interpretation of imagery and exemplifies this through apparent mutual misunderstandings in relation to the marketing and development of adventure tourism for Chinese visitors. In an invited contribution, 'Understanding Adventure Tourism by Chinese Outbound Tourists', Rong Huang reflects on this dilemma.

Chapter 11 'Paradoxes of Cruise Tourism to Greenland' addresses a number of paradoxes and raises several questions that surround the growing numbers of passengers gazing at, and visiting, Greenland from maritime cruise perspectives.

Chapter 12 comprises the invited contribution from Carina Ren and Mette Simonsen Abildgaard: 'Greenlandic Independence and Tourism Futures – Exploring *Modern* and *Ethnic* Logics'. This interweaves debates concerning the future of Greenland as a likely independent state with the future of tourism and the role of the Greenlandic people in that future. As such it provides an apposite culmination of Part 2.

Part 3, 'Conclusions' comprises a single chapter which draws the strands of the book together while reflecting on Greenland's post-Covid future of economic 'self-sustainability' and tourism's role within it.

Acknowledgements

I would like to thank again the organisers and participants of the 19 August 2020 Arctic Politics international seminar hosted by Aalborg University in Copenhagen for their insightful observations, suggestions and encouragement. In particular, I would like to thank Carina Ren for facilitating it and Ulrik Pram Gad for chairing the seminar.

My sincere thanks are also extended to the four invited contributors to this volume who responded to my requests under not always easy circumstances.

Thanks again to Claire Parfitt, Ali Thompson, Lauren Davies, James Bishop, Stuart Burke and the rest of the production team at CABI and the copy editor, Jonathan Ingoldby for their support.

Derek Hall
Maidens, Scotland
March 2021

Part 1:

Arctic Context

1

Framing the Arctic

> 'The Arctic's always changing ... Even in winter it's not quite static. You can kind of feel the planet moving all the time. It's like having an extra-terrestrial view' (Sarah Moss, 2009: 62)

> '... the Arctic is not a single place but is a wide and diverse region that is in a constant state of change' (Hansen and Johnstone, 2020: 296)

1.1 Introduction

Some two-thirds of Greenland's 2.166mn sq km lies within the Arctic Circle. Greenland (Kalaallit Nunaat) is employed in this book as emblematic of the critical stage in which the Circumpolar North's future is faced with rapid environmental, socio-cultural, economic and geopolitical change. Just as tourism can open up societies and environments to the outside world, so climate change is opening up the Arctic, and notably Greenland, to forces of globalisation and the need to adapt to dynamic environmental, economic and political conditions. And confronting those changes, Arctic societies are poised at a critical moment in their development.

Although the book's focus is on Greenland, the first three chapters attempt to offer an Arctic-wide context for what is to follow. The current chapter addresses ways of thinking about the Arctic and the people who live there, before the discussion moves on to interrogating the contemporary dynamics of Arctic development in Chapter 2, and of understanding the role played by tourism in Chapter 3.

A recent intensification of research interest in the region reflects at least three areas of deepening concern (Lee, Weaver and Prebensen, 2017a: 1):

- a growing social and political focus on climate change that is most strongly exhibited here (e.g. Barr, 2015; Broadbent and Lantto, 2016; Vincent, 2020);
- a heightening of geopolitical interest following a relative opening up of the region as a result of climate warming (e.g. Pelaudeix, 2017; Welch, 2020); and
- intensified debates over the sustainable use of natural and cultural resources, including those for tourism and travel (e.g. Bertelsen and Justinussen, 2017; Huijbens and Lamers, 2017; Bjørst, 2019).

Such emphases, however, have largely reflected exogenous influences on and interests in the Arctic, while the voices of those actually living and working there may represent considerably different perspectives – notably in relation to climate change. But, until recently, these voices have tended to be overlooked (Grydehøj, 2018b). This is a recurring theme of the book. Yet, the Arctic is home to 4 million people, over 10 per cent of whom are Indigenous – such as Inuit, Sámi, Nenet and Aleut peoples – and have long adapted to the region's severe climate and

DOI: 10.1079/9781789246728.0001

extreme environment (Durfee and Johnstone, 2019).

Eight states claim sovereignty within the Arctic (a geographical construct which is contestable: see Section 1.3): Canada, Denmark (Greenland), Finland, Iceland, Norway, Russia, Sweden and the USA (Alaska) (Fig. 1.1). These states vary considerably in terms of population numbers and density, territorial size, economy and style of government (Johnstone and Hansen, 2020a). Generalisations about the Arctic and its peoples are therefore dangerous – a strong caveat which accompanies the content and interpretations of these first three chapters.

Dodds and Nuttall (2016) talk of 'distinctive but overlapping' driving forces that contribute to the shaping and reshaping of polar regions, and which they encapsulate in six eye-catching terms (Box 1.1). These can provide markers for some of the pathways pursued in this volume.

1.2 Arctic 'imaginaries'

Much social science has been concerned with understanding the links between how we experience, understand and order the world, and by so doing shape others' experiences and understandings. In the case of the Arctic a number of (geopolitical) perspectives, conceptions or 'imaginaries' have been identified.

> The imaginary constitutes within itself a loose frame for understanding that of which we have no direct contact ... It can represent a means of meaning-making that is felt, experienced and embodied ... and whether real or otherwise, the imaginary in a general sense comes with us and ... shapes the sorts of worlds we create ... the imaginary can become so entrenched that it becomes the way of seeing ... that is difficult to dislodge.
>
> (White *et al.*, 2019: 2).

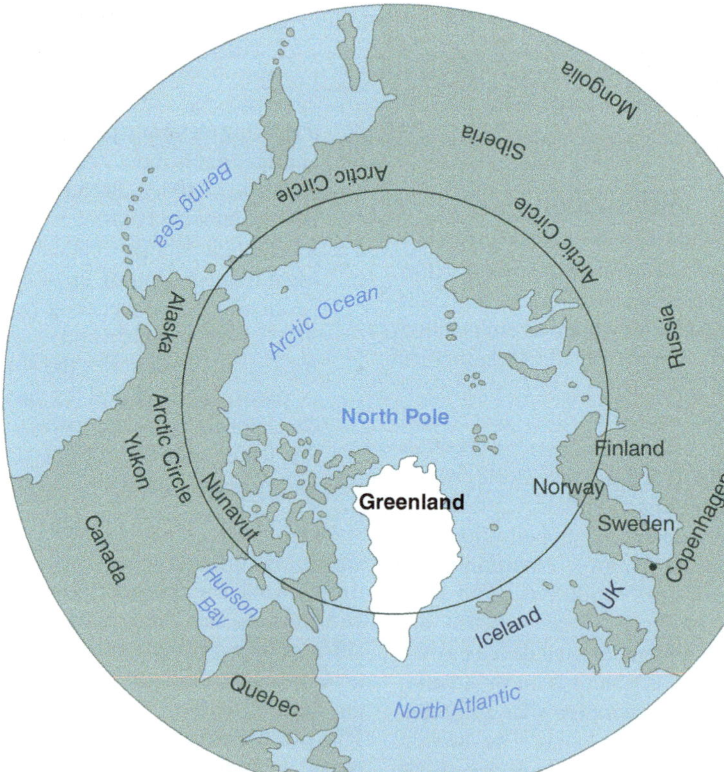

Fig. 1.1. Polar view of the Arctic. Author's own figure, modified and redrawn from Dillon, 2019: 7.

Box 1.1. Forces remaking the Arctic

1. Globalisation

Rather than being peripheral to world events, for centuries the Arctic has been tied to the global economy as well as being subject to the effects of increasing globalisation. Even by 2005 it was estimated that $230bn was being generated annually in the Arctic from intensive exploitation and export of energy resources. The growth of international tourism to and within the Arctic (Chapters 3, 6–12) is a critical generator of such flows and exchanges as the world (slowly) de-carbonises.

2. Perturbation

The Arctic plays a critical role in global climate dynamics, experiencing rapid physical changes in response to warming, with equally global consequences (Chapters 2 and 9). While ecological/physical disturbance – fate of sea ice, permafrost, stability of polar ice sheets – represents one dimension of perturbation, a perceived 'broader sense of unsettlement' includes socio-political dimensions such as forced migration policies (Chapters 2 and 4).

3. Amplification

This dimension has at least three elements: (a) an intensified shift and acceleration of geophysical change; (b) impacts generating second-order effects on marine and terrestrial environments; and (c) the creation of (unwarranted?) hope, fear or anxiety regarding future risks. As explicit geopolitical dimensions of such emotional states, the planting of a Russian flag on the bottom of the central Arctic Ocean in 2007, which inspired the Ilulissat Declaration (Box 1.3) or the Chinese aspiration to establish a 'Polar Silk Road' (Chapters 2, 8, 10) can be exemplified.

4. Polarisation

This term is intended to highlight (a) the growing activism of Indigenous peoples and circumpolar co-operation in the Arctic, involving sub-national governments and regional organisations evolving a distinctly 'northern governance' (this chapter and Chapters 2 and 4); and on the other hand, (b) the expanding interest in the Arctic of such extraterritorial actors as the European Union (EU) and campaigning environmental groups such as Greenpeace (Chapters 2 and 5). Such polarisations revolve around competing conceptions of the nature, appropriateness and sustainability of Arctic economic development.

5. Legalisation

This concept emphasises how the Arctic has become increasingly embedded in the workings of international legal regimes (this chapter and Chapter 2), creating 'an ever more complex mosaic of governed spaces' (Dodds and Nuttall, 2016: 46). Notably, as the central Arctic Ocean becomes more accessible through warming, international regulations relating to the high seas and seabed will take practical effect, just as the United Nations (1994) Convention on the Law of the Sea (UNCLOS) has played an important role hitherto in the zoning of ocean space.

6. Securitisation

As a term employed by critical security studies scholars, this refers to the manner in which countries, regions, objects, infrastructures, resources and populations are 'secured'. This can involve invocations of danger, threat and risk in order to lever political and financial resources. This highlights 'power-knowledge' relations in potentially competing claims and aspirations relating to Arctic-based resources (Chapters 2, 4, 5 and 8).

Source: Dodds and Nuttall, 2016: 32–57

Steinberg, Tasch and Gerhardt (2015) order Arctic imaginaries into two groups. These and their relevance to this volume are presented in Box 1.2. However 'exceptional' and 'extraordinary' the Arctic may be, such imaginaries are being modified or enhanced by accelerating climate change and the prospect of intensified resource-led opportunities to produce new kinds of framings.

Nevertheless, in the face of (geopolitical) change and potential threat, a desire by Arctic states to maintain the regional imaginary of status quo was embodied in a formal declaration signed in 2008 in the city now symbolic of

Box 1.2. Arctic 'imaginaries'

Group A: The Arctic as a system of bounded, sovereign territorially-defined states, albeit with specific modifications that account for the region's unique cultural and geophysical characteristics.

1. The Arctic conceived as a *terra nullius*, an unclaimed but potentially claimable space beyond the regulations of international law, where individual states are free to exercise their expansionist tendencies, whether claiming land, water, ice or seabed. This perspective, or rather the fear that others may hold it, is subliminal.
2. A moderated version of the above, where the underlying environment of the Arctic is more relevant but different, requiring different norms and legal regimes of governance. States are presented with a unique space of new and different opportunities.
3. The most conventional conception is that while the Arctic may not present opportunities for new levels of territorial formation, it does offer the arena for replicating existing forms, as in the calls for Greenlandic independence (Chapters 4 and 12).

Group B: Conceptions that highlight processes and alliances that transcend boundaries.

4. The Arctic as a resource frontier, with opportunities for states, corporations and individuals whose roots are elsewhere and who seek not incorporation of territory but extraction of natural resources (Chapters 2 and 5).
5. Indigenous nationhood: whereby an Indigenous world view, such as the pan-Inuit movement (Chapter 2), challenges the fundamental assumptions behind the modern territorial state.
6. The environmentalist imaginary: which views the Arctic as a space whose nature is pristine but endangered, and which therefore should be governed according to an ethic that transcends the prerogative and developmental ideals of the sovereign state (Chapters 2 and 9).

Source: Steinberg, Tasch and Gerhardt, 2015: 15–17

global climate concern and the focus of Greenland's tourism: Ilulissat (Box 1.3).

1.3 Arctic paradoxes

While the Arctic may be subject to contested and dynamic perceptions or imaginaries, these in their turn can generate several paradoxes.

First, some policy-makers who have attempted to direct attention to the Arctic have stressed that the region is not simply an empty space to be crossed by people from distant lands, but that it is an arena of connections that bring the world together. An early proponent of this view was the anthropologist Vilhjalmur Stefansson ([1921] 2015, [1922] 2017), who argued that the Arctic was a 'Polar Mediterranean', featuring a relatively navigable central space (Fig. 1.1) that united diverse coastal peoples in commerce and productive interaction.

This 'Mediterraneanist' view of the Arctic was reproduced some 60 years later by the then Soviet leader Mikhail Gorbachev in 1987. In his 'Murmansk Initiative' (Archer, 1988; Jonson, 1990), which came to mark the beginning of the end of the Cold War era in the Arctic, Gorbachev observed that the Arctic was the place where the Eurasian, North American and Asia Pacific regions met, where frontiers came close to each other and interests of states crossed (Åtland, 2008; Fondahl, Espiritu and Ivanova, 2020). Through this 'Mediterraneanist' lens, therefore, the Arctic is seen as a land of proximate peripheries, as a space of connection not separation. Yet it 'still exists primarily to connect others who live *outside* the region' (Steinberg, Tasch and Gerhardt, 2015: 8).

A second paradox, albeit closely related to the first, is that of the Arctic imagined as both a geographical periphery (Box 1.4; also Section 8.1) and as an experiential core in acting as an important context for the production and consumptions of tourism. In this way the role and nature of 'the ice' becomes both constraining and permissive, even seductive (Fig. 1.2).

Box 1.3. The Ilulissat Declaration

Arctic Ocean Conference: a joint Danish-Greenlandic initiative

'The Arctic Ocean stands at the threshold of significant changes. Climate change and the melting of ice have a potential impact on vulnerable ecosystems, the livelihoods of local inhabitants and Indigenous communities, and the potential exploitation of natural resources.'

In response to these perceived threats, the Declaration, signed by Canada, Denmark, Norway, Russia and the USA as the five coastal states of the Arctic Ocean (the 'A5') and adopted 28 May 2008:

- re-emphasised commitment to the principles of the UNCLOS (1994), implicitly rejecting the call for a new comprehensive international legal regime to govern the Arctic Ocean;
- recognised the UNCLOS framework to provide a solid foundation for responsible management by the five coastal states and other users of this ocean;
- highlighted the need to further strengthen search and rescue capabilities with the increasing use of Arctic waters for tourism, freight shipping, research and resource development;
- emphasised strengthening co-operation that included the collection of scientific data concerning the continental shelf and protection of the marine environment;
- affirmed that the five coastal states of the Arctic Ocean would continue to contribute actively to the work of the Arctic Council (Chapter 2) and to other relevant international forums.

Unlike the Arctic Council (Box 2.4), there was no participation in the Declaration process from Sweden, Finland, Iceland, Indigenous peoples' representatives or observer members of the Council. The Home Rule Government of Greenland, both because of its growing confidence in para-diplomacy as co-convenor and by virtue of being on home soil, did participate.

In part triggered by Russian flag planting on the seabed at the North Pole, the Ilulissat Declaration was generally considered a significant success in halting growing tensions among the five coastal states, and helped lay the foundations for Arctic co-operation that has survived confrontations elsewhere.

The Declaration did not include security and defence, and this rendered it incapable of countering the longer-term challenge perceived, especially by the USA, of China's growing presence in the Arctic.

Sources: Kingdom of Denmark 2008; Rahbek-Clemmensen and Thomasen, 2018; Viggo Jakobsen, 2019

A third paradox is that, despite the long-fixed bounding of the Arctic Circle, and universal agreement and recognition of points of the compass (a constantly shifting magnetic North Pole notwithstanding), researchers struggle to find any consistent defining view of (the) 'Arctic'. Definitions may vary depending upon scientific, environmental, geographical, political and cultural approaches, perspectives and biases, which, in their turn, are reflected, for example, in the different ways the working groups of the Arctic Council (Section 2.5) define 'the Arctic'. Further,

> ... climate change is eroding many of the physical boundaries and features, and reshaping the contours of geography that have been drawn as seemingly fixed points on maps, such as the tree line, the southern extent of discontinuous permafrost, ice shelves, glaciers and the distribution of perennial sea ice.
>
> (Dodds and Nuttall, 2016: 4)[1]

A fourth paradox relates to perceptions and representations of the Arctic as remote, pristine wilderness, whereby tourism in these latitudes appears to be implicitly defined in terms of nature – such as the *Aurora borealis* (Box 1.5) – as the dominating attraction. Such exclusivist conceptions of 'nature' – common in tourism discourses linked to colonialism and Western visions of modernity – tend to commodify, conceal or ignore Indigenous inhabitants: stakeholders who, arguably, have the most to lose or win in Arctic tourism contexts. When Arctic nature is 'scripted as a peripheral object of desire' for tourist consumption, the contemporary livelihoods of Indigenous peoples either tend to be romanticised as part of an 'authentic' primitive and threatened traditional past, or erased from the perceived landscape (de la Barre, 2013; Grimwood, 2015).

Box 1.4. Peripherality

Peripheral: '1 relating to or situated on the outer limits of something. 2 of secondary importance.'
Periphery: '1 the outer limits or edge of an area or object. 2 the less important part of a subject or group. ORIGIN Greek *periphereia* "circumference"' (*Oxford English Dictionary*).

1. An essentially contested concept, open to (re-) interpretation from spatial, economic, socio-cultural and political perspectives.
2. Often seen to be closely related to notions of marginality, a condition of apparent disadvantage that may arise from environmental, cultural, social, economic and/or political factors.
3. An internally and externally relative quality: peripheral to what?
4. The dynamism of places can highlight the temporal relativity of peripherality: what is regarded as peripheral at one time may not be at another.
5. Space within a periphery is not a homogeneous, but rather an amalgam of 'nearer' and 'farther' peripheries.
6. Peripherality may be characterised by a number of interrelated features that potentially impact on development processes:
   - geographical remoteness from mass markets, increasing transport costs for goods, services and visitors;
   - a tendency to lack effective political and economic control over major decisions affecting well-being;
   - the likelihood of weaker internal economic linkages, lower innovation rates, and fewer opportunities to benefit from economies of scale, limiting the ability to achieve high multiplier effects, with significant levels of goods and services needing to be imported;
   - the stimulation of outward migratory flows, particularly of younger age groups, reducing the vitality of a resident population while imposing burdens on social welfare provision with a resulting older demographic structure;
   - national and local state sectors may play a greater (interventionalist) role than in comparable non-peripheral locations because of the above factors; such support may serve as a means of assisting intra- and inter-generational equity that can keep open future economic, social and environmental options, not least for sustained human welfare;
   - information flows from and within may be weaker than those *to* a periphery from 'core' regions, influencing perceptions/imagery of the periphery; however,
   - there may be a greater likelihood of the retention of high amenity value, being relatively underdeveloped, but with the natural resources contributing to these requiring careful management.
7. The importance of identifying peripheral regions was underlined in the European Commission's (1997) *Cohesion Report*, and led to the calculation of an index of peripherality, albeit largely based upon economic considerations.
8. Although an inherently spatial and positivist concept, development of more holistic understandings of peripherality has seen research incorporating such aspatial dimensions as perceptual constructions, shared social norms and institutional frameworks.

Sources: Jussila, 1998; Brown and Hall, 2000; Copus, 2004; Hall, 2007; Davies and Michie, 2011; Máliková, Farrell and McDonagh, 2016; Lee, Weaver and Prebensen, 2017b. See also Chapter 8

In this way, culture in Arctic tourism discourses is often addressed as an adjunct of nature (Johnston, 1995, 2011; Smed, 2014, 2017), despite significant populations, both Indigenous and incoming, inhabiting much of the region. This apparent asymmetry is a recurrent theme of the volume, and may be exacerbated by, for example, the imagery generated by both exogenous and incomer tourism entrepreneurs.

1.4 The Arctic in popular culture: imagination and paradox

In her survey of representations of the Arctic in (largely English-language) literature and the popular imagination, Hansson (2018; see also David, 2000; Cavell, 2008), reinforces some of the above observations and articulates some of the paradoxes when arguing that the region has been often imagined as the 'last frontier', a space

Fig. 1.2. Icebergs in the North Atlantic. Author's own photograph.

Box 1.5. *Aurora borealis* and Arctic tourism paradoxes

Experiencing the Northern Lights has become a high priority on many tourists' 'bucket lists' (Box 3.1). Increased solar activity and magnetic storms have enhanced boreal activity since 2000, such that businesses offering 'Northern Lights tourism' activities are some of the fastest growing in the Circumpolar North.[2] The Lights' presentation involves a fluid, ever-changing network of actors, human as well as non-human, and their choreographies (Kivelä, 2014; Jóhannesson and Lund, 2017).

The Lights are mysterious, flickering and teasing, and sightings are unreliable due to different strengths and direction of the solar wind, and local weather conditions, notably the presence or otherwise of a clear sky. The primary condition for sightings is darkness, which has traditionally been associated with danger and fear (Edensor, 2010). This raises the paradox in that, by contrast, far north landscapes have been represented, from the work of eighteenth-century explorers and artists onwards, as pristine, majestic and white, evoking a sense of the sublime, an image that contemporary tourism image-makers continue to promote (Lund, 2016).

Creating a stage to contextualise a Northern Lights experience requires narratives that script an Arctic fantasyland that acknowledge elements of old imaginaries, although how far these are constructed from exogenous histories of exploration, research and travel and the extent to which they draw on 'authentic' Indigenous knowledge and experience is questionable (Mathisen, 2010, 2017).

A further conundrum set within a paradox is that while the Arctic as a 'peripheral' region has become a centre of contemporary experience production, like the surrounding natural environment and cultures, the central tourism product – the Aurora – is ephemeral and unstable and thus uncertain (Heimtun and Lovelock, 2017; Lee, Weaver and Prebensen, 2017b). Of course, risk is inherent in the 'supply' and pursuit of a number of tourism-related activities, and this is a central element linking Northern Lights tourism to adventure tourism (Chapters 8 and 9) and other forms of nature-based recreation, and the sustainability of small businesses that facilitate it.

of alternative livelihoods and opportunities where people – implicitly outsiders – can remake themselves.

Historically, Hansson (2018) recognises four major representations of the Arctic in popular culture (Box 1.6), some contiguous with and others cross-cutting Steinberg, Tasch and Gerhardt's (2015) Arctic 'imaginaries' (Box 1.2).

Hansson (2018) concludes that narratives about the Arctic have rarely acknowledged it as a social environment, sustaining a continued narrative absence for Arctic people. Yet her observations on contemporary fiction reflect on ambiguous processes of becoming, connection and participation, where the emergence of Arctic detective stories/crime thrillers has acted to change the character of literature about the Arctic (and the nature of the crime writing genre), through populating the otherwise 'barren landscape'. Although in Melanie McGrath's crime novel *White Heat* (2011), 'the most addictive character – both hero and villain of the piece – is the Arctic itself' (Jones, 2011), just like the main character in Peter Høeg's (1992/1993) *Miss Smilla's Feeling for Snow*, McGrath's central investigator is 'a threshold figure, part Inuk and part *qallunaaq*, or white' and the novel extends its genre by embracing the entire community where the crimes take place as victim, and not just of murder (Hansson, 2018: 54). This equally applies to the more recent Greenland-set Arctic crime novels of Mads Peder Nordbo (2017/2018, 2018/2019), whose investigating journalist Matthew Cave is again the product of Inuit and

Box 1.6. Representations of the Arctic in popular culture: exogenous imagery demoting the Indigenous presence

1. The Arctic as a site for heroic endeavour, and notably the historic goal of reaching the North Pole, with members of numerous expeditions attempting to attain that goal being portrayed as heroes. In Anglophone popular culture in particular, Hansson argues that the scientific gains of these endeavours have taken second place to the elements of competition and national honour: '... rampant hero worship in a *terra nullius* icescape'.

2. The Arctic as a futile goal. Almost the opposite of the above, Hansson notes the satirical and ironic nature to which such endeavour – particularly when it failed – was often subjected. This was notable from the later nineteenth century, with the imperialist overtones of expeditions becoming the butt of jokes that depicted the object of competitive endeavours as 'a desolate snow-desert, inviting the question ... [as to] ... the value of an uninhabitable wasteland far from civilisation'.

3. The Arctic as gendered space. The Arctic has been 'frequently represented as a woman, silently awaiting her lover-conqueror', where the North Pole becomes a boudoir, domesticating the Arctic for explorers such as Fridtof Nansen, as in a *Punch* cartoon of 1896 ('The Sleeping Beauty of the North', illustrated in Hansson, 2018: 52). 'The virginal Ice Maiden, however, had her counterpart in the idea of a formidable Ice Queen with the power to defeat the explorers daring to enter her realm ... [yet] ... her hostile nature emphasises [the explorers'] fortitude and reinforces a model of heroic masculinity.'

4. The Arctic as otherworld: as a transitional space that is disorienting. Extreme natural conditions expose the visitor to an environment where familiar ways of interacting with the surrounding world are challenged both physically and mentally. This can be exemplified by the apparently hallucinatory lead character on an isolated archaeological dig in Greenland that goes wrong during a global pandemic (no longer fanciful) in Sarah Moss' (2009) debut novel *Cold Earth*. Power relations between nature and culture are reversed, and alternative, utopian, worlds may be stumbled upon, through accident or malfunction. 'Using the Arctic as a setting for utopia relies on the same thought patterns as when the region is conceived of as available for exploitation: it is a blank space, open and ready for inscription.' We can see obvious parallels with space fiction (sci-fi) here. A more recent development, however, is *cli-fi* (climate fiction), usually dystopian stories about the effects of, and attempts to halt, climate change (e.g. Cussler and Cussler, 2009; Theroux, 2009; McEwan, 2010): '... the underlying preservation ideal perpetuates the representation of the Arctic as pristine nature under threat' (Hansson, 2018: 53). Rarely is the region depicted in social or cultural terms, as a site of modernity.

Source: Hansson, 2018: 46–55

European (Danish) parentage, and whose writing draws on and interrogates community-wide issues.

Also in relation to Greenland, somewhat belatedly, wholly 'Indigenous' voices have begun to emerge in the form of Niviaq Korneliussen's (2014) edgy *rites de passage* novel *HOMO sapienne*, which was claimed to be the first such to be first published in Greenlandic and by a Greenlandic author. It was translated into English in 2018 with the title *Crimson*. At the time of writing this had not been undertaken for Sørine Steenholdt's *Zombieland* (2015), a collection of socially critical short stories by a member of a growing community of young Greenlandic writers firmly grounded in Arctic social milieux.

1.5 Authenticity?

Heidi Hansson is professor of English Literature at Umeå University in northern Sweden, and her analysis follows in the tradition of evaluating and interpreting outsiders' perceptions and emotions when confronted by 'the Other': in this case by a forbidding physical environment that may convey the impression, and thus encourage the social construction, of being uninhabited. Prompted by the previous paragraph, therefore, we may be left asking where and who is the 'authentic' voice of such environments?

This current volume is concerned with the nature, roles and impacts of tourism development. Authenticity and commodification are central to academic debates in tourism (Cole, 2007), embracing arguments surrounding the nature, extent and consequences of the commercialisation of Indigenous peoples and their local identities. Again, most of these evaluations and interpretations have been undertaken from a 'Western' or even Euro-centric perspective, including exogenous voices suggesting that tourism development should (be seen to) promote local awareness and identity (e.g. Franklin and Crang, 2001).

Such questions that have long been raised include why does commodification, which can arise from tourism and other forms of socio-cultural economic development, result in some (not necessarily Indigenous) communities' disempowerment, while in others 'authenticity' is appropriated and becomes a powerful resource (Cole, 2007: 944)? There is, of course, no single answer given the multiplicity of variables at play and the fact that notions of 'authenticity' are socially constructed, subject to power relations, and thus malleable.

At least partly because of their perceived authenticity and cultural 'Otherness', Indigenous groups can possess an intrinsic interest for tourists and the tourism sector, an interest which, understandably, may not be reciprocated (e.g. Flora, 2019: x).

Economic development in the Arctic can symbolise modernity and global connectivity, accompanied by communities striving for 'de-peripheralization' (Kühn, 2015). Paradoxically, the development of tourism, particularly that focusing on Indigenous groups, can produce images of 'traditional' communities ascribing to them a role from which they may have struggled over decades to escape (Keskitalo, 2017). Such 'Arctification' (Müller and Viken, 2017b: 284) risks, for example, seeing northern areas that had previously been perceived as integrated parts of Norway, Sweden and Finland, being increasingly gazed upon as something 'exotic', inducing a 're-peripheralization', a process to which notions of 'Indigenous tourism' may contribute.

As Bjørst and Ren (2015) suggest in the case of Greenland, contemporary Arctic imaginaries and local people's roles within them may be seen to compete with each other. In one, the Arctic as the resource periphery is viewed offering multiple opportunities for economic development, in which Indigenous groups may be active participants. In another, the Arctic is presented as a pleasure periphery for adventure seekers in the pristine wilderness in which Indigenous groups may become a passive or vestigial element.

1.6 Voices from within (i): Arctic indigeneity

1.6.1 Indigeneity

As noted earlier, the Arctic is home for approximately 4 million ethnically diverse people, about 10 per cent of whom are considered Indigenous and around half of whom are Russian (Exner-Pirot,

2020: 307). In the Nordic north and western Siberia, some of the most populated regions of the Arctic, Indigenous populations seldom represent more than 10 per cent of the total. It is only among the least populated areas of the Arctic that Indigenous peoples are numerically the majority: in Nunavik and Nunavut in Canada and, in strictly ethnic terms, Greenland. But in the latter there is no legal distinction between Indigenous and non-Indigenous people – all are Greenlanders and possess the legal rights as such (Andersen, 2020) – although this has to be seen in the context of the nuanced complexities of Greenlandic–Danish relationships (Chapters 4–13).

Issues relating to Arctic cultures – including the (re-) interpretation, (re-) definition and commodification of culture – and historic precedence in the region can be problematic. An understanding and application of 'indigeneity' can be both challenging and challenged.

Weaver (2016) argued that Indigenous groups may not regard themselves as such, while Friedman (2008) suggested that they may not be primarily self-defining populations, but may labour under categories that have been imposed by colonial orders. By contrast, Barth (1969) affirmed that ethnic identity was a matter of self-ascription or choice.

'Indigeneity' may imply intimate connections (usually small scale) between group and locality, connoting deep-rooted feelings of belonging, attachment and identification. It therefore distinguishes 'natives' from 'others': a relational trait through which 'connections between group and locality' are negotiated (Merlan, 2009: 303). For example, any tensions that exist between Indigenous groups living modern lives and their (non-Indigenous) neighbours with whom they may share similar lives, but with differences in history, ethnicity, traditions, cultural expressions and mentality, may represent left-overs from the colonial past (Müller and Viken, 2017b).

This suggests that perceptions of indigeneity may vary significantly across the Arctic (Ren *et al.*, 2021), rendering generalisations difficult and potentially misleading (Viken and Müller, 2017). In its promotion for tourism purposes (Section 3.6) indigeneity may be intertwined with, and perhaps obfuscated, by notions of 'authenticity'.

1.6.2 Arctic Indigenous peoples' re-emergence

In recent years, Arctic Indigenous peoples have been re-asserting their rights in their traditional homelands, having for so long been marginalised or exploited. Prior to World War II, Indigenous peoples tended to be left alone by authorities who largely did not interfere with traditional economic activity (hunting, trapping, fishing, and, in Scandinavia and Russia, reindeer husbandry) (Coates and Broderstad, 2020: 13).

Indigenous living conditions were to change, sometimes dramatically, after the war. Economic development and militarisation, coupled to growing concerns about national sovereignty, saw a rapid influx of non-Indigenous workers and settlers across the Arctic. In the 1950s and 1960s, changing political priorities emphasised 'improving' Indigenous conditions, not least by providing full-time education and replacing mobile lifestyles with sedentary existences. New towns were constructed that offered a range of modern services and amenities. Infrastructure investments saw the electrification of remote communities, now rendered accessible by new highways and airfields, with commensurately upgraded communication systems (Coates and Broderstad, 2020: 14). But with such developments, an economic marginalisation of Indigenous peoples was accompanied by social discrimination exhibited by the often numerically dominant newcomers, and by negative stereotypes of Indigenous peoples.

Indigenous peoples' expressing increasing frustration with the impacts of 'development' provided the context for heightened interest in the human rights, security and development of Arctic peoples when the United Nations' *1994 Human Development Report* (UNDP, 1995) was published. This raised awareness for policy-makers and publics to a range of Arctic issues which had hitherto failed to attract serious consideration: quality of life disadvantages with such outcomes as shorter life expectancy and significantly higher rates of infant mortality, suicide and domestic abuse (Lukovich and McBean, 2009; Welch, 2020).

While Indigenous peoples everywhere have suffered cultural dislocation and a consequent sense of disempowerment, the effects have been particularly noticeable in the Arctic because of the fragile relationship between people and environment, and the combination of their often small numbers and high levels of dispersion. Emblematic of this precarious relationship in a time of rapid climate change is food and health security, since there is relatively little room for error within this ecologically vulnerable zone (Wesche and Chan, 2010; Waits *et al.*, 2018; Welch, 2020: 478).

Sámi in Scandinavia and Inuit in Greenland have been able to build upon a stronger and broader level of public support than that experienced by Indigenous peoples in other parts of the Arctic (Lantto and Mörkenstam, 2008; Coates and Broderstad, 2020: 16). But while the nature of governance has varied, sometimes considerably, across the Arctic, a number of Indigenous groups have been able to draw on transnational networks (Section 1.6.3; Box 2.5).

Climate change presents perhaps the largest challenge facing Indigenous people in the Arctic (see Chapter 2, Box 2.1). The Arctic Climate Impact Assessment (ACIA, 2004) most strongly brought the impacts of warming on Arctic life and livelihoods to global attention. With Arctic warming proceeding at twice the global average, recorded impacts on Indigenous peoples include the poleward migration of the tree line causing a loss of tundra and the invasion of alien fauna, flora and pathogens, melting permafrost causing damage or destruction to buildings and infrastructure as well as flooding, depletion of pack ice reducing opportunities to hunt with dog sleds and ice-hole fishing, and a decline in animal species that have traditionally provided sustenance and cultural identity.

In an address to the United Nations in 2008, representatives of the Sámi Council and the Inuit Circumpolar Council argued for the need to address climate change 'in an urgent manner' and emphasised that Indigenous peoples, being marginalised globally, were most at risk from the effects of climate change, the two most daunting challenges being not only the direct impacts of climate change threatening their social, cultural and economic systems, but also the prospect that the Arctic would carry the burden of poorly-conceived climate mitigation policies devised elsewhere (Stonehouse and Snyder, 2010: 67, citing *The Barents Observer* 24 April 2008).

1.6.3 Indigenous peoples' rights

International legal frameworks have increasingly protected Indigenous rights, not least as a result of the United Nations Declaration on the Rights of Indigenous Peoples (UNDRIP), adopted by the UN General Assembly in 2007 (UNDRIP, 2007). It is not a treaty: no state is formally bound to its contents, and some Arctic countries, notably Canada and the USA (Gover, 2015), even voted against it initially, while Russia abstained (Tomaselli and Koch, 2014).

International Indigenous rights law exerts certain pressures in support of policy approaches that Newman (2020: 427) refers to as the 'contextualized decolonization' of the Arctic, helping to protect against southern political majoritarianism that might otherwise treat the Arctic as a resource extraction colony. Nonetheless, the apparent reduction of such southern colonial power can take different forms in different Arctic contexts.

Key rights such as those relating to culture, land and self-determination, have implications that constrain certain types of state policies. And some rights have significant trans-boundary dimensions.

One key international legal instrument setting out commitments on Indigenous rights is the 1989 Indigenous and Tribal People's Convention, negotiated within the auspices of the International Labour Organization (ILO) and commonly known as ILO Convention 169 (ILO, 1989). This has, for example, direct legal bearing on obligations in Greenland.

Variations between states may develop further with tendencies toward regional negotiations on Indigenous rights instruments: for example, regional negotiation of the Nordic Sámi Convention (Bankes and Kuivorova, 2013), involving Norway, Sweden and Finland in making particular international treaty commitments, recognises the trans-boundary dimensions of Sápmi[3] which were agreed upon in 2017 (Newman, 2020).

Overall, several key groups of Indigenous rights can be recognised (Newman, 2020: 432–434):

(i) *cultural rights*: encompassing a range of specific rights, notably those related to traditional harvesting and relationships with certain animal species;
(ii) *land rights*: to land traditionally occupied and used, as embraced by UNDRIP articles 25–28.

Although there are complex issues relating to ownership of natural resources (Ahren, 2016);
(iii) *self-determination rights*: these are receiving gradual recognition in international instruments. For example, the 2009 Greenland Self-Government Act and future independence is partly fostered and conditioned by rights of Indigenous self-determination (Kuokkanen, 2017), although rights alone cannot determine all the practicalities of full independence;
(iv) *participatory rights* as one aspect of self-determination (Xanthaki, 2007).

In addition to their representation on the Arctic Council (Chapter 2), Indigenous groups are able to draw on support from the United Nations Educational, Scientific and Cultural Organization (UNESCO) and various non-governmental organisations (NGOs) (Viken and Müller, 2017), and Indigenous collaboration across the Arctic is not restricted to political arrangements. There are extensive cultural and artistic engagements in the region, including a variety of festivals and cultural events. A biennial event since 1970,[4] the Arctic Winter Games (AWG), for example, aims to strengthen intercultural exchange and celebrate pan-Arctic identity through sports and contests characteristic of the region and its Indigenous populations. The 2016 Games, hosted in Nuuk, Greenland's capital, was notable for its value creation (training 1,750 volunteers), capacity strengthening and identity-building, as part of Greenland's path towards independence (Ren and Thomsen, 2016; Ren and Rasmussen, 2017).

The level of international engagement extends to academic co-operation, including the University of the Arctic,[5] under whose auspices various exchange programmes for faculty and staff, numerous conferences and collaborative projects and a smaller number of business outreach initiatives are pursued. There is a growing Indigenous interest in cultural exchanges, co-operation on tourism promotion and extensive consultations on the growing environmental challenges facing the Circumpolar North (Coates and Broderstad, 2020).

1.7 Voices from within (ii): Gender and intersectionality in the Arctic

Just as Indigenous voices were long ignored, written out of the scripting of Arctic conceptions, so discourse on Arctic affairs has often been characterised by a gender-blind narrative. Bjørst (2008), for example, has notably written about the stereotyping of women in Greenland and the Arctic.

While Arctic science has been dominated by research methods and practices that reflect masculinist values of rationality, objectivity and positivism, gender and feminist research has been moving beyond these parameters to provide more comprehensive analyses of the complex relationships between the social and natural worlds. The term 'intersectionality' (Crenshaw, 1991) was designed to critically assess the intersection between race and gender, and at its core has a 'non-positivistic, non-essentialist understanding of differences among people as produced in ongoing, context-specific social processes' (Marfelt, 2016: 32). Thus the concept recognises important linkages between multiple identities of gender, race, ethnicity, class, age and other social categories, and has been able to highlight the minimal addressing of gender and indigeneity in the Arctic (Hoogensen Gjørv, 2017), despite more than two decades of a flourishing, if patchy, Indigenous feminist literature (e.g. Green, 2007).

Considering that the eight circumpolar states are among the richest and most prosperous in the world, multiple barriers to equality have been faced by women living in Arctic and northern regions. Further, with the exception of the USA, all Arctic states have ratified the 1979 Convention on the Elimination of All Forms of Discrimination Against Women (CEDAW) (OHCHR, 2006), as well as the extensive implementation obligations spelled out in the 1995 Beijing Platform for Action (UN Fourth World Conference on Women). Yet when these eight states joined together to establish the most important inter-state governing body, the Arctic Council (Chapter 2), more than 15 years after CEDAW was ratified and a year after the Beijing Platform was adopted, gender equality was not prioritised.

It was only in 2002 that the Arctic Council recognised the importance of women in developing Arctic communities in its Inari Declaration (Arctic Council, 2002: 2) by encouraging 'the integration of gender equality and women ... perspectives in all efforts to enhance human living conditions in the Arctic'.

In that same year, the Arctic Council's 'Taking Wing' conference on gender equality and women in the Arctic addressed a range of issues from women in the workplace – including the heavily male-dominated extractive industries – living conditions, traditional knowledge and self-determination for Indigenous peoples and the impacts on Indigenous women, political participation, health, trafficking and violence against women (Toka, 2002; Kohut and Prior, 2016). As the first gender-focused conference convened by the Arctic Council, this was clearly about taking an intersectional approach, even before the notion began to take a foothold in feminist analytical literature.

Much useful research has been published on such issues as domestic abuse and violence towards Indigenous women in the Arctic, questioning the dominant discourse, arguing that rapid social change does not in itself account for men's violence against women and that a focus on social agency is required (e.g. Sørensen, 1998, 2001). But such empirical research faces the risk of being exploited to stereotype Indigenous societies. It is perhaps partly for this reason that Rauna Kuokkanen has set her work within the wider frame of globalisation (e.g. Kuokkanen, 2008) and has highlighted the need to examine the ways in which gendered violence is explained, addressed and often sanctioned in Indigenous communities at higher political levels (Kuokkanen, 2014).[6] In this way Arctic gender and intersectionality issues have been set within the wider realm of Indigenous women's experiences (e.g. Sjørslev, 1998).

Events and initiatives focusing on gender since 2002 have included two Arctic Human Development Reports (AHDRs). The first (Einarsson *et al.*, 2004) was ground-breaking in addressing intersections between gender and indigeneity, with researchers taking a comprehensive overview and understanding of natural, social and political life in the Arctic. Although not explicitly highlighting intersectionality as such, it illustrated tensions between different understandings of gender and indigeneity.

In the second AHDR report, Larsen and Fondahl (2014) made an explicit decision to require all chapters to address gender as a cross-cutting issue. Not all contributions were equally successful.

In striving to contribute to sustainability and balanced participation in leadership and decision-making in both the public and private sectors, the Arctic Council launched a Gender Equality in the Arctic project in 2013, which sought to emphasise diversity within the framework of the Arctic Council's Sustainable Development Working Group (SDWG). During the first phase of this project a successful 'Gender Equality in the Arctic: Current Realities, Future Challenges' conference was held in 2014 in Akureyri (Oddsdóttir, Sigurdsson and Svandal, 2015). But such infrequent events – despite building a network of experts and creating a website to promote and expand the dialogue on gender equality in the Arctic – can still all too easily convey the impression that gender is a special interest or niche area rather than an essential mainstream area for research, analysis and understanding (Hoogensen Gjørv, 2017: 296–297). Worse, it can suggest tokenism and 'conspicuous gendering'.

Indeed, at the time of the second AHDR and the Akureyri conference, Lahey, Svensson and Gunnarsson (2014) could argue that while several international organisations had been established to support research, policy analysis, and development in the Circumpolar North, few gender and Indigenous issues had been integrated into substantive economic, social, governance or fiscal policies.

In her statement at the 2014 Akureyri conference, former Finnish president Tarja Halonen highlighted how climate change can hinder the productivity and use of land, which can adversely impact women's land ownership, inheritance, control and management over natural resources. Conversations relating to 'climate change, gender equality, ownership and control rights, and environmental protection' she argued, 'must be closely interlinked' (Ministry for Foreign Affairs, Iceland, 2015: 80).

This followed a report commissioned by the Finnish Ministry for Foreign Affairs seeking to understand how international processes and standards, such as free, prior and informed consent (FPIC), interact with self-determination and the rights of Indigenous women in the context of climate change. The report found that Indigenous women continued to face systemic violations at the intersection of gender, indigeneity and climate change (Prior *et al.*, 2013).

Yet integrating a gender dimension into domestic and foreign policy on climate change can be contentious, as when Canada's Minister of Environment and Climate Change, Catherine McKenna, received a hostile response from a 2016 G7 meeting of environment ministers, when highlighting women's social and economic vulnerability in both natural resource discussions and climate crises (Vanderklippe, 2016).

One under-reported dynamic at the intersection of gender, climate change and extractive industries (Kohut and Prior, 2016) is the movement of women across Arctic borders for the purpose of sex, particularly to remote resource extraction sites. Yttergren (2012) and Sweet (2014) have also highlighted how an increasing interest in resource extraction, resulting from climate change, can heighten the risk of human trafficking in the region. Kohut and Prior (2016) questioned whether these issues would be mitigated if women were more involved in Arctic policy-shaping and decision-making.

In attempting to link intersectional issues within a local-regional-transnational framework, in November 2014 the first ever Indigenous Circumpolar Women's gathering took place in Yellowknife, Northwest Territories, where over 80 Indigenous women shared their knowledge and approaches to programmes and ongoing projects in different regions of the Arctic. In her report *Mind the Gender Gap*, convenor Nina Larsson (2015) further examined how Arctic states approach the inclusion of Indigenous women in decision-making roles, concluding that Scandinavia's approach to gender equality translated into the appreciation of different management styles and a gender diverse workforce, which North American practice appeared to lack (Kohut and Prior, 2016: 338).

Indeed, Rachel Kohut and Tahnee Prior developed a digital platform, 'Plan A', seeking to 'weave together stories' to better inform gender-oriented policy-making across the Arctic. Both were initially part of the Tromsø-Umeå-Arkhangelsk-Rovaniemi-Kingston (TUARK) Network on Gender in the Arctic (Kohut and Prior, 2016: 339). This became Women of the Arctic (WoA), a non-profit association registered in Finland, aiming to raise awareness, support, and maintain a focus on women's and gender-related issues in the Arctic. Building on existing academic efforts, WoA has sought to carve out a non-academic space for women and girls who work on or live in the Arctic to explore the roles and contributions of women to policy-making, research, exploration, art, activism, and daily life (www.genderisnotplanb.com). Non-academic symposia have been held, often as adjuncts to larger conferences, in Helsinki and Reykjavik in 2018, Brussels and Tromsø in 2019. For the first of these, 'Bridging policy, research and lived experience' represented both the theme and philosophy of the network.

Heikkinen *et al.* (2020) have reported on recent advances in intersectionality and gender responsibility in science and technology in the co-creation of sustainable Arctic futures, while Romero-Nieva Santos *et al.* (2020) argue for the greater promotion of research and actions that promote better gender awareness in the formulation of Arctic climate strategies. There is clearly a continuing need for all Arctic states and such policy arenas as the Arctic Council[7] to ensure that recognition of the critical importance of gender, gender awareness and intersectionality is included across Arctic policy and law. The addressing of already marginalised issues in peripheralised regions remains a critical challenge; none more so than in times of rapid change.

1.8 In the next chapter

In Chapter 2, the focus moves specifically onto the role of climate change in transforming the Arctic and its hitherto perceived (multiple) peripheralities. In so doing, the chapter more starkly illuminates the dynamism of the Arctic's global roles.

Notes

[1] For a fuller discussion of definitional approaches see Dodds and Nuttall (2016: 3–8).

[2] It is claimed, for Arctic Norway at least, that the 2008 BBC television documentary *Joanna Lumley in the Land of the Northern Lights*, generated a dramatic rise in numbers of winter visitors and Northern Lights

tour operators (White *et al.*, 2019). A leading UK travel agent in the field lists 112 different available trips (Aurora Zone, 2020).

[3] Although unable to include the portion of Sápmi in the Russian Kola Peninsula.

[4] Covid-19 prevented the March 2020 Games taking place in Yukon.

[5] The University of the Arctic is a co-operative network of universities, colleges, research institutes and other organisations concerned with education and research in and about the Circumpolar North. Its international secretariat is based in Rovaniemi, Finland.

[6] Although her more localised ethnographic work among fellow Sámi is also notable (e.g. Kuokkanen, 2009).

[7] Endorsed by the Arctic Council Sustainable Development Working Group, the Gender Equality in the Arctic programme was about to publish a collective work on gender and diversity (Oddsdóttir *et al.*, 2021) as this book went to press.

2

The Changing Role of the Arctic: Transforming Peripherality

'... research into the Arctic, although often well-meaning, is also often embedded in metropolitan mindsets, which ultimately reinforce existing power structures, priorities and dependencies' (Markussen, 2017: 306–307)

2.1 Introduction: climate as a pivot of change

Following Chapter 1's framing of the Arctic, this chapter discusses the direct and indirect impacts and significance of climate change for the Arctic's changing global role. The Arctic and its peoples' double exposure to climate change and globalisation is examined in relation to notions of sustainable development and to Arctic co-operation and emerging geostrategies responding to these pressures. Climate change transformation is exemplified in the development of Arctic maritime routes and their geostrategic significance.

Increasingly, studies have been identifying the Arctic region as a 'linchpin and laboratory of climate change' (Heininen, Exner-Pirot and Barnes, 2020a: 7) requiring close examination at the local, regional and global scales and at their intersections. But such studies are also confronted with a paradox: the Arctic requirement for environmental protection – in the face of long-range air and water pollution and loss of biodiversity – and climate change mitigation, while economic activities intensify. Further, both dimensions of this paradox are encouraged by actors within and beyond the Arctic, intertwining ambiguities and contradictions.

Change, both climate-induced and economy-related, is the challenge confronting Arctic nations and cultures. But it is change played out within a global context.

2.2 Climate crisis

At no time in known history has the planet experienced a more rapid rise in greenhouse gases or a faster increase in mean surface temperature.

Because ice and snow are major features of the northern environment, small adjustments in temperature across the melting point can bring enormous change. This threshold effect takes place annually as the Arctic experiences its seasonal transition of snow melting and ice break-up. Until recently, this seasonal thaw had been kept in check by vast storehouses of ice that are contained in permanent snow-banks, permafrost, thick multi-year sea ice, glaciers and the Greenland ice sheet (NSIDC, 2020). Now, progressively warmer summers are depleting these legacies and buffers against change (Vincent, 2020: 509). Successive Arctic annual surface air temperatures from 2014 to 2018 surpassed those of any year since 1900 (AMAP, 2019).

DOI: 10.1079/9781789246728.0002

Sea ice volume and extent have decreased persistently over the last few decades, such that the area of multi-year sea ice is now some 60 per cent below that observed in the 1980s (Overland *et al.*, 2018). The total areal extent of sea ice in September has dropped by 45 per cent over the last 30 years, with more than 90 per cent loss in such areas as Hudson Bay (Stroeve and Notz, 2018). Total loss of Arctic summer sea ice will occur within the next few decades, with ice-free conditions extending ever deeper into autumn (Lebrun *et al.*, 2019).

Ice shelves are thinning and collapsing (Hamilton *et al.*, 2017), while glaciers are melting rapidly, albeit with significant regional differences in rate and extent (Noël *et al.*, 2018; Richter-Menge, Jeffries and Osborne, 2018).

A rapidly warming Arctic has the potential to:

- shift from a net carbon sink to a net carbon source (Voigt *et al.*, 2017);
- accelerate warming worldwide;
- increase the frequency and severity of wildfires and severe weather events in other climate zones (Greene and Monger, 2012), and
- both alter and amplify global climate feedback effects (Sommerkorn and Hassol, 2009) (Box 2.1).

Notably, loss of highly reflective (high albedo) snow or ice cover reduces the level of solar energy being reflected back into the atmosphere, while the energy retained acts to heat and melt, and thus further reduce albedo, resulting in a vicious circle of continued thawing and increased warmth. In particular, as reflective sea ice is replaced by heat-absorbent water, localised warming effects are created, compounding those being experienced globally. Localised warming, in its turn, leads to a reduction in land-based ice cover, as well as a further decrease in sea ice, which contributes to global sea-level rise (Steinberg, Tasch and Gerhardt, 2015: 9). Warmer air also holds more water vapour, itself a greenhouse gas, and this further amplifies the warming effect. In this way, the transfer of heat from the increasingly open Arctic Ocean to the atmosphere is likely to be one of the most critically important feedback mechanisms of climate warming (Dai *et al.*, 2019; Vincent, 2020: 508).

Melting of the Greenland and Antarctic ice sheets is of particular concern as they sit on land and are (currently) not displacing their own weight in water (Yumashev *et al.*, 2019). But even a relatively modest rise in global sea-level will submerge such island states as Vanuatu and the Maldives, and will disproportionately impact Asia, the most populous continent and increasingly the engine of the global economy (Islam and Khan, 2018; Welch, 2020).

Arctic permafrost soils contain vast quantities of carbon, which if fully mobilised by complete thawing and decomposition could more than double atmospheric carbon dioxide levels. In many areas of the Circumpolar North, some of this soil carbon is being converted into the more powerful greenhouse gas methane via microbial processes in permafrost-derived lakes, ponds and wetlands (Vincent, Lemay and Allard, 2017). Greater warming is a serious concern for release of this carbon and associated feedback effects (Fischer *et al.*, 2018). Models of greenhouse gas emissions suggest that on a current trajectory northern permafrost lands could be transformed from being a net sink to a net source of carbon beyond 2100, indicating the vital importance of mitigation actions to forestall this critical feedback effect on climate (McGuire *et al.*, 2018). New factors are also likely to accelerate permafrost thawing and methane production, such as warm rainfall events in spring (Neumann *et al.*, 2019).

The intrusion of warm Atlantic currents into the east Arctic was seen as the likely cause of the release of frozen methane hydrates observed on the east Siberian slope of the Laptev Sea in October 2020 that could indicate the triggering of a new climate feedback loop of critical significance (Watts, 2020).

The consequences of accelerated warming have significant development policy and geopolitical implications at all levels, from dog-sled hunting, local development and conservation plans to mineral resource accessibility, shipping routes and safety provisions (Vincent, 2020) (Section 2.6). This is to emphasise that climate change is not just about mechanical changes in air and sea temperatures, melting ice and shrinking glaciers, but about people and other living entities in relationship with both their local and the global ecosystem, with all the geopolitical, economic, cultural, ecological and human rights implications that spring from this (e.g. Huntington *et al.*, 2019).

For example, much of the Circumpolar North's infrastructure was built during the twentieth century at a time when permafrost was considered truly permanent: solid and concrete-like such that it could act as the foundation for homes, major buildings and transport infrastructure. But permafrost is warming, with fastest rates in the Arctic at around +0.39°C over the decade 2007–2016 (Biskaborn *et al.*, 2019). Now, built structures and infrastructure are increasingly at risk (Vincent, Lemay and Allard, 2017), with potentially a third of all infrastructure across the Circumpolar North subject to thaw instability over coming decades (Hjort *et al.*, 2018). Stark evidence of this was exhibited by the collapse of an oil storage tank at a Norilsk power station in May 2020, following 'abnormally' warm weather, with the subsequent escape of more than 20,000t of diesel fuel into two nearby rivers (Shapovalova, 2020).

Total precipitation amounts in the Arctic are increasing and transitioning to rainfall rather than snow (Bintanja and Andry, 2017). This is hastening land erosion and snow melt, resulting in further flooding hazards and permafrost subsidence. Arctic soils containing high concentrations of ice in fine sediments are especially susceptible to thawing and subsidence (Vincent, 2020: 512–513).

Current and likely climate change impacts in the Arctic are summarised in Box 2.1.

Such human and cultural security challenges clearly offer both threats and opportunities, not least in relation to biosecurity and food supply. The prediction of future requirements demands delicate policy decisions based upon well-calibrated science: '... climate change models are notoriously sensitive to assumptions, specifications, and inputs – but in the Arctic there is relatively little room for error. It is a particularly delicately balanced ecological zone' (Welch, 2020: 478). One manifestation and consequence of such vulnerability is the fact that Arctic ecosystems have a lower diversity of fauna and flora than temperate zones; the loss of just a few species can completely disrupt food chains. With continued warming, many species of plants, animals and microbes will be pushed to the upper limit of their thermal tolerances, which will increase their sensitivity to other stressors. These effects are compounded by increasing human presence: new and existing protection zones require strengthening (Vincent, 2020: 516).

Arctic protected areas will experience increasing economic and political pressure if the drive to extract resources and the development of access infrastructure continues to accelerate. In anticipatory response, the World Wide Fund for Nature (WWF) has been planning a 'Last Ice Area' where the diminishing ice may still be in place in 2050 and could provide a refuge for ice-dependent marine species (Fig. 2.1). This area extends across the Canadian Arctic Archipelago to northern Greenland, encompassing several areas already protected, such as Northeast Greenland National Park (the world's largest), Quttinirpaaq National Park in Nunavut, and a planned marine conservation area, Tallurutiup Imanga, at the eastern end of the Northwest Passage (NWP) (WWF, 2018).

The prospect of the central Arctic Ocean opening up to future exploitation has generated an international agreement to prevent unregulated fisheries in this 2.8mn km^2 region. A 2018 moratorium, 'Agreement to Prevent Unregulated High Seas Fisheries in the Central Arctic Ocean', signed in Ilulissat, Greenland[1] (European Union, 2019), was seen as the best precautionary approach to fisheries management here, given that, until recently, ice has perpetually covered the high seas portion of the Central Arctic Ocean, rendering commercial fishing impossible. There have been calls for this international conservation policy to be extended into other shipping activities. With the development of permanent transpolar sea routes by the end of the century (Section 2.6.2, Fig. 2.1) and concerns about oil spillage, noise and pollution, this area could be designated as a 'Particularly Sensitive Sea Area' under international law (Stevenson *et al.*, 2019).

Unsurprisingly, several countries are investing significantly in Arctic climate research, including those in Asia and Europe that lie well outside the Circumpolar North (Vincent, 2020: 508). The International Arctic Science Committee (IASC), the umbrella organisation for coordinating Arctic research, was inaugurated in 1990 with eight participating Arctic nations. By 2015 it was composed of government-nominated delegates from 23 nations, including strong representation from Asia and Europe (Rogne *et al.*, 2015). With five working groups (IASC, 2020a), the IASC now oversees several large-scale initiatives such as the MOSAiC (Multidisciplinary

Box 2.1. Climate change impacts in the Arctic

Immediate local impacts:

- sea ice loss: causing, on the one hand, modifications to hunting traditions, particularly a reduction in the use of dog sleds on pack ice (see Chapter 6); while on the other, rendering fish stocks more accessible (such as Greenland halibut);
- thinning ice threatening traditional migration routes;
- ice-shelf collapse;
- northern coastline erosion and potential loss of coastal villages into the sea caused by permafrost thaw and the greater wave exposure caused by extensive water conditions;
- melting permafrost causing roads to buckle, buildings and other fixed structures to be damaged or to fall;
- shortened ice-road seasons compromising industrial development;
- physical and seasonal extension of agriculture on the region's southern periphery;
- migration northward of certain fauna and flora; warming and the arrival of invasive species may prompt harmful algal blooms in coastal regions, making shellfish dangerous to eat, while some toxins in harmful algae may be passed up the food chain and have direct effects on the health and reproduction of marine mammals.

Global economic significance:

- increasing commercial viability of exploiting sub-sea hydrocarbons and subterranean mineral ore deposits;
- increasing the commercial viability of the Arctic's fishing grounds;
- opening up new inter-continental transport corridors: increasing the number of days each year that Arctic sea routes are open, potentially rendering both extraction and transport of oil and gas (particularly bulk liquid gas, LNG) easier and more attractive, although in a de-carbonising world, demand for these fossil fuel products should, notionally, shrink;
- raising potential environmental threats: to recover from an oil spill or similar disaster much more time would be required for biotic adjustment in the Arctic than in more temperate or tropical latitudes.

For Arctic peoples this represents:

- possible further opportunities for agriculture, fisheries and tourism;
- potential investment in resource extraction and transport bringing the opportunity for employment, improvements to infrastructure and better living standards;
- a threat to accelerate the erosion of traditional cultures and to heighten the danger of environmental catastrophe;
- a modification, at best, to traditional modes of hunting and fishing and associated ways of living: Inuit leader Sheila Watt-Cloutier has argued that the effects of climate change on northerners constitute a violation of international human rights, including the rights of Indigenous people to hunt on the snow and ice;
- an impact on life chances as climate change affects the marine and continental parts of the Arctic in different ways, and within the region differently.

Sources: ACIA, 2004, 2005; Sejersen, 2015; Watt-Cloutier, 2015; Nuttall, 2016; Aksenov *et al.*, 2017; Barry, 2017; Fritz, Vonk and Lantuit, 2017; Roeben and Azubuike, 2020; Shapovalova, 2020; Vincent, 2020; Welch, 2020

Distributed Observatories for the Study of Arctic Connections) expedition, in which researchers from 20 countries spent a year, 2019–2020, in the Arctic sea ice. As T-MOSAiC (Terrestrial Multidisciplinary Distributed Observatories for the Study of Arctic Connections), research findings and land-based developments from the project were extended into 2021 (IASC, 2020b).

Indigenous experience and understanding are a source of knowledge that can prove invaluable for incorporating into climate-related policies; for example, in relation to impacts on marine ecosystems (Falardeau and Bennett, 2020). In supporting national and international calls for closer partnerships between local communities and research programmes,

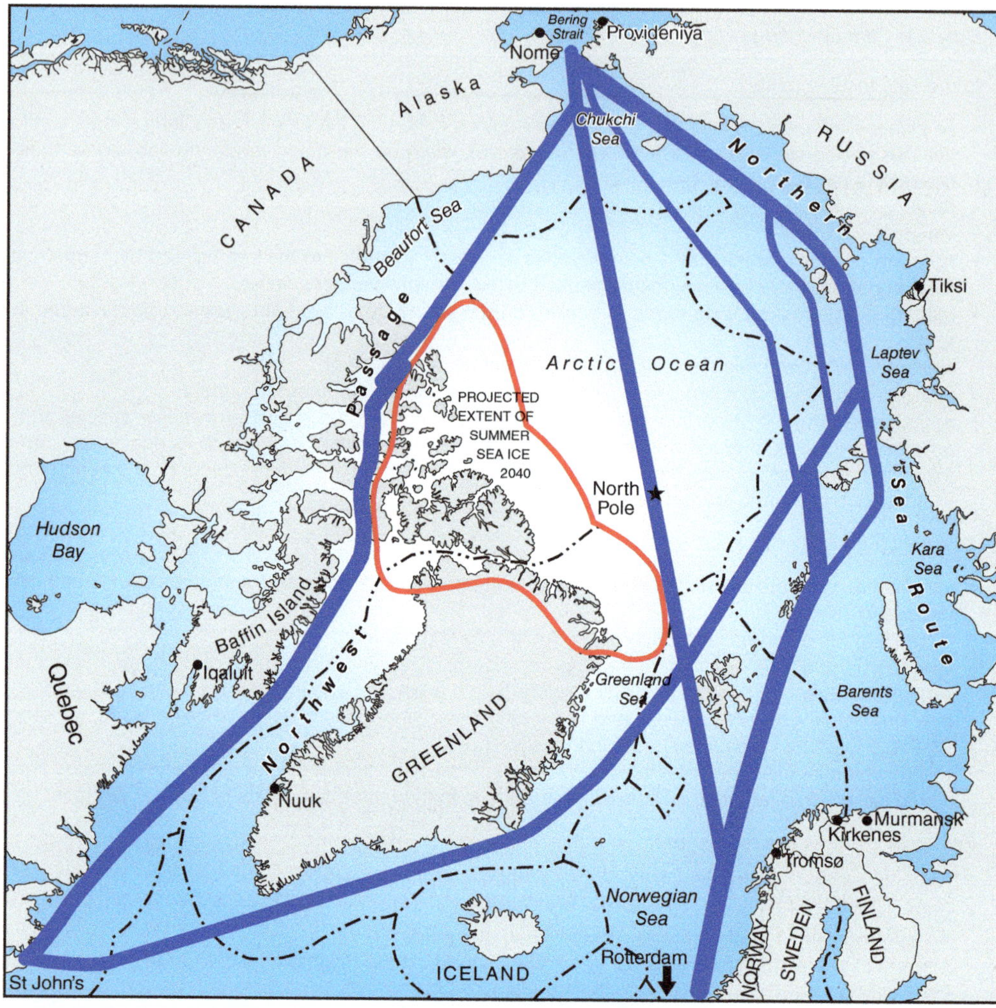

Fig. 2.1. Optimal September Arctic maritime navigation routes, 2040. Author's own figure, modified and redrawn from an amalgamation of Smith and Stephenson, 2013, and WWF, 2018.

including Indigenous-led research, the second Arctic Science Ministerial (ASM)[2] (2018: 3) declared that

> Indigenous Peoples should be involved as appropriate ... in the assessment and definition of Arctic research priorities ... [and that there is] ... the necessity for all States and the European Union conducting research in this region to work together, in collaboration with Arctic Indigenous Peoples and local communities.

The increasing involvement of northern communities in research will assist incorporating Indigenous knowledge into Arctic policies (Vincent, 2020: 521), not least in contributing to an understanding of 'sustainable development'. Roeben and Azubuike (2020) also emphasise that climate change is a human rights issue where the life chances of Indigenous people are particularly affected (e.g. OHCHR, 2020). Yet, while co-operation to mitigate and adapt to the adverse effects of climate change distinguishes much of the strategy and policy documents of Arctic states and Arctic declarations, there is political hesitation, even ambivalence, and certainly a lack of consensus on the part of Arctic states in the actual implementation of climate change mitigation (Heininen,

Exner-Pirot and Barnes, 2020a: 8), as exhibited in Greenland (Chapter 5).

2.3 Double exposure: sustainable development in the face of climate change and globalisation

Much research in the Arctic has focused on assessments of vulnerability and adaptive capacity, both in human and non-human living communities. In a study of Canadian Inuit, vulnerability was conceptualised as a function of exposure to climatic stresses and of the adaptive capacity to cope with such stresses (Ford and Smit, 2004; Sejersen, 2015: 165). Stressors were seen to be derived from a number of factors that included not only climate and loss of biodiversity, but also population pressures, urbanisation, neo-liberal market reforms and political institutions.

From such situations, Leichenko and O'Brien (2008) postulated a 'double exposure' whereby communities, not least Indigenous peoples, are subjected to pressure from both global environmental change and globalisation – the human world is being 'compressed, shrunken, interlinked, connected and subjected to similar systems ... and values' (Sejersen, 2015: 167). This 'double exposure' views global change as being deeply entangled with questions of equity, and considers that these interacting processes are exceeding the capacity of both humans and ecosystems to adapt, thus undermining sustainability. Those who are experiencing the negative consequences of one of these dimensions are also most likely to be experiencing negative qualities of the other, thereby succumbing to their multiple interactions.

While the study of exposures, adaptive capacities, vulnerabilities and resilience may help to identify 'areas, sectors, peoples and communities in special need of assistance' (Sejersen, 2015: 166), the 'double exposure' model can highlight pathways generated by, for example, the processes of climate change accelerating industrialisation in the Arctic which in feedback loops is contributing to, and likely to exacerbate, climate change. With explicit (short-term) winners and losers emerging from these interrelationships, issues of equity and sustainability are foregrounded.

If sustainable development in the Arctic aims at improving the well-being, health and security of Arctic communities and residents while preserving ecosystems' functions and resources (Petrov *et al.*, 2017), strategies need to consider the economic, social and environmental linkages within the Arctic and between the Arctic and other global regions.

While, according to Barnes (2019: 198), sustainable development has become a globally recognised term for 'responsible growth', opinion within the Arctic about what this actually means, and how it should be measured, remains diverse and often complex. Huebert *et al.* (2012: 17) suggest that several policy statements by Arctic states elide 'sustainable development' with the need to maintain 'environmental security' (Sam-Aggrey and Lanteigne, 2020) (Box 2.2). Dalby (2013) also acknowledged the link between sustainable development and environmental security, and recognised the parallel international discussions surrounding resource management, environmental degradation and development that led to the institutionalisation of 'sustainable development' in the United Nations' World Commission on Environment and Development report *Our Common Future* (the Brundtland Report) (WCED, 1987).

But what constitutes 'sustainable development' (or 'responsible growth') can vary at the local, regional and state levels of governance, creating significant environmental as well as socio-cultural policy issues (Government of Canada, 2018; Kikkert and Lackenbauer, 2019). Indeed, the ability of local Indigenous communities to adapt to climate change in their own way and to develop their communities in the way they perceive as being 'sustainable', can be

Box 2.2. Environmental security

In 1995 the World Bank tied the notion of environmental security to the concepts of inter-generational and intra-generational equity by arguing that environmental sustainability is closely connected with both these concepts. It suggested that when the wealthy consume more resources overall, the poor tend to rely on the direct exploitation of natural resources and may have no choice but to engage in unsustainable uses of environmental resources (Barnes, 2019: 198).

compromised by climate change policies and strategies pursued at the state level, thereby emphasising the need to extend effective participation in decision-making processes (Dalby, 2013; Barnes, 2019: 199–200).

While climate change may 'threaten' certain aspects of life in the Arctic, it is possible that a changing relationship with the land is prompting a different perception of what constitutes a threat to the sustainability of the community (Barnes, 2019). Expressed views that economic growth (alone) can provide solutions to the various issues facing Indigenous communities may highlight the limited options available, and a consequent future that could be characterised by economically dependent development.

In this way, climate change can highlight the potentially contradictory juxtapositions of the Arctic resource- and traditional subsistence-based economies, as two of the three main economic sectors (Box 2.3), albeit economies that represent a diverse range of local, regional and national systems of different size and resource geographies.

Interest in the Arctic increased with the release of US Geological Survey reports on offshore oil and gas potential consequent upon greater accessibility with climate change (USGS, 2007, 2008). Subsequent offshore oil and gas exploration projects generated expressions of concern from Arctic residents and concerned campaigns from Arctic communities and environmental groups, resulting in Greenpeace's emotive 'Save the Arctic' initiative.

The formal sector of the Arctic economy has a relatively narrow base of renewable and non-renewable natural resources, the exploitation of which has created significant wealth inequalities (Duhaime, 2004). Resource development frequently requires the import of capital equipment, human and financial resources, entailing substantial economic leakages and potential local social disruption (Huskey, Mäenpää and Pelyasov, 2014), with the introduction of different social norms and practices emphasising such projects' semi-colonial impact.

While large-scale resource projects can be an important engine of growth,[3] they can exert severe environmental impacts, with only limited benefits remaining locally, and extra pressures placed on existing facilities and infrastructure. Where change and uncertainty is brought to communities (Wilson, Hansen and Rowe, 2017), a key element of sustainable development should be ensuring that investments made in research and development provide longer-term benefits for the region's residents, especially its Indigenous peoples.

Demands for high quality, informed and influential public participation in large-scale extractive industries – for minimising environmental risk and social impacts, and for strategies of sustainable development – have been growing. As noted in Section 1.6.3, the UN Declaration on the Rights of Indigenous Peoples (UNDRIP) includes the right to free, prior and informed consent (FPIC), also recognised in the ILO Convention 169 (Wilson, 2016). While only Norway and Denmark have ratified the latter, all Arctic states have ratified UNDRIP. FPIC recognises the rights of Indigenous peoples to make free and informed choices about development projects on their lands, water, and territories. It aims to ensure that they are not coerced or pressured, and that their consent is freely given prior to the start of a project, that they have full information about the scope and impacts of proposed activities, and that their choice to give or withhold consent is respected and upheld (Heinämäki, 2015). But even in countries where consultation is federally mandated, questions remain about how to implement FPIC in practice (Johnson, 2020: 99, 112).

Box 2.3. Main Arctic economic sectors

1. The **formal market-based economy** founded on large-scale resource extraction that accounts for a large share of regional income generated, and related industries that often provide smaller, yet important, contributions to overall income.
2. The **traditional subsistence-based economy**, including small-scale family production, which is commonplace in smaller and more remote settings, and in predominantly Indigenous communities. Based on traditional pursuits such as hunting, trapping, gathering, fishing and crafts, it is increasingly part of a mixed economy with close connections to the local market economy.
3. The **transfer (public) sector**, which includes provision for direct transfers to households, public sector jobs and provision of services generally.

Sources: Poppel, 2006; Petrov, 2016; Larsen and Petrov, 2020

Disputes about the ownership, use, management and conservation of traditional lands and resources are central to the Inuit Circumpolar Council's 'Declaration of Resource Development in Inuit Nunaat' (ICC, 2011), intended to guide policy and decision-making while emphasising the rights of Inuit to be involved in all decisions that affect their lands and waters, and to sustain their hunting and harvesting activities (Johnson, 2020: 100). In Greenland, the settlement of land claims under self-government has meant that some Indigenous communities have entered into resource development projects through joint ventures with industry and government, alongside impact benefit agreements and environmental monitoring projects (Dodds and Nuttall, 2016: 127–128):

> While companies can move on to other projects when reserves are exhausted or if mistakes are made, a community generally only has one chance at development of extractives, and therefore it is of utmost importance to get it right the first time.
>
> (Hansen *et al.*, 2016: 25)

Further research needs to be targeted at improving planning and decision-making processes to include more effectively local stakeholders, and to improve administrative capacity within communities when dealing with extractive industries (Dale, Bay-Larsen and Skorstad, 2018). An understanding of local aspirations is essential for securing local benefits, which are not achieved by means such as job creation alone. Residents are not necessarily interested in working in industry if it compromises traditional activities.

There are several enduring challenges that need to be overcome to ensure more effective and meaningful impact assessment processes for Arctic communities (Hansen and Tejsner, 2016). Not least, oil spills represent a 'low probability, high impact' risk for local ecosystems, and are a continuing concern for coastal communities (McDowell and Ford, 2014), as oil disperses and degrades very slowly at cold temperatures. With a heightened level of accident risk consequent upon increased shipping and cruise tourism activity, the need for improvements in Arctic marine disaster and response policies is vital (Mileski *et al.*, 2018).

Other forms of contamination occurring from industrial activity include seismic testing, drilling, and operation of ships, helicopters and planes, all of which can affect Arctic wildlife profoundly. Underwater noise from a drillship operation off the coast of Greenland was detected at a level that would be audible to whales 38km from the source (Kyhn, Sveegaard and Tougaard, 2014). Noise from underwater seismic testing can travel 4,000km, causing habitat avoidance of fish and marine mammals, as well as hearing impairment and even death in some species of marine invertebrates (Denchak, 2018).

Benefits may not automatically accrue from development projects, but are increasingly negotiated through impact and benefit agreements (IBAs). These facilitate the distribution of benefits from resource extraction between companies, states and local communities, and help contribute to a 'social licence to operate' for energy companies (Tysiachniouk and Petrov, 2018).

Formal sector activities such as ecotourism ventures, commercial fish harvesting, or employment in resource extraction may be combined with traditional or subsistence activities. Climate and environmental change are necessitating longer travel distances to reach hunting grounds and are raising demand for more modern harvesting and transport equipment (Larsen and Petrov, 2020: 87–88). The livelihoods of significant numbers of Indigenous people – and some non-Indigenous residents – continue to depend on harvesting and the use of living terrestrial, marine and freshwater resources (Poppel, 2006). Many of these resources are used as food, clothing and crafts, and make important contributions to the mixed cash economy of local communities and households.

Not all regions are well positioned to nurture strong non-industrial economies, but they can draw on traditional knowledge, and their growth can contribute to moving towards locally sustainable, or at least, less unsustainable, development (Bjørst and Ren, 2015).

2.4 Frontier to fulcrum?

World War II and the Cold War more firmly entrenched the Arctic into states' foreign policies because of its strategic location, and the region became progressively more militarised. Gorbachev's Murmansk Declaration of 1987, calling for a zone of peace (Åtland, 2008: 294) (Chapter 1), brought into focus the critical importance of northern environmental challenges. In the waning years of the Soviet Union these became the focus of efforts to engage the West in initiatives to improve scientific and environmental co-operation, and to establish new political institutions such as the Barents Euro-Arctic Region (Box 2.5) and the Arctic Environmental Protection Strategy (Hønneland, 2010): 'The Murmansk speech set in motion a new normative structure for the post-Cold War Arctic in which states and Indigenous peoples committed to a co-operative and rules-based regional order organised through consensus-based institutions' (Greaves, 2019: 157).

As communist structures in Central and Eastern Europe were dismantled during 1989–1991, the Russian half of the Arctic opened up to international co-operation, and it could be said that '[a]n international region was born' (Exner-Pirot, 2020: 309), 'characterised primarily as an apolitical space of regional governance, functional cooperation, and peaceful co-existence' (Käpylä and Mikkola, 2015: 4), even if the Arctic, in practice, has never ceased to be characterised by sovereign states and the pursuit of their interests.

But it was clear that environmental issues such as climate warming, transnational pollution, marine risks and ocean management demanded Arctic-wide attention as potential threats to human security (Greaves, 2016, 2019), and Indigenous co-operation could assist in establishing a framework.

Arctic Indigenous communities have generally worked within national and regional boundaries, but have also found inspiration and guidance in the work of other Indigenous organisations. The Sámi Council (initially named the Nordic Sámi Council), was established in 1956, and there was Arctic involvement in various Fourth World organisations, such as the World Council of Indigenous Peoples, which the Sámi played a significant part in establishing.

Although Arctic Indigenous peoples may not have greatly interacted with southern organisations within their host countries, they have effectively engaged with other Arctic peoples. This process began with an Arctic Indigenous conference held in Copenhagen in 1973, and has continued with the Inuit Circumpolar Council (ICC), established in 1977 – representing 160,000 Inuit from Canada, Greenland, Russia and the USA – and the Arctic Athabaskan Council (AAC),[4] thereby generating a high level of intra-regional and international interaction (Gerhardt, 2011; Coates and Broderstad, 2020: 17–18).

2.5 The Arctic Council: articulating or mitigating change?

In the face of climate change, environmental monitoring and governance has demanded greater international, Arctic-wide collaboration.

As one collaborative framework evolving out of the 'Murmansk process', the Arctic Environmental Protection Strategy (AEPS), was established in 1991 as a multilateral agency with four working groups:

(i) Arctic Monitoring and Assessment Programme (AMAP);
(ii) Conservation of Arctic Fauna and Flora (CAFF);
(iii) Emergency Prevention, Preparedness and Response (EPPR), and
(iv) Protection of the Arctic Marine Environment (PAME).

Arctic states shared responsibility for each core area with international governmental organisations and NGOs.

Within this remit, research projects were launched whose findings, including dangerously high levels of pollution and reduced life expectancy for Indigenous peoples, stimulated environmental protection groups, human rights activists and politicians to advocate for greater intervention in the Arctic. In response, Indigenous peoples' organisations called for an increased voice and role in Arctic governance. Most notably, the ICC took a leading role in lobbying the Canadian government to form a multilateral Arctic body. With Canadian and US support for transforming the AEPS into a formal institution, the Ottawa Declaration of September 1996 brought the Arctic Council into existence (Wiseman, 2020: 441) as a multilateral forum for promoting mutually beneficial co-operation on Arctic issues (Box 2.4).

Since the Council's founding, two further 'environmental' working groups have been generated – the Arctic Contaminants Action Programme, and the Sustainable Development Working Group – highlighting the importance and range of issues acknowledged as arising from climate change.

In addition to the Council's eight permanent Arctic member states and six permanent Indigenous peoples' organisations (Box 2.4), Observer status has been an area of some contention, being sought by states and international institutions wishing to contribute to the governance of climate change-related issues and to gain a political voice in Arctic economic development processes (Chater, 2016). Although Observers have a minimal role within the existing framework, their presence and co-operation arguably enhances the international legitimacy of the Council (Wiseman, 2020: 443–444). Adopted in 1998 and revised in 2013, the Arctic Council's (2013) *Rules of Procedure* govern conditions for admission as an Observer. These include agreeing to respect both the sovereignty of Arctic countries and the rights of the five Arctic coastal states as specified under the United Nations (1994) Convention on the Law of the Sea (UNCLOS).

Questions are continually raised over the sustained effectiveness of the Council's existing structures in overcoming the evolving complexity of inter-governmental challenges, not least in responding to Arctic climate change. In an attempt to clarify the Arctic Council's evolving position in this endeavour, Kuusama (2020) developed a linear, five-stage model (Box 2.6).

Box 2.4. The Arctic Council

Overall purpose: A high-level forum for promoting mutually beneficial co-operation among Arctic states; a forward-thinking framework for multilateral policy making and governance.

Legal status: Its mandate as an inter-governmental forum facilitates discussion and action on all Arctic affairs except military security issues, and emphasises environmental protection, sustainable development and Indigenous peoples' rights. The Arctic Council does not have the authority to make binding decisions or develop policies.

Composition: Eight permanent member states (Canada, Denmark/Greenland, Iceland, Norway, Russia, Sweden, USA); six permanent Indigenous peoples' organisations (Inuit Circumpolar Council, Sámi Council, Russian Association of Indigenous Peoples of the North, Aleut International Association, Arctic Athabaskan Council, Gwich'in Council International); 13 non-permanent, non-Arctic Observer states (China, France, Germany, India, Italy, Japan, the Netherlands, Poland, Republic of Korea, Singapore, Spain, Switzerland, UK); and various appropriate NGOs and other non-state actors (Box 2.5). Observer status provides a direct window into multilateral Arctic negotiations as well as offering access to the Arctic Council's Members and Permanent Participants.

Role: To facilitate scientific research assessments and binding agreements for the protection and promotion of Arctic peoples, lands, waters, and resources; and to address social and environmental issues in the Arctic through a multilateral mandate.

Development: A permanent Secretariat became operational in 2013, enabling the Council to expand its research capacity and evolve into a policy-making forum able to produce influential scientific assessments relating to environmental, ecological, economic and cultural change in the Arctic.

Impacts: Has advanced policy decisions and brought increased visibility to global issues through an Arctic lens. Ongoing change highlights the need for multilateral co-operation over a range of issues.

In arguably the only international political region of significance to evolve since the Cold War, the Arctic Council 'has not so much blazed a trail as invented and occupied a unique space in international relations' (Brigham *et al.*, 2016: 9). Collaboration has created a high level of trust between the eight Arctic states, a multilateral relationship that has been referred to as 'Arctic Exceptionalism' (Käpylä and Mikkola, 2015).

The question remains over whether a new/separate organisation is required to address the challenges of climate change in the Arctic (Box 2.6).

Sources: Koivurova, 2010; English, 2013; Kankaanpää and Young, 2012; Chater, 2016; Poto and Fornabaio, 2017; Riddel-Dixon, 2017; Schaller, 2019; Heininen, Exner-Pirot and Barnes, 2020a; Exner-Pirot, 2020; Wiseman, 2020

Box 2.5. Main categories of non-state actors in the Arctic

1. Acknowledging the importance of traditional knowledge systems, the Council's structure integrates Indigenous peoples' organisations (IPOs) – the ICC, the Sámi Council and the Russian Association of Indigenous Peoples of the North – all of which held the status of permanent participants (PPs) in the AEPS, and have retained their privileged status (Hossain and Maruyama, 2016). The Aleut International Association (AIA)[5] received PP status in 1998, as did the Arctic Athabaskan Council and the Gwich'in Council International (GCI)[6] in 2000. In this way, the Arctic Council has provided an internationally recognised forum for widely dispersed populations, enabling meaningful engagement and co-operation among a number of Arctic Indigenous groups (Wiseman, 2020: 443). The symbolic importance of a major international organisation that has permanent, circumpolar-wide Indigenous involvement is considerable (Koivurova and Heinämäki, 2006).

2. Scientists: the scientific community has been a participant in Arctic affairs since at least the International Polar Year 1882–1883. For example, according to the Svalbard Treaty of 1920, although Norway has sovereignty over the territory, it cannot be used for military purposes, and citizens of any country can live and work there. Consequently, ten states have been able to establish research stations at Ny-Ålesund, a research town on Svalbard's Spitzbergen Island (Exner-Pirot, 2020: 311). Transnational relations have strengthened as a result of scientific networks, several of which were formalised following the end of the Cold War: the International Arctic Science Committee (IASC) (https://iasc.info/), the International Arctic Social Science Association (IASSA) (https://iassa.org/) and the University of the Arctic (UArctic) (https://www.uarctic.org/about-uarctic/organization/). Co-operation negotiated under the auspices of the Arctic Council[7] includes the 2017 'Agreement on Enhancing International Arctic Scientific Cooperation' (https://iasc.info/ASA) (Tsui, 2019).

3. NGOs, especially environmental ones: have played a prominent role in regional agenda-setting. The large NGOs, such as Greenpeace and WWF, have been significant in shaping outsiders' perceptions of the Arctic. WWF, as an Observer member works with stakeholders to inform and advocate for environmental conservation policies. Greenpeace, not an Observer, pursues high profile, populist campaigns, such as criticising the seal hunt or protesting against oil exploration (Exner-Pirot, 2020: 316).

4. Several northern sub-national polities – territories (Canada), states (US), autonomous constituents (Greenland) and counties (Nordic states) are represented by the Northern Forum, a pan-circumpolar NGO established in 1991 (https://www.northernforum.org/en/). Other sub-regional forums, include the Nordic Council, the Barents Euro-Arctic Council (https://www.barentscooperation.org/en/) and the Pacific Northwest Economic Regions Northern Caucus (PNWER) (Alaska, Yukon and NWT) (http://www.pnwer.org/arctic-caucus.html) (Exner-Pirot, 2020: 315–316).

The Arctic Climate Impact Assessment report (ACIA, 2004) establishing the Arctic as a region vital for studying and highlighting climate change, brought attention to the fact that warming in the Arctic was increasing at twice the rate of any other global region.

The impact of climate change in shifting external conceptions of the Arctic and its ocean, from the image of a hostile environment to a dynamically changing region with economic possibilities, has required stricter governance measures. This need was emphasised in the Arctic Council's assessments of the social and environmental consequences of hydrocarbon exploration and increased shipping in the region, and the implications of climate change for biodiversity (Arctic Council, 2009; CAFF, 2010, 2019).

AMAP was tasked to 'produce information to assist local decision makers and stakeholders in three pilot regions in developing adaptation tools and strategies to better deal with climate change' (AMAP, 2017: 4). These included the Baffin Bay Davis Strait Region (AMAP, 2018), and identified tools to aid local adaptation confronting uncertainties in climate prediction and global carbon emission trajectories (Vincent, 2020: 516).

Since the Arctic Ocean, like other oceans, is subject to UNCLOS, there is a large body of jurisprudence for its regulation. The delineation of extended continental shelf in the Arctic Basin is occurring under its provisions. Several international environmental agreements, such as the Minamata Convention on mercury or the Stockholm

Box 2.6. Kuusama's five-stage model of the growing climate awareness of the Arctic Council

I: Pre-problem stage: Climate change is a global environmental problem (1991–1992)
In the AEPS founding document (1991: 7), the Arctic states committed to international co-operation to ensure protection of the environment and sustainable development. The focus was on trans-boundary human-induced long-range contaminants. Climate change was initially classified as a problem best suited for global environmental governance, although there were signs it would become a significant factor for the Arctic: it was important for AMAP to remain aware of international research.

II: Specific Arctic interest in climate monitoring (1993–1999)
Climate change was now identified as significant for the Arctic region, spawning a call for more specific Arctic monitoring, reinforced by the Intergovernmental Panel on Climate Change (IPCC) confirming the Arctic to be 'very vulnerable to projected climate change and its impacts' (IPCC, 1997: 8). The first comprehensive *State of the Arctic Environment Report on Arctic Pollution Issues* (AMAP, 1997: 169) highlighted 'climate change is likely to be more pronounced in the Arctic than in other areas of the world', and climate change was included in the first Arctic Council Ministerial Declaration, 1998.

III: Arctic climate awakening (2000–2005)
The dramatic effects of climate change were highlighted by the Arctic Climate Impact Assessment (ACIA, 2004, 2005) process. ACIA is the world's most comprehensive regional climatic and ultraviolet radiation assessment to date. This work resulted in an increased visibility of the Arctic Council at international forums.

IV: Climate change as 'game changer' (2006–2017)
The Arctic Council fully incorporated impacts of climate change into its work. Arctic states became more aware of climate warming's opportunities and challenges, and the Arctic Council identified itself in a lead role regarding these. The Council's enlargement to include Asian Observer countries saw the Kiruna Declaration (Arctic Council Secretariat, 2013: 3; Steinberg *et al.*, 2014) alluding to major Arctic and 'Arctic-minded' greenhouse gas-emitting countries, urging them to commit to common climate efforts.

The Iqaluit Declaration (Arctic Council Secretariat, 2015: 4–5), called for global and national action to fight climate change, prior to the Paris Climate Conference. Resilience and adaptation to climate change were stressed. The Arctic Contaminants Action Programme (ACAP) was one result (Arctic Council, 2015).

While the role of the Arctic Council as an environmentally aware standard setter was strengthened, not much changed in relation to addressing the major causes of climate change largely because of vested interests in the hydrocarbons development sector.

V: Time of uncertainty (2017–present)
The Arctic Council celebrated its twentieth anniversary in Fairbanks in 2017, where the Trump administration tried to weaken phrasing of the Declaration, particularly regarding climate change and greenhouse gas emission reduction (Sevunts, 2019). Shortly after, the US withdrew from the Paris Agreement. Climate collaboration became more complicated, but an innovatively delivered Rovaniemi Statement by the Chair indicated that the Arctic Council was not ready to reverse course: '[a] majority of us regarded climate change as a fundamental challenge facing the Arctic and acknowledged the urgent need to take mitigation and adaptation actions and to strengthen resilience' (Arctic Council, 2019: 7).

While climate change is an omnipresent, cross-cutting issue, differing national interests impinge on international co-operation, yet the Arctic Council depends on consensus. It remains to be seen how far the Biden administration is willing to ameliorate the environment he inherited.

Source: Kuusama, 2020: 482–489

Convention on persistent organic pollutants, and the 2017 IMO Polar Code on shipping regulations have been strongly influenced by Arctic efforts (Exner-Pirot, 2020: 314). Such agreements emphasise that states cannot tackle governance of the Arctic Ocean independently.

A tenth anniversary meeting of the Ilulissat Declaration (Box 1.3) in 2018 re-affirmed the geopolitical status quo established ten years earlier which had provided the foundation for responsible and shared management of the Arctic Ocean, and had precluded the requirement for a new international legal regime (Rahbek-Clemmensen and Thomasen, 2018; Viggo Jakobsen, 2019). Nonetheless, increasing economic opportunities in the Arctic could draw the Council's priorities away from its environmental protection focus (Chater, 2016). The importance of the Arctic Council can only increase in coming decades with a heightened prominence of the Arctic in global affairs. Specifically, the region's inhabitants require an active and effective platform for addressing challenges to the region's environmental, social and cultural sustainability (Wiseman, 2020).

2.6 Climate change and geopolitical perturbation in the Arctic

2.6.1 Parallel pathways?

An important, if challenged, hypothesis, promoted by Wilfrid Greaves (2019), in contrast to some of the more optimistic projections, is that the transformation of the Arctic environment wrought by climate change is *undermining* the material basis for co-operation and security in the Arctic at the pan-regional level. If the natural environment provided a shared foundation for Arctic security in the post-Cold War period, then it logically follows, Greaves (2019: 159) claims, that as the environment is changing so too are the conditions for, and dynamics of, regional co-operation and security. Climate warming has transformed the context for human (economic) activity in the region in ways that are encouraging participation by a wider range of actors with distinct, sometimes conflicting, interests. In this way, Greaves (2019: 160) contends, climate change has encouraged a resurgence of geopolitical competition as Arctic and non-Arctic states seek to maximise their interests in the region. At stake – in crude economic terms – are shipping lanes, fisheries, minerals, and an estimated 13–30 per cent of global undiscovered hydrocarbons (Gautier *et al.*, 2009). This has coincided with Arctic states' need to submit claims to their extended continental shelves within ten years of ratifying UNCLOS.

Greaves goes on to draw parallels between Arctic maritime warming processes and evolving Arctic geopolitics. The terms 'Atlantification' and 'Pacification' have been employed to refer to the northward intrusions of warm water, nutrients and fauna from the Arctic's neighbouring oceans (Katz, 2018). More fish and animal species have been sighted at higher latitudes than ever before, taking advantage of milder conditions caused by these large volumes of warm water, a process that will intensify in coming decades. From this observation, Greaves (2019: 160) contends that the Arctic, for long perceived as distinct from the rest of the globe because of its unique environment, 'appears certain to increasingly resemble other ecosystems', and goes on to suggest that this ecological phenomenon is also proceeding geopolitically as Arctic co-operation and security dynamics become transformed as a result of climate change.

While the Arctic's vast natural resource wealth has been identified as a potential source of conflict, the majority of resources are believed to lie in undisputed sovereign territory. Nonetheless, Greaves (2019: 167) argues that the post-Cold War era of pan-Arctic governance may be weakened as issues are negotiated bilaterally, and as Arctic sub-regions become incorporated into adjacent blocs of regional politics with their own inter-governmental institutions. In this way, just as climate change is rendering the Arctic environmentally less distinctive, so, as a region of the world hitherto ascribed as 'apolitical', the Arctic will be drawn closer to the fragmented nature of the world's centres of political influence.

Greaves concludes, however, that any conflict is more likely to be caused by outside effects spilling into the circumpolar region than by overt competition generated from within the Arctic itself. He notes two major interrelated climate change-induced geopolitical trends contributing to an Arctic fragmentation: the Arctic's opening up, notably for shipping and resource exploitation, and the growing role of Asia, led by China, in assuming more influence in Arctic developmental processes. These two trends are addressed below.

2.6.2 Opening up: invigorated navigation routes

Until the 1980s, transits through either the NWP on the North American side to the west of

Greenland, or the Northern Sea Route (NSR) off northern Russia, were perilous and rare due to the preponderance of sea ice. As climate warming has critically reduced this constraint, preparation for, and use of, Arctic shipping routes has intensified (e.g. Choi *et al.*, 2015; Aksenov *et al.*, 2017; Schach and Madlener, 2018).

In attempting to understand how projected geophysical changes in Arctic sea ice would impact on the viability of navigation through the region, Smith and Stephenson (2013) simulated optimal September routes for Polar Class 6 (PC6) ships – those with moderate ice-breaking capability – and common open-water (OW) ships. They found that by 2040 the region's overall navigation potential would increase substantially (Fig. 2.1), with the ability of OW ships to complete voyages along the NSR through a number of optimal routes shifting away from the Russian coast. Unprecedented navigation routes for PC6 vessels would be available through the central Arctic Ocean and NWP by 2040. Such a dramatic northward shift of optimal PC6 routes well away from the NSR, which currently dominates optimal PC6 traffic, to the North Pole (for transits to/from Europe) and the NWP (for transits to/from eastern North America) is a stark indication that, with a predicted

> ... robust PC6 corridor directly over the North Pole ... sea ice will become sufficiently thin and/or diffuse enough to surpass a critical technical threshold – and the shortest possible great circle route will thus become feasible – for ships with moderate ice-breaking capability.
>
> (Smith and Stephenson, 2013: 4872)

Further, by 2040, Smith and Stephenson (2013) calculated that September sea ice conditions in the NWP will have changed sufficiently such that trans-Arctic shipping to/from North America will often be able to capitalise on the 30 per cent distance savings that this route offers over the NSR. Indeed, much has been written about savings that could be gained for OW ships using a shorter Arctic route compared to the existing Panama and Suez canal routes, between Asia and North America (up to 7,500km shorter) and Asia and Europe (up to 7,635km shorter) respectively, with commensurate reductions in fuel consumption and emissions.

But constraints remain considerable, with high environmental and safety regulatory requirements, as enshrined in the IMO (2017) Polar Code, and limited infrastructure available in terms of ports, ice-breaking services and search and rescue capabilities (Struzik, 2018). Most importantly, ice and weather conditions remain unpredictable.

Such factors suggest that in the short- to medium-term Arctic corridors are unlikely to compete with the more established southern routes. Yet the existence of at least seasonally ice-free passage through Arctic waters possesses implicit global economic and strategic significance (Exner-Pirot, 2020: 310).

The potential impacts of new shipping routes and increased vessel traffic on endemic Arctic marine mammal (AMM) species have been largely un-researched despite such creatures' critical social, economic and ecological roles and widely recognised susceptibility to ice loss.

Hauser, Laidre and Stern (2018) developed a vulnerability assessment of 80 sub-populations of seven AMM species to vessel traffic during the ice-free season. Their estimates of level of vulnerability varied considerably both between species and locations. For example, narwhals were estimated to be most vulnerable to vessel impacts, given their high exposure and sensitivity, while polar bears had low vulnerability because of their low exposure and sensitivity. Geographical bottlenecks, such as the Bering Strait and eastern Canadian Arctic, were characterised by levels of two to three times higher vulnerability than more remote regions. These pinch points are necessary route-ways for both vessels and migratory AMMs, and so represent areas of potentially high conflict but also opportunities for conservation-informed planning. Hauser, Laidre and Stern see their research as providing a necessary first step towards developing best practices for maritime sectors moving into this rapidly changing seascape.

How realistic are the navigation predictions? Any transition to improved Arctic maritime accessibility is tempered by the fact that prospects for the NSR are more realistic than for the NWP, as the former is now ice-free for a larger proportion of the year, while the latter is vulnerable to persistent icing at crucial choke points. The NSR also offers a shorter journey for a higher volume of shipping, and has superior infrastructure. But the unpredictability of conditions even on this route is a deterrent to just-in-time oriented container shipping, in that speed uncertainties have knock-on effects for scheduling, fuel efficiency and insurance costs; and further, this route requires shippers to take on expensive Russian pilots (Lasserre, 2018; Welch, 2020: 479).

The most notable example of taking advantage of such emerging opportunities is the 'Polar Silk Road' (PSR), a component of China's 'Belt and Road Initiative' (BRI).[8] Both are intended to reduce China's dependence on the US market and to increase China's global foothold (Kopra and Nojonen, 2020: 15). To these ends, the PSR involves the development of the NSR in co-operation with Russia (NDRC and SOA, 2017; Lim, 2018; Moscato, 2018). A subsidiary of China's largest shipping company COSCO started regular use of this route in 2017, and transport activity is growing: cargo shipping rose to 31.5mn t in 2019 from 7.5mn in 2016, with the number of voyages increasing by 58 per cent (Gunnarsson, 2021). But the overall scale of activity remains small in global terms (Vincent, 2020: 513), and year-round shipping currently only takes place in the western-most part of the NSR.

Nonetheless, as Rong Huang explains in Section 10.4, China is developing a comprehensive national ocean strategy to develop and protect its 'blue economy' (Erokhin, Gao and Zhang, 2018), and clearly desires to be a great sea power (Brady, 2017). In the Arctic, China's special interest in the maritime opportunities has been stimulated with improved access to markets and to Russian energy supplies. Hitherto, apart from domestic coal, China's economy has been heavily dependent upon potentially unpredictable deliveries of Middle Eastern oil (Dubois and Gagaridis, 2018).

These developments are contributing to a reconfiguring of international (i.e. Western) security concerns: Russia being provided with a vast Asian market for its western Siberian gas reserves, and China exploiting a trade route to Europe outside the influence of the USA (Deng, 2018; Vincent, 2020: 514) (see Section 2.6.3). Indeed, China's current impact on Arctic security concerns is closely related to its access to Russian energy supplies (Bertelsen and Gallucci, 2016; Stepien *et al.*, 2020). The cornerstone of this relationship is a $27bn project to ship liquified natural gas from Russia's Yamal Peninsula to China via the NSR: 'China's investment in Russia's Yamal Arctic LNG project ... demonstrates the complementarity and mutual reinforcement between the use of Arctic shipping routes and the development of Arctic resources ...' (Deng, 2018: 59).

The foreign investment entering Russia through this arrangement – in which Chinese corporation Sinopec has a 20 per cent stake – has been critical in mitigating the impact on Russia's economy from Western sanctions imposed following the annexation of Crimea (Staalesen, 2018). The need for refuelling, surveillance, and search and rescue infrastructure to support increased traffic along the NSR has also provided the justification for Russian investments in military infrastructure along its northern coastline (Greaves, 2019: 165–166), while providing the maritime 'hard core' for China's 'PSR' and a strong reason for China's interest in preserving stability in the region.

Similarly, the vital contribution of the Arctic to its economy has led Russia to insist on its peaceful intentions and desired co-operation with its polar neighbours, and this has proceeded, despite relational problems in other spheres.

2.6.3 Partnership and paranoia: Asian participation

As reflected throughout this volume (Chapters 3, 5, 8 and 10), China's role in the Arctic (not least in Greenland), has become both significant and contentious.

China's application for Arctic Council Observer status in 2008 triggered concerns both within and outside the Arctic. Arctic states and the forum's Indigenous representatives, known as Permanent Participants or PPs, were concerned that the introduction of global interests in such areas as resource development, shipping and climate change science, could squeeze out regional priorities of Indigenous self-determination, sustainable development and environmental protection. Some states were alarmed by statements from high-ranking Chinese officials that the Arctic should be considered a common heritage of mankind, beyond the jurisdiction of any one state (Xinhua, 2018).

A number of other Asian countries followed in China's wake and submitted their own applications for Observer status: Japan, Republic of Korea and Singapore – all of whom have a particular interest in shipping and shipbuilding – and India.

After several years of deliberations, all five Asian states, as well as Italy, were accepted as Observers in 2013. The Arctic Council reconciled its concerns by including a stipulation that Observers 'recognise Arctic states' sovereignty, sovereign rights and jurisdiction in the Arctic' and 'respect the values, interests, culture and traditions of Arctic Indigenous peoples and other Arctic inhabitants' (Arctic Council, 2018). On becoming Observer Members, all five Asian states have invested in natural resource extraction in the Russian Far North (Lunde, Yang and Stensdal, 2015), deepening the political and economic connections between the Russian Arctic and the Asia-Pacific region.

Whereas Russia has been an Arctic power for centuries, China – recognising the economic and strategic potential of a warming Arctic – only recently signalled its commitment to developing Arctic capabilities. Beijing has made significant investments in Arctic science, research, co-operation, resource extraction and tourism, and the *China's Arctic Policy* white paper, released in 2018[9] (European Parliament, 2018; Lim, 2018; State Council, 2018; Liu, 2019), declared the country to be a 'near Arctic state'. On this basis Beijing launched the 'PSR' extension of its 'BRI'.[10] Significantly, the white paper was followed by the announcement of a substantial restructuring of state bureaucracy, whereby the Ministry of Land Resources, which had traditionally overseen the country's polar activities, was enlarged and renamed the Ministry of Natural Resources (MNR). Together with the creation of a Polar Navigation Committee these developments signalled a step change in the ongoing process of institutionalising Beijing's planned PSR, developing and integrating the country's polar assets to achieve greater connectivity with the wider BRI framework (Eiterjord, 2018).

In terms of physically facilitating policy in the Arctic, China launched a state-of-the-art ice breaker (*Xuelong 2*) to join a re-purposed heavy ice breaker (Brigham, 2018). The symbolism was clear: China was taking Arctic shipping seriously.

As a consequence, China's Arctic aspirations and actions have attracted much scholarly and policy-making attention and debate, including a special supplement to the *Arctic Yearbook 2018*, as it has established itself as the foremost non-Arctic state active in the circumpolar region (e.g. Brady, 2017; Gad *et al.*, 2018; Kobzeva, 2019). This has been met with concern from other circumpolar states. The Canadian government has invoked national security to prevent Chinese companies from acquiring private corporations, even though Chinese investment could help fund sorely needed infrastructure and natural resource projects (Lackenbauer *et al.*, 2018). Relations were soured further with the Canadian arrest of Huawei's chief finance officer, wanted in the US for allegedly violating trade sanctions against Iran (see section 3.7 for the consequences of this for tourism).

The short-lived Xi–Trump honeymoon of April 2017 resulted in multiple memoranda of understanding (MOU) between US and Chinese oil and gas corporations, including an agreement to purchase Alaskan liquified natural gas (LNG). Subsequently this relationship morphed into a trade war (Feng and Saha, 2018; Persily, 2020).

Within the 'Polar Silk Road' artifice, Chinese interests are involved in a number of Arctic infrastructure projects, highlighting tensions between local and regional infrastructure demands and state-level security concerns (Greaves, 2019: 165–166; Chater, Greaves and Sarson, 2020).[11]

Beijing 's overtly state-level strategy to turn the global pandemic crisis into a geo-economic opportunity (Kopra and Nojonen, 2020: 21) has included accelerating digitalisation in the Arctic, where northern-most areas lack 'reliable, accessible and affordable broadband' (Arctic Council, 2017: 10), with the enhancement of 'connectivity' seen as a regional priority (Finnish Chairmanship, 2017; Lanteigne, 2020). China's 'Digital Silk Road' initiative has seen Chinese investors involved in the Arctic Connect project seeking to connect Europe and Asia,[12] while Huawei had planned to deploy high-speed internet in northern Canada's remote regions (Levinson-King, 2019). As the expansion of 5G was viewed as a critical element of China's economic recovery after Covid-19, it was expected that Beijing would intensify its 'Digital Silk Road' efforts in the Arctic (Blanchette and Hillman, 2020; Kopra and Nojonen, 2020).

Bislev, Gad and Zeuthen (2018) offer a number of useful perspectives on the mutual resource interests of the Arctic and China:

- debates are often about imagining the future and hypothesising motivations rather than about existing and concrete facts on the ground;
- resources that are sought, may be ascribed different values at different times and by different actors;
- such 'resources' may be diverse, ranging from raw materials and finance, know-how, leisure and tourism-related, and human resources, to international recognition, security reinforcement and geostrategic positioning;
- since China and Arctic communities until recently had little direct contact with each other, pre-conceived mutual images may not be well founded, exposing substantial mutual knowledge gaps (this is explored further in Chapter 10 in relation to Visit Greenland's expectations of Chinese 'adventure' tourists' requirements).
- 'Chinese-Arctic co-operation' and interaction is characterised by a variety of actors: diverse Chinese state and non-state actors engaging with the fragmented nature of Arctic governance, which, in most cases, emanates from capitals located outside the Arctic itself – neither 'the Arctic' nor 'China' may be the most appropriate scale for analysing such relationships.

Further, Foley (2018: 99) has argued: '... that political discourse can take on a life of its own, contributing in material ways to perceptions of China as a security threat in the absence of evidence suggesting that these perceptions are warranted.' He cites Johnston's (2013) notion of an 'echo chamber effect' in international relations, a source of information distortion that would have material consequences for how states define threats and set security policy (Foley, 2018: 100).

2.7 In the next chapter

The hitherto general perception of Arctic peripherality and quiet co-operation is being transformed by a warming climate and the possibilities it is opening, in economic, geopolitical and geostrategic terms, to generate some critical perturbations extending beyond the Arctic, while requiring Arctic populations and institutions to adapt to new and uncertain circumstances. In the next chapter, issues of sustainable development and relationships between the local and the global are exemplified in a focus on the nature and role of Arctic tourism.

Notes

[1] The ten signatories were: Canada, China, Denmark (in respect of the Faroe Islands and Greenland), the EU, Iceland, Japan, the Republic of Korea, Norway, Russia and the USA.

[2] Intended to enhance international co-operation in Arctic science, a first meeting of Arctic Science Ministers (ASM1) was convened in Washington DC in September 2016, with science ministers from 25 governments, the EU and representatives from Arctic Indigenous peoples' organisations. A subsequent meeting was held in Berlin (ASM2) (October 2018) (co-hosted by Germany, Finland and the European Commission); while a postponed ASM3, co-hosted by Iceland and Japan, was to take place in Tokyo in May 2021 (Arctic Science Ministerial, 2020).

[3] Between 1900 and 1950 less than 5 per cent of the Arctic was affected by infrastructure development. Some estimates have suggested that by 2050, 50-80 per cent will be disturbed by development (Dodds and Nuttall, 2016: 126).

[4] The Arctic Athabaskan Council is an international treaty organisation representing 45,000 Indigenous peoples of Athabaskan descent spanning 76 communities in Alaska and Canada's Yukon and Northwest Territories, speaking some 23 different languages.

[5] The AIA represents Indigenous peoples of Aleut descent in the USA (15,000) and the Russian Federation (350).

[6] The GCI represents 9,000 Gwich'in in Alaska, the Yukon and Northwest Territories. Gwich'in is one of 47 Athabascan languages.

[7] Specifically through the Council's Task Force for Enhancing Scientific Cooperation in the Arctic (SCTF).

[8] More than 70 countries are involved in the BRI across Eurasia, Africa and the maritime Asia-Pacific, covering more than two-thirds of the global population and over one-third of the world's GDP. But the standard of Chinese companies' corporate social and environmental responsibility in this 'Initiative' is considered 'rather low' (Kopra and Nojonen, 2020: 15).
[9] 'The white paper is not a strategy document, and is more interesting for what it omits, such as the national security dimension that is a major driver of China's Arctic ambitions' (European Parliament, 2018: 1).
[10] Some Chinese scholars suggested that this should have been named 'One Belt, One Road, One Circle' (Huang, 2017).
[11] As noted in Section 8.3, in 2019 Denmark apparently intervened to prevent Chinese companies from securing a contract to (re-)construct international airports in Greenland, again citing national security.
[12] A 13,800km sub-sea cable project led by the Finnish state-owned company Cinia.

3

Arctic Tourism: Sustainability, Resilience and Identity

> '... the crossing of the Arctic Circle takes on ritual proportions, and the border between the Arctic world and the ordinary, everyday world south of that border is filled with symbolic meanings, narratives and histories ...'
>
> (Mathisen, 2017: 70)

3.1 Introduction

This chapter aims to briefly view the nature of Arctic tourism and the research that has been undertaken into it. As such, it acts as a context within which subsequent chapters addressing tourism in Greenland can be placed. However, sustaining the caveats of the last chapter, generalisations about Arctic-wide characteristics and processes can be misleading and should be treated with caution. This is further emphasised by the often fragmentary and/or conflated nature of data available.

Academic interest in Arctic tourism has increased significantly in recent decades, with the Circumpolar North being constructed as an international tourism space. Hitherto, it could be argued that, unlike the Antarctic, the Arctic had been subject to an intensified territorialisation of space – on both land and sea – representing expressions of state power. Such power, through subsequent economic activity, reinforced permanent occupation that would later strongly influence patterns of tourism development as an element of economic diversification and geopolitical repositioning (Gelbman and Timothy, 2010; Horejsova and Paris, 2013). Nonetheless, as Ioannides (2019), among others, has emphasised, the Arctic contains some of the least visited places on the planet, not least because of their often extreme isolation, sparse populations and lack of visitor infrastructure, with often poor and highly expensive air and sea connections, if any.

The precursors of contemporary visitors included 'knapsack adventurers' (Williams, 1859; Snyder, 2007): independent mountaineers, sport anglers and hunters who were sufficiently wealthy to be able to engage in the recreational possibilities offered by 'wilderness, unique cultures, unexplored mountains and seasonally abundant wildlife' (Stonehouse and Snyder, 2010: 26). Indicating a literary interest in such recreation, during the first decade of the nineteenth century a guidebook was published for Scandinavian Arctic travellers (Viken and Jørgensen, 1998). At this time visitors employed local transport and their own two feet, while either camping or calling upon the hospitality of local people. Staying in Arctic homes and sharing local lifestyles, for however short a time, brought new dimensions to travellers' experiences.

Mountain climbing for recreation had become popular by the middle of the nineteenth century (Dent, 1892). Norway was the initial focus of interest, and by the end of the century this had extended to Canada, Alaska, Spitsbergen and Greenland, with activity being publicised by both individual accounts and the journals of newly-founded mountaineering clubs and associations (e.g. Conway, 1897).

DOI: 10.1079/9781789246728.0003

Sport hunting and fishing increased with knowledge of abundant sources of fish such as salmon and trout, and exotic trophy wildlife. Close partnerships often developed between Indigenous people and their recreational visitors, such that by the end of the nineteenth century there had evolved both an economic interdependency and mutual cultural tolerance. The value of Indigenous guides for these activities was widely circulated in new recreational magazines (Stonehouse and Snyder, 2010: 27).

By the 1850s, the Industrial Revolution in North America and Europe had generated a wider wealth base and stimulated further opportunities for travel. Commercial steamship tourism began in Norway in 1850, and by the 1880s Arctic tourism by this means and by other forms of sea and river transport had become extremely popular, to destinations including Norway's fjords and North Cape, Spitsbergen, Greenland, Baffin Bay, Iceland, Alaska's Glacier Bay and the Canadian Yukon. With the aid of on-board explorers and

Table 3.1. Estimates of tourist numbers to the Arctic: c. 2007/8 and c. 2017. Author's own table derived from Maher, 2017: 215–216; OECD, 2020; and the author's additional calculations.

Country/region/province	c. 2007/8	c. 2017	2018
Canada			
Yukon	8,049	255,000	
Northwest Territories	62,045	93,910	
Nunavut	9,323	16,750	
Nunavik (Northern Quebec)	25,000	1,000	
Nunatsiavut (Northern Labrador)	565	19,840	
National total			*21.15mn*
Finland			
Finnish Lapland	2.11mn	2.52mn	
National total			*5.63mn*
Greenland	c. 55,000	80,862	
National total			*c. 100,000***
Iceland	277,800	1.29mn	*2.34mn*
Norway			
Finnmark	2.42mn		
Nord Norge: 3 northernmost counties		1.05mn	
Svalbard	29,813	118,614	
National total			*5.69mn*
Russia	c. 30–40,000 ('A few tens of thousands')	500,000	
National total			*24.55mn*
Sweden			
Norrbotten	1.70mn	2.15mn	
National total			*10.75mn*
USA			
Alaska	1.63mn	2.07mn	
National total			*79.75mn*
Approximate totals	8.4mn	10.2mn	

Note: These 'estimates' should be treated with extreme caution. They are based upon different data collection methods: whole number arrivals and/or departures, and by varying modes of transport; registered overnights/overnights by non-residents/overnights in tourist accommodation; data not necessarily from the appropriate year but within spans of up to five years; and from secondary even tertiary sources. These shortcomings highlight the need for much more precise and comparable data sets upon which to base forecasting, strategic planning and development.

**Combining numbers of land-based visitors with cruise ship 'arrivals', who are unlikely to stay overnight, is no more than a crude indication of visitation levels. This is exemplified for Greenland in Chapter 7.

naturalists, little-known or recently discovered glaciers, bays, wildlife and Indigenous communities attracted much tourist interest (e.g. Collis, 1890; Stonehouse and Snyder, 2010: 28).

Canada established the world's first national park system, and the first in Europe was created in Sweden. Railway systems grew and national awakenings saw governments of Arctic countries encouraging their citizens to visit the natural attractions of the north on the iron roads. Tourism and travel activities were further encouraged by a burgeoning of personal accounts, guidebooks and periodicals devoted to travelling to, and experiencing the benefits of, the great outdoors (e.g. Scidmore, 1885; *National Geographic Society Magazine*). Twentieth-century transport innovations – from snowmobiles and zodiacs to helicopters and cruise liners – allowed visitors to pursue a wider range of both group and individual activities and to gain access to more affordable localities, such that a number of Arctic tourist destinations assumed mass market characteristics. Nonetheless, accounts of 'pioneering' journeys and experiences still found a ready market (e.g. Ritter, 1938/1954/2019), and good travel writing on experiences in the Arctic continues (e.g. Lopez, 1986; Campbell, 2018).

Indeed, prior to Covid-19, increasing volumes of visitors were travelling to the Arctic (Table 3.1), both reflecting and further stimulating a growing awareness of Arctic issues, not least climate change and its associated impacts (Chapter 2). In terms of tourist motivation, the 'threatened polar bear' and other tropes of climate change imagery have generated what some depict as voyages to experience disappearing environments: 'last chance' and 'bucket list' tourism.

As deduced from Table 3.1, tourism activity across the Arctic is, of course, far from evenly distributed, and, with the exception of Iceland, relatively low in terms of its economic contribution to Arctic countries as a whole (Table 3.2). Again, such figures are not necessarily helpful, given the relative areal extents, degrees of economic activity and population sizes of each country's Arctic region. Although tourism in the Arctic is by no means a new phenomenon, the penetration of northern space by infrastructure is in many cases relatively recent. Where accessibility is difficult and such infrastructure costly, public sector involvement is often significant, not least if tourism development, as a policy instrument, is viewed as a potential remedy for local economic and social problems.

While some locations are expensive and difficult to access, others, such as the North Cape (Norway), Kiruna ice hotel (Sweden), Rovaniemi Santa Park (Finland) and the Gulf of Alaska, have become mass tourism destinations (Hall, 2014a; Müller, 2015; Müller and Viken, 2017a), raising issues of carrying capacity and 'overtourism'.

Perhaps understandably, few comparative studies in Arctic tourism development have been undertaken, reflecting the disparities between the territories and the limited range of comparable data available. Work by Maher and colleagues (e.g. Maher, 2007; Maher *et al.*, 2014) is exceptionable in this respect. Table 3.3, albeit now somewhat dated, draws on this; Greenland (Chapters 4–13) and Norway are notable omissions from Maher *et al.*'s compilation. It is within such a context that this chapter addresses the nature and roles of tourism in the Arctic as setting the sectoral context for, and perhaps contrast to, tourism development in Greenland (Chapters 6–13). It addresses the direction of Arctic tourism research, developing Arctic tourism markets and, in the face of climate change, the persisting and contested debate on sustainability.

3.2 Paradoxical dichotomies

The consumption of tourism in the Arctic has been characterised as a paradoxical dichotomy (Lee, Weaver and Prebensen, 2017a: 1), in that the 'experiential core' of Arctic tourism is sought in a context of geographical peripherality (Chapter 1). Given both the fragility and extremes of Arctic environments, compounded by the impact of climate warming, it is not surprising that much Arctic tourism research has been concerned with the environmental, socio-cultural and economic repercussions of the tourism production/consumption nexus.

The challenges and opportunities associated with the Arctic's insularity and peripherality reflect the appeal of remoteness, 'otherness', distance, distinctiveness and scale. Such characteristics, while attractively appealing, can also be the source of constraints that may be difficult to

Table 3.2. Tourism's economic contribution to Arctic countries. Author's own table derived from Árnadóttir, 2019; Lee, 2020; OECD, 2020; Resource Development Council, 2020; and the author's additional calculations.

Country	Indicators of economic contribution
Canada	2000 Can$26.6bn 2019 Can$39.7bn (€24.6bn/£22.7bn/US$30.2bn)
Finland	2.5% of GDP 2017 €4.4bn (£4.1bn/US$5.4bn)
Greenland	DKK500mn (€65mn/£60mn/US$80mn) 11.7% share of the economy Around 750 direct and indirect jobs
Iceland	Tourism is the major export earner 2017 8.6% of GDP, 30,000 employed: 15.7% of total 2018 ISkr 520bn (€3.3bn/£3.0bn/US$4.0bn): 39% of export value
Norway	Total tourism 2017 NOK120.3bn (€11.7bn/£10.8bn/US$14.4bn): 3.6% of GDP Inbound tourism 2018 NOK55.3bn (€5.4bn/£5.0bn/US$7.2bn) 2016 162,400 employed in tourism
Russia	2019: 4.04mn jobs, of whom 1.12mn directly employed Value: Rbl5,443bn (€59.9bn/£55.3bn/US$73.7bn)
Sweden	2018 Tourism SEK337bn (€33.0bn/£30.5bn/US$40.6bn): 2.6% of GDP, 6.4% of export value 172,000 employed (3.4% of total)
USA: Alaska	2018 value: US$4.5bn (€3.7bn/£3.4bn) c.10% employment in tourism.
Global average	10% GDP
OECD countries	4.2% GDP, 21.7% service exports

overcome. Issues such as inaccessibility, inadequate infrastructure, tourism marketing, education, training and support, climatic disruption and extreme seasonality may be common across Arctic 'destinations' (Timothy and Olsen, 2001).

Consequently, Arctic tourism has tended to be associated with sentiments of 'edge of the world' adventure (Gyimóthy and Mykletun, 2004) and unique nature-based experiences. Increasingly, adventure in nature has become a central element of Arctic land-based tourism. Skiing, mountain hiking and snowmobiling (Eckerstorfer, Christiansen and Neumann, 2009) embody the individualistic 'challenging the elements' recreation (Chapter 9) that, as an emblem of Arctic imagery, and as highlighted in Chapter 1, tends to marginalise the human dimension of those who live and work in the Arctic.

With increasing awareness and alarm concerning the climate crisis, there has arisen the second paradoxical spectacle of increased tourism to destinations threatened – and thus perceived more valuable to pursue as a scarce resource – through 'last chance' visitation (Lemelin *et al.*, 2010; Lemelin, Dawson and Stewart, 2012). With its rapidly changing geography, fauna and flora, the Arctic has risen significantly on the list of such 'disappearing' destinations (Grenier and Müller, 2011; Johnston, Viken and Dawson, 2012) (also Box 4.6 in relation to Greenland).

This 'disappearing Arctic' syndrome, attributable to increased development pressures as well as climate change, is clearly paradoxical in the way that the very concern for the environmental vulnerability of the Arctic generates growing numbers of tourists to visit the region, thereby exacerbating those same pressures (Box 3.1). While media reporting often contributes to a sense of urgency by amplifying narratives of imminent disappearance (Piggott-McKellar and McNamara, 2017), it rarely acknowledges contested debates on sustainable development and the (un)sustainability of tourism in fragile physical and cultural environments (Scott *et al.*, 2016).

3.3 Arctic tourism research

Although visits to the Arctic by tourist-outsiders can be traced back to at least 1795 (Guissard

Table 3.3. Comparative characteristics and issues in Arctic tourism. Author's own table derived from Maher *et al.*, 2014 and the author's additions and interpretations.

Arctic country/territory	Main tourism characteristics	Major tourism challenges
Alaska	1. Offers a rare non-extractive economic option	1. Cruise ship pollution and onshore disruption
	2. Economically valuable but threats of 'overtourism' in specific areas	2. Non-resident hunters infringing Indigenous territory
	3. As a tool for economic development in far north communities	3. Lack of infrastructure in the far north, including search and rescue capacities for ship-based tourism
	4. Marine access to remote coastal communities is expected to improve as Arctic sea ice shrinks	4. Terrestrial access has been predicted to *decrease* as permafrost thaws
	5. Focusing more on non-wildlife types of tourism, such as historical and cultural may be important for the future of the industry	5. Retreating glaciers and shifting ranges of wildlife are causing visitors and tourism operators to modify their activity patterns to secure viewing opportunities
		6. High costs of getting to Alaska – global fuel prices could play a large role in future trends, highlighting how the fortunes of Indigenous entrepreneurs are tied to the health of the global economy in an unequal dependency relationship
Arctic Canada	1. Rapid social, economic, and environmental change as a result of climate change and the growth of northern governments and institutions	1. Strategies for natural environmental and Indigenous cultural protection need stronger enforcement
	2. Tourism numbers are uneven across the region	2. Inadequate transport infrastructure, scarcity of local products, a lack of skilled labour, and insufficient marketing resources
	3. Growth of expedition cruise tourism	3. Negative cultural and environmental impacts of cruise tourism: onshore carrying capacity excesses, the sale of marine mammal parts for souvenirs, and increased refuse. Guides need to act as educators
	4. More empirical research into the impacts of climate change is required	4. Cost structures and taxes render destinations expensive
Arctic Finland	1. Nature tourism important since the 1980s, also snowmobile excursions to reindeer or sled dog farms, wellness centres, Santa Park and ice hotel	1. Winter activities not seen as adventurous
	2. Hierarchical categorisation of destinations	2. Seasonality issues with the need to provide more summer attractions
	3. High quality	3. Limited accessibility from the south
Iceland	1. Natural environment attractions: Þingvellir waterfalls, Geysir hot springs system; volcano climbing	1. Tourism policy 2. Social and environmental impacts 3. Infrastructure, and 4. Research, education and training. 5. Reflecting the above, tourism activity was seen to be constricted in both spatial and temporal terms

Continued

Table 3.3. Continued.

Arctic country/territory	Main tourism characteristics	Major tourism challenges
	2. Culture–nature combinations: Althing, site of first parliament, 930, in a rift valley; Blue Lagoon open air thermal baths; Iceland pony riding	Since Maher *et al.*'s research, tourism activity has expanded significantly with both seasonal and geographical extensions, exploiting the international cruise market and trans-Atlantic flight stop-overs. Over-tourism in the 'Golden Circle' excursion itinerary from Reykjavik is notable
Arctic Russia	1. Limited number of different attractions: hunting and fishing, ethnographic tours, adventure tourism	1. Variable quality with questionable health and safety regimes
	2. More expensive yacht excursions, helicopter and fixed-wing flightseeing	2. Increasing pressures on fragile environments and infrastructure are inadequately monitored
Arctic Sweden (Norrbotten (Swedish Lapland) and Västerbotten)	1. 34% of the area of Sweden, with only 5.5% of the Swedish population	1. Potential conflicts of interests concerning property rights, tourism and the right of public access to nature
	2. Nature-based tourism is the main attraction with a high concentration of ecotourism companies. But also ski resorts, Sámi culture with reindeer herding, and the ice hotel	2. Conflicts between tourism development and forestry, mining and reindeer herding
	3. Good transport accessibility	

and Lee, 2017), compared to other disciplines, concerted research into Arctic tourism is a relatively recent pursuit, appearing in the mid-1990s (Maher, 2007), and given impetus by designation of the International Polar Year (IPY) 2007–8. Although tourism was barely included in the large IPY research projects of the time, increasing arguments for more and better Arctic tourism research were voiced. Lee, Weaver and Prebensen (2017a: 2) offered two reasons for focusing on the Arctic:

- with notable exceptions (Mason, 1994, 1997; Mason *et al.*, 2000), until quite recently research and discussion of Arctic tourism tended to be subsumed as an element of a broader polar tourism construct that often paid more attention to the Antarctic (Hall and Johnston, 1995; Stewart *et al.*, 2005; UNEP, 2007): the context for both is, of course, very different;
- the production of a more systematic knowledge base on Arctic tourism was needed: a database that would track social changes brought about by the expansion of Arctic tourism particularly resulting from climate change and its impacts (Fay and Karlsdóttir, 2011) (subsequently pursued in the EU Interreg Nord cross-border co-operation projects in the high north of Scandinavia: https://interreg.eu/programme/interreg-sweden-finland-norway/).

Grenier and Müller (2011; also Stewart and Liggett, 2018: 363–364) identified a complex network of factors – several of which have been outlined in previous chapters – that coalesced to stimulate the maturing of Arctic tourism research:

- political shifts, such as the ending of the Cold War and a new impetus to develop Arctic resource economies;
- emancipation and empowerment of Indigenous peoples;
- emergence of the Arctic as an entity for governance and jurisdiction;
- a strengthening of environmental movements;
- increasing attention paid to climate change issues reinforced by the fear of losing species and landscapes;

Box 3.1. The ethical paradox of 'last chance' tourism

'Last chance' tourism (also known as LCT) has generated research into travel to destinations where (usually iconic) features, species or landscapes, such as glaciers and polar bears, are perceived to be at risk of disappearing, often as the result of climate change (e.g. Dawson *et al.*, 2010; Welling, Árnason and Ólafsdóttir, 2015; Groulx *et al.*, 2016; Lemieux *et al.*, 2017). Such destinations are not wholly confined to remote areas, but in practice are mostly in this category, as visitors seek to experience 'before it's too late' the decline/disappearance of pristine nature and/or of 'traditional' Indigenous groups, their values and belief systems (e.g. Walker and Moscardo, 2016).

A second type of LCT, although not necessarily recognised as such, is the individual 'last chance' opportunities of older or less-abled people who may face limited or decreasing opportunities to travel. Some aspects of this dimension may be seen as 'bucket list' tourism – things to do before you die (e.g. Bosangit, Hibbert and McCabe, 2015) or to impress social media friends with (Thurnell-Read, 2017). Arguably, the polar cruise tourism sector has benefited from this dimension (e.g. Eijgelaar, Thaper and Peeters, 2010), given the ageing demographic structure of the economically more developed world (although the age profile of cruise tourists has been falling in recent years: Mariani *et al.*, 2014; Domenèch and Gutiérrez, 2019).

Both forms of LCT represent the acquisition of meaningful experiences through travel as a central element through which identity and selfhood (selfie-hood?) is formulated, performed and mediated: a process of 'experientialisation' – travel experience as fashion accessory – that has been identified as one of the central trends in contemporary tourism (Wang and Alasuutari, 2017).

Thus LCT markets appear to be established around an ethical paradox (Dawson *et al.*, 2011) in that 'last chance' visitors appear to convey a lack of awareness or an ability to self-justify the environmental damage caused by their own consumptive travel behaviour. That such a disconnect exists between valuing a place and contributing to the climate (and development) impacts that threaten that place, suggests a weak ethic of care within the LCT marketplace (Groulx *et al.*, 2019). There remains the need for more nuanced understandings of how last chance tourists' values and motivations are underpinned by self-interest, in order to inform policies and messages to encourage less unsustainable behaviour (Hindley and Font, 2018).

Although LCT has gained popularity as a contemporary travel trend, it reflects a longer history of viewing contemporary consumption as a mutual source of harm (Harrison, Newholm and Shaw, 2005), and is just one part of the broader range of ethical dilemmas arising from multiple forms of tourism consumption on an increasingly threatened planet (e.g. Hall, 2014b). Although studies on LCT have evolved, the tensions and ethical conflicts between its promotion and the balance of benefits derived from it, continue (Lemelin and Whipp, 2019), and calls to better appreciate host community perspectives have gained support as a result (e.g. Schweinsberg, Wearing, and Lai, 2020).

- changes in tourism demand, with tourists' tendency to be more active, seeking out alternative and remote destinations, while avoiding politically sensitive regions;
- a proliferation of online technologies – blogs and online diaries by Arctic tourists themselves, media platforms and the ability to access satellite tracking data on vessels in Arctic waters.

In the Arctic, discussion of a research agenda for tourism began with the work of Stewart, Draper and Johnston (2005). Creation of the International Polar Tourism Research Network (IPTRN), embracing both the Arctic and Antarctic, emerged from the 2006 Canadian Association of Geographers meeting at Lakehead University, Thunder Bay. The need for inter-disciplinary and inter-institutional collaboration was highlighted, and following a further planning meeting in Rovaniemi, Finland, in 2007, the network was formalised. The first IPTRN conference and community workshop was held in Kangiqsujuaq, Nunavik, in 2008. Since then, each biennial conference has expanded the Network's focus. Some conferences have had a greater industry presence, while others have embraced more community representation. These IPTRN conferences have led to joint research projects and publications (e.g. Stewart, Liggett and Dawson, 2017; Chimirri, 2020). The sixth IPTRN conference, convened Whitehorse and Dawson City, Yukon, Canada in June 2018, focused on heritage, culture, environment and community economic development as they intersect with tourism.

At much the same time, the University of the Arctic (UArctic), headquartered in Rovaniemi and established as an international bridge for teaching and research in 2008, developed a Thematic Network on Northern Tourism. It now has more than 20 partners, with a joint masters curriculum (Maher, 2017: 217). The university's objectives include: connecting traditional and Indigenous knowledge systems with contemporary academia; partnering with northern communities and Indigenous populations to respond to their educational aspirations and requirements; and broadening the voice of the North in the world (UArctic, 2020).

Complementing these developments, in their review of (mostly English-language) tourism research published in peer-reviewed journals between 1980 and 2016, Stewart, Liggett and Dawson (2017) identified eight overlapping 'key themes' which, they claimed, assisted in characterising the focal areas for polar tourism researchers and aided an understanding of how those areas of focus might have changed and evolved (Box 3.2). Unsurprisingly, they found that, according to their categorisation, 'tourism development' and the 'management of tourism' dominated the literature in representing 62 per cent of the total peer-reviewed journal publications. This replicated

Box 3.2. Identified themes of published Arctic-related tourism research

1. Development: an increasing focus on the Arctic (as compared to an early predominance of interest in the Antarctic) coincided with the growth of sustainability discourses in the 1990s.
Sub-themes: Greenland; Svalbard; LCT; sustainability, ethics and nature conservation; notions of place – protected areas; wilderness; rural locations; nature-based activities.

2. Management: while earlier dominated by the development of guidelines and codes of conduct and the assessment of management models, increasing attention was being paid to the management of growing numbers of cruise ships and their passengers. Marketing and branding could be added as a sub-theme linking this and the previous theme.

3. (Tourist) experience:* place-related experiences; experience 'consumption' and wildlife viewing.

4. Global change:* a theme to have emerged since 2006. Tourism–climate feedback loops would appear to be an implicit area of further fruitful investigation here.
Sub-themes: environmental costs of tourism; perceptions and effects of climate change; implications of and for cruise tourism. A later addition could be: responding to the Chinese outbound market.

5. Governance:* seen as an expanding area of research interest.
Sub-themes: appropriateness and efficacy of governance structures for cruise ship tourism; regulatory frameworks required for safe and responsible visits; requirements of local communities.

6. Impacts: as research has become more sophisticated, so impact-related publication has fragmented into various sub-themes subsumed within other focus areas.
Sub-themes: wildlife related: economic; 'biological'; ways of life dependent upon it.

7. Community
Sub-themes: relationship between Arctic tourism and resident perspectives; involvement of 'host' community representatives in research; subsequent work on indigeneity merits inclusion here, and relates this theme directly back to 'governance'.

8. Research reviews
Sub-themes: taking stock of Arctic tourism research; gateways; biodiversity.
More recent work to be added here could include the selective collection of reviews celebrating 20 years of the *Scandinavian Journal of Hospitality and Tourism* (e.g. Hall and Saarinen, 2021; Ren *et al.*, 2021).

Note: * indicates themes regarded by Stewart and Liggett (2018) as having become critical areas of research activity.

Sources: Stewart, Draper and Johnston, 2005; Stewart, Liggett and Dawson, 2017; Stewart and Liggett, 2018; author's additions

a predominance identified some 12 years earlier (Stewart, Draper and Johnston, 2005).

That earlier review had identified two areas that required further attention: (i) the nature of tourist experiences, expectations, knowledge and satisfaction; and (ii) global trends and large-scale influences such as climate change. Box 3.2 incorporates the – not necessarily coherent – themes and sub-themes of these reviews, with additions, and with specific reference to the Arctic.

Stewart and Liggett (2018) highlight the importance of a 'global change'–'cruise ship tourism' nexus reflecting important issues related to accessibility and safety (as well as dependency issues – see Chapter 11 in relation to Greenland). Changes in sea ice extent and thickness, greatly influenced by atmospheric and hydrologic conditions, have obvious implications for landing sites and navigational safety. These continue to raise questions regarding the sufficiency of existing governance structures for Arctic cruise tourism, not least in relation to high levels of cruise ship emissions of sulphur dioxide (further taken up in Chapter 11). And, of course, such emissions contribute to climate warming which is permitting the increasing accessibility of ships in Arctic waters. Such feedback inter-relationships emphasise the need for tourism research perspectives to adopt a more holistic approach to climate change issues.

While Lee (2020: 106) has reasonably argued that 'Research outputs on Arctic tourism ... to date might be assessed as being rather haphazard in terms of areas of emphasis as well as disciplinary sources that are drawn from', he might have added that research has often lacked a critical edge in addressing the climate crisis.

Interestingly, Weir (2017) contends that in much of the literature, contemporary climate change is often presented as a new phenomenon, one creating new issues and challenges for researchers and for the tourism sector. Yet, he argues, climate change in some form has been a constant presence through the pre-history and history of humanity, and that 'confusion' about climate change being a 'new' issue, is perhaps the result of focusing only on the anthropogenic dimension. Weir goes on to suggest, therefore, that current research (and, one assumes practice) could benefit from examining past experience in (tourism) adaptation to climate change.

Understanding the impacts of climate change (Hall and Higham, 2005; Gren and Huijbens, 2014) and developing effective mitigation (e.g. Scott *et al.*, 2008) and adaptation (e.g. Dubois and Ceron, 2006; Becken and Hays, 2007, 2012) measures, and their policy outcomes (Gössling *et al.*, 2010, 2012; Scott, Hall and Gössling, 2012; Saarinen, Rogerson and Hall, 2017), have been subjects of tourism research for some time. But the balance of effort between investigating impacts of climate change on tourism and research into mitigating tourism's contribution to anthropogenic climate change has not been comparable (e.g. Hall, 1999a; Peeters, 2007).

Recognising the complex relationships between the cultural, economic and environmental realms, Kaján (2014a) therefore argued that increased collaboration between climate and social scientists was required for a more holistic comprehension, to better inform policy for minimising negative impacts and identifying potential opportunities. In the context of Arctic tourism, the overall negative impacts of climate change have been viewed, somewhat superficially, to be balanced by possible opportunities, such as easier access for cruise ships (notwithstanding the contribution of cruise ships to greenhouse gas emissions and climate feedback effects). On the other hand, caution has been advised by Stewart *et al.* (2007) who pointed out that the warming effect alters the distribution and character of sea ice and may actually have negative implications for cruise tourism in the region by generating more icebergs, at least in the short term.

3.4 Sustainable tourism goals: a Nordic governance perspective

Although sustainability – both the nature and development of 'sustainable tourism' and the actual sustainability of tourism in Arctic circumstances – may not appear explicitly in many of the sub-themes in Box 3.2, it is nonetheless implicit in much Arctic tourism research and practice.

Political acknowledgement of the importance of sustainability issues in the North included a 'special emphasis' being placed on sustainable tourism during the Icelandic presidency of the Nordic Council of Ministers in 2019. An accompanying paper on tourism policy analysis, as one of a series of publications addressing environmental, tourism and tourism-related issues, highlighted 'Nordic' political thinking, with observations on, and a wish-list for, tourism sustainability within the pursuit of sustainable development goals (Árnadóttir, 2019). While this paper was not aimed specifically at the Arctic regions of the Nordic countries, it embraced policies and practices to be applied there.

In the paper, sustainability is claimed to be a central issue in Nordic tourism, and 'environmental integrity' is seen as a key component for destination attractiveness, regional branding and the overall economic viability of communities and regions (Øian *et al.*, 2018). Such integrity embraces both nature and culture, and Nordic destinations are encouraged to aspire to direct their marketing to selected 'high yield–low impact' groups and sway away from overcrowding and mass tourism. This should reinforce Nordic imagery of being a destination for unspoilt nature experiences.

As an increasingly scarce global resource, the value of nature experiences in Nordic destinations is assumed to increase.

All Nordic countries except Denmark emphasise sustainability in their national tourism policies, and all are committed to the 2015 Paris Agreement and its targets on combating climate change. All have incorporated the UN Sustainable Development Goals (SDGs) Agenda 2030 into national policy objectives, and a common Nordic Strategy for Sustainable Development is in place until 2025 (Nordic Council of Ministers, 2013, 2017).

Nevertheless, despite being among world leaders in being closest to achieving the UN SDGs by 2030 (Larsen and Alslund-Lantén, 2017), Nordic countries stand out as overconsumers of natural resources and substantial producers of waste (Bauer, Watson and Gylling, 2018), and meeting UN SDG12 ('Responsible Production and Consumption') targets (United Nations, 2015) is challenging.

Further, despite the importance placed upon sustainability at the national and Nordic levels, it is suggested that there continues to be a lack of co-ordination, objectives remain unclear and monitoring tools are often not in place (Bauer, Watson and Gylling, 2018). The challenges posed by sustainability need to be a driver for better Nordic co-operation, particularly in the areas of shared information, practices and technology.

Until the Covid-19 pandemic, tourism's carbon footprint had grown rapidly, estimated at about 8 per cent of global greenhouse emissions, four times greater than previously thought (Lenzen *et al.*, 2018). Unsurprisingly, the World Tourism Organization's estimate of carbon emissions from tourism is lower, at 5 per cent of global emissions, of which 75 per cent is said to originate from the transport sector (UNWTO, 2018).

Has Covid-19 helped to better consider such questions as the degree of beneficial effect of tourism for the planet? And, particularly critical for the more isolated and globally peripheral Arctic destinations, in respect of the UN SDGs and commitments to the Paris Agreement, in a post-Covid world, is it morally justifiable to continue encouraging tourism growth, and, especially, to be targeting long-haul markets and destinations?

In Bauer, Watson and Gylling's (2018) critique of Nordic progress in this regard, they argue that a more holistic approach to the development of tourism as a positive force in economic, environmental and social development is required. They emphasise the need to develop and implement tools to monitor tourism's sustainable development impacts in creating jobs and promoting local culture and products.

Nordic countries have taken different measures to tackle the environmental challenges of tourism. While Norway and Sweden have implemented aviation taxes intended to reduce emissions, Denmark has been opposed to such measures, which it fears will undermine competitiveness (Bauer, Watson and Gylling, 2018: 35). Surprisingly, the Nordic countries have *not* been notably proactive in terms of innovation and technology application in the tourism sector, and Nordic ministers have called for new networks of knowledge-sharing and best practice to be developed. Yet there have been exceptions.

The development and implementation of tools to monitor the impact of tourism to assist sustainable destination management includes an ambitious project initiated by the Icelandic Ministry for Industries and Innovation and led by the Icelandic Tourism Task Force, on developing impact indicators for Icelandic tourism (EFLA Verkfrædistofa, 2018), taking into account economic, environmental and social parameters. Iceland has also been working towards developing a tourism carrying capacity assessment methodology (Árnadóttir, 2019: 53).

In Norway, the development of a sustainable destination standard 'road map' led by Innovasjon Norge (2017) and key industry organisations (De Historiske *et al.*, 2017) incorporates criteria and indicators that can be measured, registered and monitored, covering aspects of nature, culture, environment, social values, community involvement and economic viability, with a long-term view towards encouraging 'sustainable destinations' (Hillgrén *et al.*, 2016). It sets out to define what characterises sustainable travel and tourism and to identify the challenges involved. A key objective is to provide direction towards sustainable travel and tourism by 2050. The 'road map', which is part of Norwegian government policy, could set an example – for collaboration between public and private actors, technology developers and investors – of how to tackle issues involved in working towards sustainability in Nordic Arctic tourism.

3.5 Tourism, sustainability and the local voice

As noted in Chapter 1, perceptions and representations of the Arctic that reify 'pristine' nature can obscure and even erase the lives and livelihoods of Arctic Indigenous peoples by privileging a particular, largely exogenous version of nature, along with its attendant socio-cultural rules, norms and structures. In the tourism context this is perpetuated in visions of the Arctic as a spatially remote 'experience frontier' and/or 'wilderness', as a place to be travelled *across*, and consumed temporarily, rather than being inhabited. This suggests that tourism to and in the Arctic needs to more firmly embrace an understanding and a demonstration of the wider influences of culture and history (Stewart, Draper and Dawson, 2011).

Arctic tourism is situated within fields of power that, historically, operate to favour certain knowledges, practices and values, while disenfranchising others, and 'sustainable' tourism development in the Arctic needs to embrace and complement the views and aspirations of local people, alongside environmental considerations (Johnston, 2011): a more inclusive approach is required with respect to the plurality of voices, values and knowledge systems layering the terrain (Grimwood, 2015).

Further, while Arctic tourism literature tends to emphasise the importance of enhanced intergovernmental and inter-institutional co-operation for co-ordinated and effective governance across the Circumpolar North (Maher, Stewart and Lück, 2011) (see Chapter 2), improved co-operation at interactional scales between tourists, local tourism promoters and local inhabitants is also required in the light of well documented 'incompatibilities' between tourism and Indigenous communities (e.g. Hinch, 1998; Buckley, 2005; Johnston, 2006; Stewart, Draper and Dawson, 2011).

Two decades ago Franklin and Crang (2001) argued that much tourism research addressing 'nature' was 'uncritically confused' and/or conflated the term with 'environment'. Such obfuscations mirror those tourism-related perceptions and promotions that situate nature as something 'pure', distinct from, and external to, social and/or cultural realities (Castree, 2001). 'Empty wilderness' is perhaps the most extreme expression of this Cartesian division between nature and society in that it has been employed to justify the appropriation of Indigenous lands for preservation and/or 'natural resource' exploitation, disregarding human ecological connectivity (Grimwood, 2015). And 'nature' remains a contested political terrain, both spatially and ideologically (e.g. Franklin, 2004; Olafsdóttir, 2013).

Being both an exotic enterprise of escape and an everyday performance that embraces relations with other actors, mobilities and objects (Grimwood, 2015), tourism can help demystify the Arctic as a 'primitive' and authentic periphery, and engage residents as active participants rather than passive recipients of the tourism machine (Stewart, Draper and Dawson, 2011), or subjectivities (e.g. 'host community') determined by visiting researchers. Thus conceived, Arctic tourism can be located within broader spheres of human experience and social science.

For example, food tourism has received increasing attention, intertwining experiences of environment and its products (ingredients for local food production). Consuming food in the Arctic, while part of the quotidian for local people, may be presented as exotic for tourists and set within storytelling that can enhance the consumption experience (de la Barre and Brouder, 2013), especially when it involves local food that is perceived as disagreeable (such as decomposing whale blubber: Section 6.5). The promotion of food in Arctic tourism as a central consumption experience is one of those dimensions that can help 'demistify' Indigenous culture and the quotidian.

3.6 'Indigenous tourism'

Indigenous tourism, a perhaps dated concept, has been defined as a tourism activity in which Indigenous people are directly involved through agency and/or by having their culture serve as the essential element of attraction (UNBC, 2020).

From Britton (1982) to contemporary evaluations (Linehan, Clark and Xie, 2020), critiques of tourism in 'less developed' and minority population destinations have maintained that tourism can perpetuate (and encourage?) colonial attitudes and power structures such that tourism development processes in such areas may remain in the hands of incomer entrepreneurs and/or non-Indigenous national and regional destination marketing organisations (DMOs), generating exogenous images of indigeneity (Viken and Müller, 2017).

Indeed, part of the criticism of writings about Indigenous peoples is that, even in the era of post-colonialism, authors may appear to observe and analyse through colonial lenses, and debates on the need for a real de-colonisation of academic writing, and for a more prominent place for alternative voices, continue (Hall, 2013; Viken, 2013; Chambers and Buzinde, 2015; Linehan, Clark and Xie, 2020).

The Arctic has been socially constructed and influenced by contemporary discourses in which indigeneity has had varying positions, including a fuzziness and randomness of Indigenous self-categorisation (Section 1.6), Arctic areas being presented and represented as exotic and 'Othered' (Müller and Viken, 2017a). Keskitalo (2017) asks whether the 'aboriginalisation' which tourism may enhance, is a blessing or a curse. However, '... northern native people are the best mediators between southern visitors and their homeland – not a last frontier, not an untouched wilderness, but a homeland which has shaped and sustained their communities, their culture, and their economy' (Notzke, 1999: 60).

Indigenous people are far from being automatically passive victims of such processes: they may actively participate to gain economic benefit (Tuulentie, 2006; Cole 2007), and 'Otherness' can be a unique selling point.[1] In the Arctic, many Indigenous people view tourism as a path to modernity and economic development (Viken and Müller, 2017: 21). Encountering Indigenous cultures can be a way of connecting with history, traditions and alternative global perspectives. They can make a difference for those performing tourism, perhaps helping to revitalise and regain ethnic consciousness and pride, and as a means of empowerment (Weaver, 2016; Viken and Müller, 2017).

Arctic Indigenous tourism has often been a reaction to the decline in other industries, offering employment for both Indigenous and non-Indigenous northerners. It should be seen in the context of a changing Arctic vacillating between traditionalism and modernity (see Chapter 12 for a Greenlandic context). Hence members of Indigenous populations may have alternative roles in terms of employment, education and lifestyles not contingent upon traditional activities (Leu and Müller, 2016). Tourism may not be the only option for a livelihood, but may be a deliberate economic decision embracing the desire to disseminate knowledge about culture and history (Tuulentie, 2006; Müller and Kuoljok Huuva, 2009).

'Progress' can mean the loss of traditions and cultural distinctiveness, as well as cultural pollution and commodification, potentially conflicting with the images of 'Otherness' promoted by tourism marketing. Although most groups wish to retain their identities, they require strategies to cope with the pressures of commodification, stereotyping and marginalisation (Castañeda, 2012). For example, the exoticisation of costumes, performances and mass reproduction of 'unique' cultural artefacts for tourist consumption can not only distort and falsify the image and representation of a people and its culture, but can also stall evolution, imposing a form of societal stasis (Kirshenblatt-Gimblett, 1998) that may result in 'frozen images' of a culture (MacCannell, 1992).

Inevitably, the nature and roles of 'Indigenous tourism' vary across the Arctic.

Challenges may not be solely related to indigeneity but also to the geographical characteristics of the product (Müller and Viken, 2017b). That tourism development may be feared by some as an intruder and a threat, while being embraced by others as an opportunity to support

Indigenous livelihoods, is a too simplistic dichotomy, as such decisions are taken within a complex ethno-political context (Müller and Pettersson, 2006; Müller and Viken, 2017a).

Research with a Sámi community in Norway identified three main reactions to tourism from within the community (Viken, 2006):

- irritation – by the way the tourism industry handles Sámi culture, and by the fact that the most profitable parts of the business remain in the hands of non-Sámi (an emotion expressed in Greenland: Chapter 12);
- 'reflexive rejection' – where tourism is claimed to be of minimal importance for cultural and identity issues; and
- 'discursive awareness' – people admitting that tourism is a significant institution and as such is part of the contexts that over time form their views of themselves, their culture and the outer world.

Arctic transformation through climate change, globalisation and land development pressures are generating both threats and opportunities across local communities (Kaján, 2013, 2014a, 2014b). The impacts of climate change are being felt more intensely in communities that are dependent on natural resources. This is important for Arctic tourism development as it is often based on nature, and can be acting as an economic complement to traditional livelihoods (Kaján, 2014a): the subsistence/adaptation model of socio-economic change (Notzke, 1999: 62). The interplay between tourism and community-based adaptation can be significant, with Indigenous knowledge generating invaluable insights concerning the impacts of, and strategies for, mitigating climate change impacts on community well-being, sense of place attachment and potential future challenges.

In their suggested trajectories for future Arctic and Indigenous tourism research, Ren, *et al.* (2021) propose three 'trails' to help further diversify and strengthen accounts of how tourism and Arctic Indigenous cultures interrelate:

- Continue developing approaches and conceptual tools that acknowledge difference and multiple tensions and which move beyond such binary distinctions as modern/traditional and culture/nature so as to better learn from and about contemporary Indigenous lives and challenges (Kramvig and Førde, 2020). In other words, to engage with hybridity (Chapter 12).
- Continue engaging with questions concerning the tensions of co-existence between tourism development and Indigenous land governance, where the former is an agent in the human reshaping of Arctic landscapes, not least through the construction of infrastructure that may raise disputes and conflicts over land management and land claims that can be crucial elements for Indigenous communities.
- Reconsider and expand ideas on the interplay betwen tourism and culture by engaging with creativity, design and cultural innovations (such as story-telling) that are important in many Arctic and Indigenous communities (Guttorm, Kantonen and Kramvig, 2019).

The irony of 'Indigenous tourism' for this volume is that, exceptionally, in Greenland the 'Indigenous' population represents by far the majority (Chapters 4 and 6), yet, legally 'Greenlandicness' ignores indigeneity, and consequently does not overtly seek to exploit an 'Indigenous' dimension for tourism purposes (Chapter 12).

3.7 Geostrategy and Arctic tourism: the Chinese market

> The exponential growth in Chinese outbound tourism and the increased sophistication and diversification of Chinese tourist preferences have meant that Chinese tourists have become an interesting resource for destinations around the globe including of course Arctic destinations. The pristine Arctic nature is likewise seen as a valuable resource by the experienced Chinese tourists.
>
> (Bislev and Smed, 2018: 29).

As noted in various sections of this volume (2.6, 8.3, 10.3–10.4, 11.6.2) the multiple roles of China and Chinese actors in the Arctic, and the perception of those roles from within the Arctic and from the sources of Arctic governance outside, have generated debates and narratives along a number of often intertwining dimensions. The role of China and Chinese citizens in Arctic tourism development is the source of several of those.

After extractive industries, maritime transport is one of the key economic growth areas in, if not necessarily for, the Arctic (Gauthier, 2017) (Section 2.6.2). The accruing of maritime experience in expanding Arctic waters and the development of cruise tourism are not unrelated, with China's role in the former becoming increasingly important, while numbers of Chinese tourists on cruise ships in the Arctic are also increasing. A substantial if variable growth in Chinese tourists to Arctic countries (Tables 3.5–3.9) can therefore be viewed both within the context of growing Chinese economic interest in the Arctic (Section 2.6) and as part of the enormous overall growth of Chinese outbound tourism undertaken in recent decades (Table 3.4).[2]

In 1995 an 'Approved Destination Status' (ADS) policy was formalised, permitting Chinese citizens to travel for leisure purposes on group tours to countries with which an ADS agreement had been negotiated. By 2011 there were 70.25mn outbound tourist trips, rising to 162mn by 2019, spending $US255bn (€207bn/£191bn): 20 per cent of the global total (Table 3.4). Pre-Covid, the Chinese outbound market had become by far the world's largest, and its relatively high spending power represented a seductive attraction for tourism promoters and governments alike. Globally, China has sought access to tourism destinations on favourable terms alongside its trade and foreign relations policies. For example, in January 2018, a collaboration programme was established with the European Union – EU Ready for China – for the travel and tourism sectors. The EU's goal was to achieve a 10 per cent annual increase in Chinese visitors, worth €1bn, as part of a range of commercial and cultural exchanges. This included EU-China Tourism Year (ECTY), funded by the EU, which sought to assist travel and tourism sectors to attract and better welcome Chinese visitors. Such visitors would include educational tourists (e.g. see Huang, 2020).

The interface of tourism and geopolitics and the consequent relative fragility of (pre-Covid) Chinese outbound tourism was more starkly exhibited towards the end of the Canada–China Year of Tourism 2018. When planned in 2015, the aim of the Year was to assist doubling the number of Chinese visitors to Canada by 2021: in practice, 2018 saw a slow-down in growth (Table 3.5). Late in the year Canada arrested Meng Wanzhou, Huawei chief financial officer, wanted in the US for allegedly violating trade sanctions against Iran. As China–Canada relations swiftly deteriorated, Canadian tourist agencies suspended their marketing efforts in China, including those on *Wechat*; the federal tourism agency suspended its *Weibo* account, and a joint closing ceremony to be held in Beijing was cancelled (Dangerfield, 2018).

As noted earlier, the availability and comparability of reliable Arctic statistical data is

Table 3.4. Peoples' Republic of China (PRC): key travel indicators. Author's own table derived from Worldometers, 2019; CIA, 2020; National Bureau of Statistics, 2020; OECD, 2020.

Population	1.4bn 18% of world total
2030 forecast (OECD)	3bn
Population median age	37.3
GDP per capita PPP (US$)	18,200
2019 GDP growth over previous year (%)	6.1
Annual average GDP growth over previous decade (%)	9.0
Population in possession of a passport (UK figure is 76%) (2019)	9.0% (120mn)
Outbound trips ('for private purposes')	2015 121.7mn 2019 162.1mn
2015–2019 increase	33%
Projected outbound trips	2030 >400mn
International tourist expenditure (US$bn)	2008 36.2 2013 128.6 2015 250.0 2018 277.3 2019 254.6

Table 3.5. Chinese visitor arrivals in Canada, 2016–2019. Author's own table derived from Chinese Tourist Agency, 2017; Lat Multilingual, 2018; Dangerfield, 2018; Destination Canada, 2020.

Year	No. of Chinese visitors	Annual % change	*Notes*
2016	610,139	24	
2017	682,000	12	
2018	737,000	8	
2019	571,000	–22.5	*Air arrivals only*

problematic, both in terms of the difficulty in abstracting solely 'Arctic' regional tourism figures from individual national statistics, and the lack of comparability of collecting methods, definitions and data presentation between countries (as evidenced in Lee, 2020) (Tables 3.6–3.9).

To place these tables in some perspective, Wang and Nan (2012) claimed that only around 500–800 Chinese tourists annually visited the Arctic *and Antarctica* before 2011. Fu (2012) suggested that formal Chinese *polar* tourism began around 2008 with one tour agent selling tickets for a European Arctic tour company. Online Arctic cruise ticket auctions began in 2015 (Huang, Tang and Weaver, 2017: 91). *China Daily* (2017) reported that between 2011 and 2016 numbers of Chinese visitors to Iceland had increased by 660 per cent to 67,000 (see also Huijbens and Alessio, 2015).

Even in 2020 Lee could still argue that despite the growing importance of Asian/Chinese tourists in (European) Arctic destinations, researchers and industry practitioners were yet to fully understand the reasons, motivations, conditions or factors behind the apparent 'sudden rush' of Asian tourists to European Arctic destinations. (Such a comprehension deficit is reflected in the marketing of adventure tourism to Chinese tourists in Greenland: Chapter 10.) Huang, Tang and Weaver (2017: 97) suggest that it seems probable that Chinese explorer-tourists will move into deeper peripheries as the Arctic becomes more accessible to conventional tourist segments.

In seeking a wider theoretical frame for China–Nordic Arctic relations and encounters, Jørgensen and Bertelsen (2020) argue that tourism, perhaps as a precursor to Chinese initiatives such as mining projects, trade deals or political alliances, offers both challenges and opportunities for Arctic businesses and communities in terms of economic, community up-skilling and human capital benefits. In this way Chinese tourism to the Nordic Arctic is seen to create incentives to acquire global skills and knowledge necessary for effective engagement in an increasingly Asian-centred global economy. The connection between Asia and the Arctic, mediated by tourism, can thus be viewed as a prelude to the pathways of different development sectors to be taken in 'the Asian century' (Asian Development Bank, 2011). In other words, tourism can be viewed

Table 3.6. Attempts to enumerate Chinese visitation to Nordic countries. Author's own table derived from Lee, 2020 and sources cited in the table.

Country	Observations	Sources
Denmark	Hosted China–Denmark Tourism year, 2017 China-Ready programme devised for the sector	Ministry of Foreign Affairs, Denmark, 2017a
	Accounted for 17% (235,138) of Chinese overnights in Nordic countries in 2018: an increase in number of 325% compared to 2008	Visit Denmark, 2020 Visit Sweden, 2019
Greenland	To attract more Chinese visitors, Visit Greenland hosted a promotional workshop 'Chinese Arctic Travel Summit' in 2015	Por, 2020
	Around 1,000 Chinese have visited Greenland each year 2013–2019	See section 10.3
Finland	China is fourth largest incoming market; Finland accounted for 25% (332,546) of Chinese overnights in Nordic countries in 2018: an increase of 282% compared to 2008	Visit Sweden, 2019
	Lapland has the largest number of Asian visitors (Table 3.7). Finnair helped to make Helsinki an important European air hub for Chinese travellers	Visit Finland, 2017
Norway	There were 44,213 Chinese overnight stays January–April 2017: an increase of 87% over the previous year. Compare with Table 3.8	NHO Reiseliv, 2017
	In 2018 Norway accounted for 32% (437,767) of Chinese overnights in Nordic countries: an increase of 856% compared to 2008	Visit Sweden, 2019
Sweden	Accounted for 26% (347,254) of Chinese overnights in Nordic countries in 2018: an increase of 244% compared to 2008 (see also Table 3.9)	Visit Sweden, 2019

Table 3.7. Finland: percentage of foreign overnights taken by visitors from PRC and Hong Kong, 2000–2020. Author's own table derived from www.visitfinland.stat.fi.

Visitors from PRC and Hong Kong	% of foreign overnights	
	Whole of Finland	Lapland
2000	1.0	0.4
2005	1.7	0.6
2010	1.7	0.9
2014	2.6	1.9
2015	3.8	2.6
2016	4.7	4.3
2017	5.4	5.5
2018	5.6	5.7
2019	6.3	6.1

Table 3.8. Norway: numbers of overnights taken by visitors from PRC, 2019, compared to other market sources. Author's own table derived from NHO Reiseliv, 2020.

Visitors' source country	Total number of overnight stays of visitors in 2019	% change from 2018
China	471,822	+5
'Rest of Asia' (excluding Russia)	207,004	0
Russia	154,189	+15
UK	649,981	+4
USA	956,174	+17

Table 3.9. Sweden: ten-year growth of Chinese and total foreign visitor overnights by season, 2017/18 compared to 2007/08. Author's own table derived from Visit Sweden 2019: 14.

Season	Percentage increase in overnights	
	Chinese	Total foreign visitors
Winter (December–April) 2017/18 compared to 2007/08	172	50
Spring (May) 2018 vs 2008	220	60
Summer (June–August) 2018 vs 2008	373	37
Autumn (September–November) 2018 vs 2008	200	66

as being at the forefront of emerging interactions between the Arctic and Asia, in areas of transport accessibility, industrial activity, trade, cultural exchange and geopolitics. These multi-dimensional aspects of interaction are expected to be played out in the context of Arctic tourism as Asia sustains its role as the biggest contributor to the world's economy (Nomura, 2017). With this global outlook, Asian tourism, largely led by China, and not least in the Arctic, is expected to be one of the most significant arenas to reflect the changing global order (Huang, Tang and Weaver, 2017; Lee, 2020: 106).

3.8 Conclusions

Arctic destinations will be required to adapt to a predicted (post-Covid) 'boom' in tourism. But according to Wang *et al.* (2020: 9), Arctic tourism sites are scattered and the responsibilities, powers, and interests of different tourism stakeholders are unclear. The Arctic lacks tourism framework conventions and operational institutions, making inter-regional co-ordination and communication more difficult. Land-use competition is also quite strong, presenting further obstacles to Arctic tourism development (Dubois, Ceron and Peeters, 2013).

Therefore, according to Wang *et al.* (2020: 9), there is an urgent need to establish an Arctic tourism framework convention to regulate inter-regional co-operation and create further opportunities for the sector. It is necessary to build a community of tourism in the Arctic region, strengthen protection of the natural environment and Indigenous culture, and build a platform for exchange between Arctic and non-Arctic countries.

The Arctic has no body equivalent to the International Association of Antarctic Tour Operators (IAATO), which is concerned about Antarctic heritage and conservation and subscribes to what has been referred to as 'environmental stewardship' with the intention of advocating, promoting and practising safe and environmentally responsible private-sector travel to the Antarctic (IAATO, 1992–2019). While Wang *et al.* (2020: 9) bemoan the fact that the Arctic has no equivalent body, and argue that it should, nowhere in their paper do they acknowledge the role of the Arctic Council.

There continues to be a need for good quality empirical research and for solid, comparable data, as a foundation for appropriate policy, planning and management in the face of the dynamic consequences of climate change and potential growth in (post-Covid) visitation.

3.9 In the next chapter

In no context does tourism exist in isolation from other sectors and forces, both 'natural' and 'human', as will be explored in subsequent chapters focusing on Greenland. Such focus opens in Chapter 4 with an evaluation of the cultural, political and economic 'evolution' of Greenland.

Notes

[1] Of course, non-Indigenous 'local' cultures can also provide tourists with 'Otherness'.

[2] Chinese roles in Arctic tourism also extend to running tourism business operations, such as hotels (Kirchner, 2018).

Part 2:

Dynamic Greenland

4

Evolving Greenland (Kalaallit Nunaat)

'While Greenland is the world's largest island, in many respects it makes no sense to think of it as an island place, but rather as a series of more or less isolated yet interconnected locations'
(Nielsen, 2020: 46)

'... the territory's political, economic, and cultural development over the past seven decades has been a series of crossroads...'
(Grydehøj, 2020a: 230)

4.1 Introduction

This chapter addresses Greenland's emergence from a *terra incognita*, through disparate migrations from North America and northern Europe to become a contemporary symbol of geopolitical and climatic uncertainty. It briefly illustrates how Greenland's international position has been shaped by a history of colonialism and contested sovereignty, sustaining the territory's sometimes ambivalent position linking northern Europe and North America while sustaining a pan-Arctic role. Its basic characteristics (Box 4.1) emphasise Greenland's exceptionalism relative to those Arctic Indigenous territories without autonomy (e.g. Sápmi) and Arctic Indigenous territories with autonomy but without a clear method of achieving political independence (e.g. Nunavut) (Grydehøj, 2020a: 217).

This chapter therefore focuses on Greenland-specific processes addressed more generally within an Arctic framework in previous chapters, and provides a political–historical context for subsequent Greenland-focused chapters.

4.2 Indigeneity?

As the location for some of the world's most northerly settlements, Greenland has been inhabited discontinuously for 4,500 years (Grønow, 2000). The earliest known residents were Inuit who originated from Central Asia and arrived via Siberia, Alaska and Canada into the north-west of the island. The earliest settlers have been named after locations where similar cultures have been recorded, all in North America (Box 4.2). The Thule culture was materially distinctive from those that preceded it. Tools and technology were radically advanced compared to the Dorset culture, both groups co-existing in north-west Greenland until the Dorset peoples disappeared. Some 250 years later, possibly because of a cooling climate, Thule culture Inuit habitation had dispersed. This included migrating to south Greenland, where Norse had settled. Hunting tools became highly developed during this period, and Thule technology perfected the kayak (*qajaq*), the boat (*umiaq*), the bow and arrow (*pisisseq qasorlu*) (Fig. 4.3), the harpoon (*unaaq*) and the dog sled (*qamutit*). These provided much greater mobility permitting the hunting of large sea mammals as well as

DOI: 10.1079/9781789246728.0004

Box 4.1. Greenland: the basics

- An autonomous sub-national island jurisdiction (SNIJ) of the metropolitan state of Denmark, one of three constituent nations of the Danish realm (*Rigfællesskabet*).
- The territory extends 2,700km over three climatic zones from high Arctic climate in the north through low Arctic to relatively mild sub-Arctic climates in the south, providing a wide variety of terrestrial and aquatic ecosystems. Two-thirds of the territory lies within the Arctic Circle (Fig. 4.1).
- 81% of its 2,166,000km^2 is covered by an ice sheet.
- With a population of just over 56,000,[1] Greenland has the lowest population density of any self-governing geopolitical entity: 0.026 persons per km^2 or 38km^2 per person. In 2020 87.3% of the population were classified as living in towns.
- After a period of rapid growth from the 1950s to the 1970s, Greenland's population is declining, with government estimates suggesting that by 2040 it will be below 50,000 (Section 6.6).
- When Greenland left the European Economic Community (EEC) in 1985, the Community's land area was reduced by more than half (from 3.8mn km^2 to 1.7mn km^2).
- By virtue of being citizens of Denmark, paradoxically Greenlanders have remained EC/EU citizens.

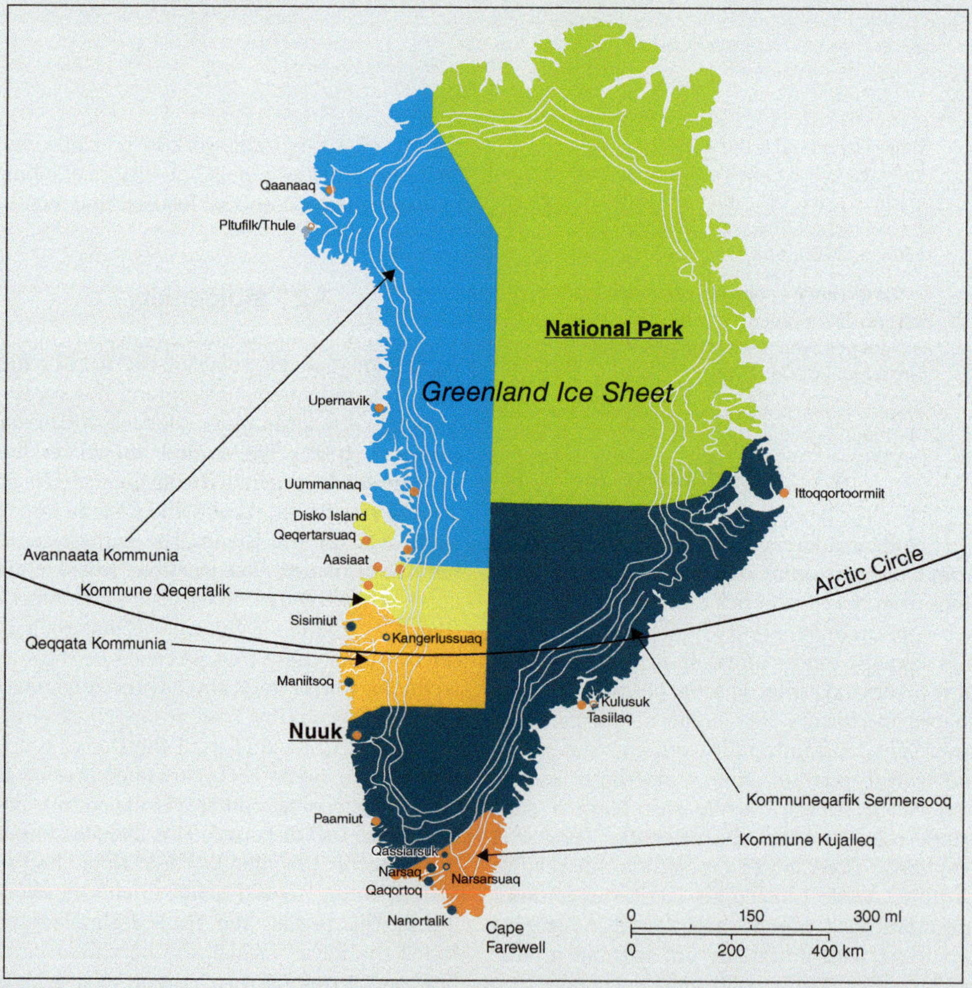

Fig. 4.1. Greenland: communes and major locations. Author's own figure, modified and redrawn from Statistics Greenland, 2018: 4 and *Enclyclopaedia Britannica*.

(*Continued*)

Box 4.1. Continued.

- Claimed as the largest island in the world (Australia being viewed as a continent), Greenland may actually be composed of three islands, the reality of which may eventually be revealed with continued ice sheet melt.
- No roads (nor railway) connect the 77 coast- and fjord-hugging towns, villages and small coastal settlements (Fig 4.2).
- Most emigrants move to Denmark, where there are more than 16,000 Greenland-born residents together with some three thousand descendants.

Sources: Kaae, 2006; Weaver and Lawton, 2017; Qaqartoq Museum, 2019; Grydehøj, 2020a; Statistics Greenland, 2020.

Fig. 4.2. The village centre and harbour of Narsarmijit (population approximately 65) on the western shore of Ikerasagssuaq (Prins Christian Sund). Author's own photograph.

land mammals such as musk ox and caribou, following their seasonal migrations to ensure adequate supplies for long winters in more permanent settlements. By contrast, the Dorset culture people mainly hunted seals at their breathing holes. Distinctive cosmologies were developed by these groups (Sonne, 2017).

Most of the prehistoric remains to be found in Greenland originate from the Thule culture (Gad, 1967; Gulløv, 2000; Kaae, 2006; Flora, 2019: 13–14) (also Section 6.3).

Norse involvement in Greenland possibly dates from as early as AD 930, when it is recorded that Gunnbjörn Ulfsson, making for Iceland, was blown off course, discovered land, but did not settle. In 978 a settlement established by exiled Snæbjörn Galti and associates on the east coast at Bláserk Fjord was not sustained. Permanent settlement came with another exile in 982, when Eiríkr Rauði Þorvaldsson, or Erik the Red, sailed around southern Greenland to reach the west coast. In 985 Erik convinced many Icelanders to join him in a land he called *Grænaland*, choosing the name carefully so as to disguise its true nature. Out of 25 ships that set sail, 14 survived to make 'Greenland' their home. These

Box 4.2. Greenland's earliest settlers

- The Independence culture arrived around 2,400 BC and may have lasted for 800 years.
- The Saqqaq culture covered more territory, and existed at around the same time, but lasted for perhaps 200 years longer.
- The Dorset culture – named after Cape Dorset in Nunavut, Canada, where the first evidence of their existence was found – may have sprung from either of the previous two cultures, arriving in two waves between 500 BC and AD 200.
- The Thule culture arrived from Canada between AD 1200 and 1300, eventually spreading around the coast of Greenland. It is their descendants who are the contemporary Inuit of Greenland.

Sources: Gad, 1967; Grønow, 2000; Kaae, 2006; Lincoln, 2020; Mason, 2020

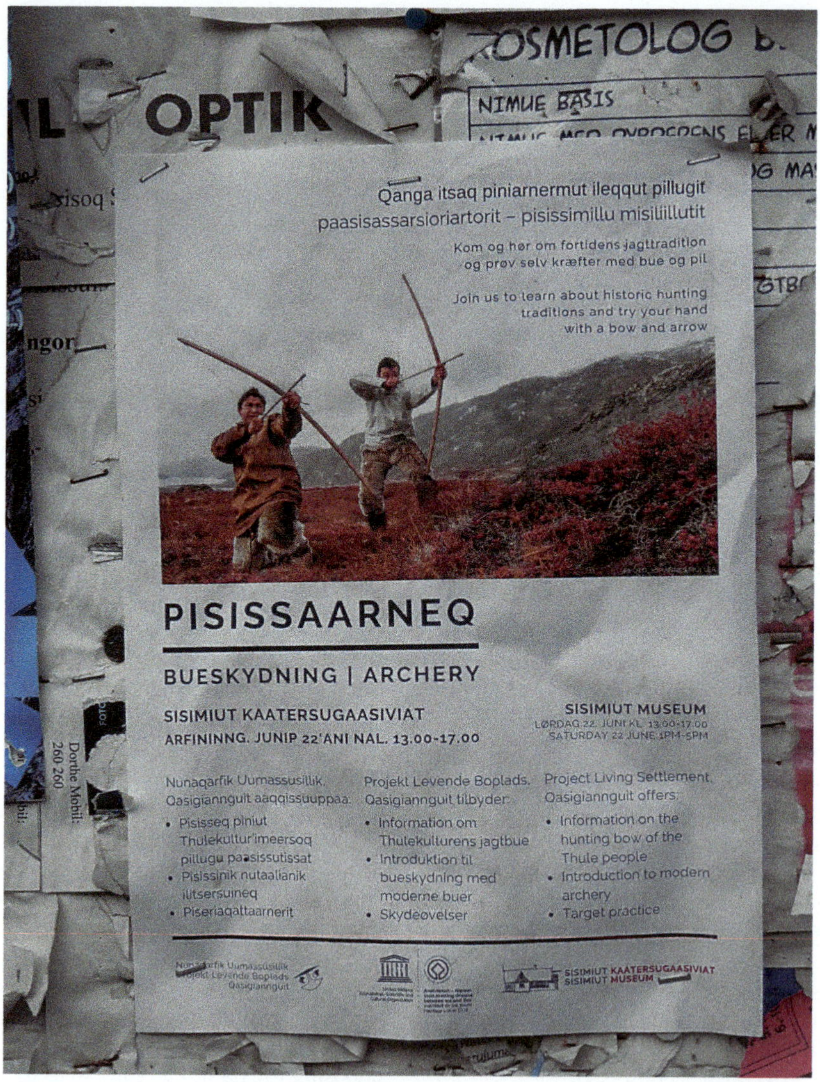

Fig. 4.3. Trilingual poster for 'Thule culture' archery lessons, Sisimiut. Author's own photograph.

Norse farmers settled in the fjords of southern Greenland and lived from hunting and fishing as well as farming (Arneborg, 2000).

Erik the Red's son, Leif Eriksson (*Leif den Lykkelige*/Leif the Happy), visited Norway in the year 1000 and converted to Christianity. Later, losing his way home to Greenland, he apparently discovered North America (Seaver, 1996). After settling briefly and calling the place Vinland, he returned to Greenland and set about converting the population to Christianity, in the process constructing several churches. Despite all the hard work the Norse settlers put into raising sheep, goats, cattle and horses, and growing winter fodder, they were dependent upon supplies from Norway. When Norway became embroiled in European conflict, losing the port of Bergen by the end of the fourteenth century, it was no longer in a position to supply such a far-flung outpost as Greenland.

This supply chain problem was one likely contributory factor to the much debated disappearance of Norse settlers around this time. Other claimed contributory factors include a fall in wool prices, the demise of demand for narwhal tusks and seal skins, and a deterioration in climatic conditions from the mid-thirteenth century which required the settlers to re-orient their main sources of food from the land to the sea.

The final Norse written record in Greenland is of a wedding at Hvalsey (Whale Island/ Qaqortukulooq) in 1408. There is no historical account of their subsequent disappearance, although archaeological evidence suggests it was an orderly departure (Arneborg, Lynnerup and Heinemeier, 2012; Stockinger, 2013). Today, Norse ruins and building reconstructions (Fig. 4.4), commemorations (Fig. 6.2) and story-telling (Fig. 6.1) are significant tourist attractions.

For several generations, Inuit people once again had Greenland to themselves (Arneborg, 2003; Dugmore, Keller and McGovern, 2007). There was no continuity between Greenland's medieval Norse population and the Dano-Norwegian colonisers of the eighteenth century (Gulløv, 2000). Nonetheless, today, Greenlanders occupy many of the same locations along fjord arms in the south-west of the country where the Norse settled a thousand years ago, now also raising sheep for meat and wool (Thomson and Thomson, 2006: 172) (Fig. 4.5).

Fig. 4.4. The partly reconstructed, and now much visited, Norse settlement at Brattahlíð (Qassiarsuk) overlooking Eiriksfjord (Tunugdliarfik). Author's own photograph.

Fig. 4.5. A pastoral scene at Qassiarsuk where Norse farmed a thousand years previously at Brattahlíð. Author's own photograph.

4.3 European colonialism

As a 'product of colonial cartography' this 'Inuit territory boxed in with Denmark and separated from its immediate neighbours in Nunavut to the West' has only been imagined as a collection of towns and settlements, joined together in a national consciousness, within the past century (Grydehøj, Kelman and Su, 2020: 294).

Initially in the European colonisation process, circumpolar regions were not objects of speculation, but places to venture into and to cross en route to other places as destinations of trade and commerce (Dodds and Nuttall, 2016: 121). European influence began in earnest in the sixteenth century, when vessels travelling northwards – including Spanish, Portuguese, Basque and Dutch whalers, traders and explorers – dominated the voyage to Greenland. As early as 1500 the Portuguese Gaspar Corte-Real reached southern Greenland.

When King Christian II of Denmark and Norway ordered an expedition to be sent there, sea ice confounded the attempt. In the 1570s the Company of Merchant Adventurers of England financed three expeditions to the Arctic led by Martin Frobisher. Frobisher's return in 1576 generated stories of mineral wealth – mountains of silver and gold in Greenland and elsewhere – and these tales were sustained in Europe for the best part of 200 years.

In the following decade John Davis made three voyages to West Greenland and managed to map the entire area between Nuuk and Upernavik: an exercise not just important for scientific field knowledge, but also one of great political significance. English place-names on the map of north-west Greenland, such as Sanderson's Hope – the passage just south of Upernavik (Fig. 4.1) – named after William Sanderson, a London brewer and one of the expedition's sponsors, mark Davis' progress and aspirations (Dodds and Nuttall, 2016: 122).

It was such European interest in Greenland that moved King Christian IV to assert Danish ownership, and in 1605 three Danish ships succeeded in landing there. In search of silver, the Danes traded bearskins and tusks with Greenlanders but also took two Inuit men captive, who were later displayed for public entertainment in Copenhagen (Gad, 1967; Lidegaard, 1991).

European activities in the seventeenth and early eighteenth centuries were mostly dominated by whaling. Lamp oil from whale blubber was sold throughout Europe and proved to be a valuable resource for the Dutch, Danish and Norwegian whalers. By the beginning of the nineteenth century, the industry consisted of hundreds of vessels and employed several thousand people, and whaling missions combined their work with scientific and mapping purposes. But barter and trading were accompanied by conflict with and abduction of Greenlanders, some of whom, repeating earlier humiliations, were subject to scientific experiments and employed as a source of public entertainment in Europe (Flora, 2019: 14–15).

In 1721 the Danish–Norwegian Lutheran Hans Egede (Box 4.3) headed a missionary expedition to Greenland. He sought to check if the earlier Norse inhabitants had given up their faith, or worse, had clung to pre-Reformation Catholicism. Finding no trace of the old colony – which had possibly not existed for 300 years – Egede turned his attention to the Inuit, who, unsurprisingly, were none too accommodating to the idea of proselytism. After pleas for support, Egede's mission in Greenland became the financial and political responsibility of the Danish crown in 1728 (Gad, 1969; Fleischer, 2003). It became a hub for missionaries, fishing and whaling (Kristensen and Rahbek-Clemmensen, 2018). Egede's 'mission' initiated a process of colonisation, which rendered Greenland a frontier of cultural and natural scientific exploration and expedition. Colonisation also brought disease (Box 4.4), a cruelly dark element of global connectivity that resonates today.

Danish colonial rule in Greenland concentrated on the establishment of two institutions: (i) the mission, which sought to convert all Greenlanders and to translate the Bible into Greenlandic; and (ii) the trading company Den Kongelige Grønlandske Handel (the Royal Greenlandic Trade) (KGH), established in 1774 to monopolise trade with Greenland (Emmerson, 2010: 303–304; Flora, 2019: 15). Small 'colonial towns' were established along the coast and some fjords to support these religious and mercantile pursuits, trading the Bible for seal skins.

By the middle of the nineteenth century, KGH had established 13 colonies: outposts of European empire, although in practice managed as individual enterprises at a local level. These rarely made a profit but acted as instruments of government, and the trading monopoly remained unquestioned (Sørensen, 2007).

But this was never a settler colony project, neither was it grounded in physical violence or forced labour. A Danish trade monopoly and the colonial desire for marketable products did encourage seal hunting at the expense of other traditional economic activities (Rud, 2017), ultimately serving to entrench a colonial division of labour. When seal populations collapsed in the early twentieth century, the colonial administrators encouraged a transition to another export-oriented primary sector activity, fishing. Unlike seal hunting, a formalised fishing industry required Greenlanders to live in permanent settlements, yet it could not sustain a subsistence economy. Thus by the 1920s most Greenlanders had been drawn into a dependency relationship with the Danish colonial administration (Grydehøj, 2020a, 2020b).

Norway and Denmark were separated politically in 1814, and Denmark asserted full control over Greenland. This position derived essentially from the Treaty of Kiel of that year, on the conclusion of the Napoleonic wars. By that treaty, the defeated Danes had ceded to the victorious Swedes the entire territory of Norway, excluding her dependencies of Greenland, Iceland and the Faroe Islands, which 'passed' to the Kingdom of Denmark.

The concept of native sovereignty over Greenland did not seriously occur to nineteenth-century imperial decision-makers. For them, the transfer of Greenland from one European people to another was entirely appropriate, even though the nature and extent of Greenland was uncertain at this time. A single parenthesis in the Treaty of Kiel and the presence of a few small Danish settlements on the west coast, inhabited by no more than a couple of hundred Danes, were considered sufficient to confer sovereignty over this huge land mass (Emmerson, 2010: 103).

Box 4.3. The contested legacy of Hans Egede

In 2020, on the eve of 21 June, Kalaallit Nunaat national day, the prominently positioned statue of Egede, looking out across the sea from old Nuuk, was splattered with red paint to make the figure's robes appear bloodied, while the staff carried in the figure's right hand was transformed into a whip. The statue's pedestal was tagged with Inuit symbols together with the word 'DECOLONIZE' spray-painted in red (Fig. 4.6).

Debate concerning the future of Egede's statue had taken on new impetus and relevance in the wake of worldwide 'Black Lives Matter' protests following the death of George Floyd at the hands of Minneapolis police four weeks earlier. Protesters elsewhere had removed statues of a number of slavers and colonial rulers, such as that of Thomas Colston in Bristol and of King Léopold II in Antwerp. These actions prompted international debate, and in an anonymous statement, the group responsible for daubing Egede argued (via Greenlandic artist Aqqalu Berthelsen), 'It is about time that we stop

Fig. 4.6. Statue of Hans Egede in Nuuk daubed with red paint, Inuit symbols and 'DECOLONIZE', June 2020. Approving sources of the photograph wish to remain anonymous.

(*Continued*)

Box 4.3. Continued.

celebrating colonisers and that we start taking back what is rightfully ours. It's time to decolonise our minds and our country. No coloniser deserves to be on top of a mountain like that. We need to learn the truth of our history.'

The name of Hans Egede, a proselytising Lutheran priest whose Bergen Greenland Company was able to exploit abundant natural resources in Greenland, remains attached in Nuuk to the city's main hotel, museum, a church, and, at least in its Danish language version, the road in which Visit Greenland's main office is located.

Unsurprisingly, opinions in Greenland on the subject of the Egede statue have been divided: in July 2020 the citizens of Sermersooq region (population 23,000), which includes Nuuk, voted on-line in a non-binding poll on the future of the statue. Some 921 supported its retention while 600 voted for the statue to be demolished.

The statue – erected in 1922 through local public subscription – was first targeted by paint in the 1970s, and more recently, was tagged in 2012 and 2015.

Other statues of Egede include those located outside the Marble Church in Copenhagen, by the Trinity Church in Oslo, outside the Harstad Church in Harstad, Norway, and in front of the Lofoten Cathedral in Vågan, Norway.

Sources: Birkhold, 2019; Berthelsen, 2020; Enge, 2020; McGwin, 2020c; Schøler, 2020; Shamonova, 2020; Walsh, 2020

Box 4.4. Disease: one pervasive consequence of colonisation

In 1733/34, smallpox brought from Copenhagen killed an estimated 1,000–3,000 Inuit, with the disease spreading rapidly as survivors fled from its epicentre in the areas surrounding Nuuk. Recurrent smallpox epidemics saw:

- almost half of Nuuk's population killed in the early 1780s;
- in an 1801 outbreak, the population of Aasiaat was reduced by 42% and that of Sisimiut by 89%.

The earliest census, undertaken in 1789, suggested a total population of about 5,122, indicating that such epidemics of the early colonial period killed a significant proportion of Greenland's population, and doubtless incapacitated many more.

Sources: Grydehøj, Kelman and Su, 2020: 291, drawing on earlier works

Yet Norway would continue to lay claim to parts of the territory. While Greenlanders supported the Danish colonial power against Norway, they also criticised the Danes for paying the Inuit insufficient respect (Langgård, 2014).

The Norwegian challenge resurfaced in the 1920s. From a legal point of view, the fact that Greenland had been 'discovered' by Norse sailors (Norwegians rather than Danes), and that Hans Egede, the missionary who re-established a European colony on the island in 1721 was Norwegian (now often referred to as 'Danish-Norwegian'), was of limited relevance. Although the Treaty of Kiel had transferred Greenland to Denmark, Norwegians had long accessed the east coast of Greenland for commercial purposes, principally hunting.

The Danish administration had not attempted to exercise jurisdiction over this area, and despite pressure from Copenhagen, Norway refused to issue a declaration of disinterest. Instead, when Denmark unilaterally declared sovereignty over the whole of Greenland in 1921, Norway strenuously objected, claiming East Greenland was a 'no man's land' (Preuss, 1932), and supported the actions of five trappers, who, in June 1931, raised the Norwegian flag, proclaiming occupation of part of East Greenland. The dispute was finally settled by the International Court of Justice in The Hague finding in favour of Denmark in April 1933 (Emmerson, 2010: 103–105).

Local district councils had been established in the mid-nineteenth century, although real authority remained in Copenhagen (Gad, 1984). But also at this time, two teacher training colleges were established, in Godthåb (Nuuk) and

Jacobshavn (Ilulissat), primarily to complement the missions (Lidegaard, 1991; Thuesen, 2016). The Ilulissat college focused on the teaching of Danish, while that in Nuuk also functioned as a publishing house. Inuit customs had begun to be documented, and most notable among early publications was Hinrich Rink's ([1875] 1997) collection of Greenlandic myths, and the first Greenlandic grammar, by Samuel Kleinschmidt.[2] These institutions were to become important in nurturing an educated Greenlandic elite (Flora, 2019: 16–17).

4.4 American neo-colonialism

In the flood of diplomatic and political activity between 1867 and 1880 which ushered the USA and Canada into the Arctic, two European colonies – Greenland and Iceland – were subject to close scrutiny. This was not for the first time, as Greenland had been identified within the American sphere of influence in the 1823 Monroe Doctrine, albeit with little effect.

But in 1868, US Secretary of State William Seward, recognised as a leading opponent of slavery and largely responsible for the US acquisition of Alaska, requested a State Department report on the economic resources of Greenland and Iceland and their potential value to the US. This report indicated that the territories were worthy of consideration as American investments. Besides seals, whales and the products of the sea, Greenland's geology offered 'much promise'. In 1868, the only substance mined in any exportable quantity was cryolite – a mineral used in making soap and, later, more importantly, in producing aluminium. It was considered that Greenland's interior could offer further minerals.

Acquisition of Greenland could extend Seward's grander northern designs: US purchase of Alaska had 'flanked British America on the Arctic and Pacific'. It was assumed that purchase of Greenland would effectively force Canada to join the USA. But the report came too late. Seward soon left his position, and US–Danish relations had been damaged by Seward's failed attempt to purchase the Danish Virgin Islands. After a bruising congressional battle to purchase Alaska, buying Greenland and Iceland was simply a step too far for a country still recovering from the Civil War. Yet the issue would rear its head again in 1916, 1946 and, notoriously, in 2019 (Emmerson, 2010: 91–94; BBC, 2019).

In the early twentieth century, the American challenge concerned north-west Greenland, which had been unknown to the drafters of the Treaty of Kiel. It had only been discovered, and (briefly) occupied, by Charles Francis Hall, an American explorer, in the early 1870s. Denmark had only established a permanent settlement there in 1909, when a Greenlandic church mission was set up on the shores of North Star Bay (close to Thule). Denmark needed an American declaration that it did not entertain a claim for northern Greenland, and it became a condition of the sale of the Danish Virgin Islands to the US in August 1916. A few years later Britain tried, and because of American opposition, failed, to obtain a right of pre-emptive purchase of Greenland should Denmark ever consider a sale (Johnston, 1933: 36–37).

During World War II, after Denmark had been occupied by Germany in April 1940, civil servants within Greenland declared the island a self-ruling territory (Birkhold, 2019). But with Nazis threatening from Norway, the Danish ambassador in Washington, Henrik Kauffmann (Christiansen, 2020), invited the US to administer Greenlandic affairs, and during the protectorate that followed some nine air bases were established. The subsequent post-war defence treaty between Denmark and the US signed in 1951 gave the US military considerable rights (https://avalon.law.yale.edu/20th_century/den001.asp).

> The defence agreement *de facto* created a system of shared sovereignty over Greenland, where the United States military was allowed to operate whenever and wherever it needed, while staying out of Greenlandic affairs and allowing the Danes *de jure* sovereignty over the island ... (Kristensen and Rahbek-Clemmensen, 2018: 4).

The island was considered to be strategically important for the US military, being situated on the shortest route from North America to Europe (and the Soviet Union). It enabled Denmark's otherwise meagre contribution to NATO to appear substantially enhanced: as the Alliance's third largest member by area (Ackrén and Jakobsen, 2015). Indeed, for Denmark, sovereignty over Greenland has been critical in allowing it to play a greater role within NATO: both in terms of size and in its strategic position with the continued presence of the American base of Thule (Rahbek-Clemmensen, 2016). As the last remaining base, located 1,200km (750 miles) north of the Arctic Circle in north-west Greenland (Fig. 4.1), Thule remains part of a US ballistic missile early warning system (Box 4.5). But the legacy of the other bases lives on (Box 4.6).

> Denmark took the blame for several scandals that surfaced during or after the Cold War as a consequence of the American activities, including the presence of American nuclear weapons in Thule in the 1960s in contradiction with [*sic*] official Danish policy.
> (Kristensen and Rahbek-Clemmensen, 2018: 5)

Indeed, what became known as *Thule-sagen* or 'the Thule Problem' was to symbolise the changing relationship between Greenland and Denmark.

Box 4.5. Thule and the symmetry of Greenlandic–Danish relations

1. Initially a meteorological station, construction of what was to become the largest airforce base in the Arctic actually began in the month prior to the 1951 Danish–US agreement being concluded.
2. In 1953, the Danish state carried out the forced removal of the population of Pituffik settlement to make more room for the air base.
3. From the second half of the 1950s, the Distant Early Warning Line (DEW Line) – a line of interlinked radar stations extending from the Faroe Islands and Iceland in the east to northern Canada and Alaska in the west – was established with the intention of securing early detection of bombers flying from the Soviet Union, and the later Ballistic Early Warning System (BMEWS) was introduced to counter the strengthening of the Soviet ICBM force.
4. Takahashi *et al.* (2019) postulate that Denmark's transforming Greenland's status in 1953 from a colony to an integrated part of the state (Section 4.5) was partly to limit Denmark's vulnerability and to invalidate any opposing voices in Greenland to the military base.
5. As noted above, although Denmark obtained early membership of NATO, this was achieved in exchange for providing Thule, as Denmark had little else to offer. In this way, in terms of national security, Denmark was, and has continued to be, dependent upon Greenland, thereby exhibiting an intriguing power relationship reversal, the implications of which Copenhagen has clearly wished to minimise.
6. With increasing confidence following Home Rule in 1979 (Section 4.6), the Greenlandic parliament approached Denmark to seek participation in negotiations if the US sought to include Thule in that country's National Missile Defense plan. In the early 1990s the Nuuk authorities managed to exercise some influence in the conversion of Kangerlussuaq and Kulusuk air bases into civil airports, and in 2003 a joint declaration of the Danish government and the Home Rule government affirmed the participation of Greenland in the fields of diplomacy and national security. This Itilleq Declaration re-affirmed Greenlandic commitment to Danish security, and permitted Greenland to conclude a complex agreement with the US and Denmark – the Igaliku Agreement – the following year.
7. Towards the end of the Cold War, US financial considerations saw discussion of a potential withdrawal of the US military from Thule.
8. In the context of multiple cold wars and hot local conflicts, Thule's continued role today is of significant strategic importance: as a US missile defence shield hub, as part of the Air Force Satellite Control Network, and as infrastructure for the US space programme, as well as for scientific research.
9. With climate change – melting of the ice sheet and the possibilities of exploiting commercial sea lanes in the Arctic – the various facilities at Thule – port, tanks and oil storage facilities, factories, hospitals and accommodation facilities – could serve as a hub or a platform for intensifying economic co-operation in the region.
10. In relation to Thule, as in other contemporary areas of development, the para-diplomacy that Greenland is able to bring to bear in its horizontal relationships with state-level entities – sometimes beyond its formal capabilities – is a source of increasing confidence, as evidenced by the 2020 tripartite agreement (Section 4.6).

Sources: Archer, 1988, 2003; US Arctic Research Commission, 1990; Department of Defense, 2011; Udenrigsministeriet/Udenrigsdirektoratet, 2011; Petersen, 2011, 2013; Takahashi *et al.*, 2019; Grydehøj, 2020a; Haycox, 2020; Humpert, 2020

Box 4.6. Legacy of the US presence: a heritage tourism of dark geopolitics?

A number of dark dimensions of the US military presence in Greenland persist, offering opportunities for equally dark heritage and 'last chance' tourism.

1. To study the feasibility of working under ice, a highly publicised US military project, known as 'Camp Century', was launched in 1960, at an elevation of 2,000m (6,600 feet) some 240km (150 miles) from the Thule air base. The 'official purpose' of this project, as explained by the US Department of Defense to Danish officials in 1960, was to test various construction techniques under Arctic conditions and to explore practical problems with a semi-mobile nuclear reactor, as well as supporting scientific experiments on the ice sheet (Petersen, 2008). This 225-person nuclear-powered army base was 'used to project popular images of techno-scientific control, nuclear containment, and American values into Cold War American culture' (Nielsen, Nielsen and Martin-Nielsen, 2014: 443).

Under the code name 'Project Iceworm', the actual purpose of this Cold War site was to construct a system of tunnels 4,000km (2,500 miles) in length, to deploy up to 600 medium-range nuclear missiles, covering, or rather, underlying, an area of 130,000 km^2 (52,000 sq miles), some three times the size of Denmark (Dodds and Nuttall, 2018). The launch complex floors would be 8.5m (28 feet) below the surface. Within three years of the start of construction, ice core samples demonstrated that the glacier within which the project was burrowed was moving faster than anticipated and would destroy the tunnels and missile launch stations in about two years. By this time 21 trenches had been cut and covered with arched roofs to contain prefabricated buildings. With a total length of 3,000m (1.9 miles) excavated, these tunnels also contained a hospital, a shop, a theatre and a church, with electricity provided by the world's first mobile nuclear reactor.

The facility was finally evacuated in 1965, and the nuclear generator removed. Project Iceworm was cancelled, and Camp Century closed in 1966.

When the camp was decommissioned, its infrastructure and waste were abandoned under the assumption they would be entombed forever. Ironically, the Americans failed to anticipate continued climate warming. Colgan *et al.* (2016) found that the portion of the ice sheet covering Camp Century would start to melt by 2100 if current trends continued. Since that research was undertaken, Arctic warming has accelerated. When the ice melted, the camp's infrastructure, as well as remaining biological, chemical and radioactive waste – including 200,000 litres of diesel and polychlorinated biphenyls (PCBs) – would re-enter the environment and potentially disrupt nearby ecosystems (Henley, 2016), although melt-water percolating down through the ice sheet could pick up and wash abandoned waste downstream decades sooner (Rosen, 2016).

2. On 21 January 1968, 12km (7.5 miles) from the US.Thule air base, an American B-52 bomber on a Cold War 'Chrome Dome' alert mission from Plattsburgh NY air force base crashed into the sea ice with four thermo-nuclear bombs on board, and, melting the ice thereabouts, sank to the sea bed. It was claimed that most of the components for three of the bombs were recovered; controversy has persisted surrounding the belief that components for the fourth remain lost. Much contaminated ice was removed to the USA for disposal. In 1995, a report revealed the Danish government had given tacit permission for nuclear weapons to be located in Greenland, in contravention of Denmark's 1957 nuclear-free zone policy (Christensen, 2009). The potentially still dangerous site of the crash is too remote for most dark tourism visitors.

3. When new surveillance technologies made Greenland less critical to NATO security, the Americans began to close down its bases (Emmerson, 2010). US Bluie West One Narsarssuak (*sic*) was the most important of the US air bases and the command centre for Greenland Operations. It was transferred to the Danish government in 1958 (Guldager, 2016). Cabin offices of the air base now act as the Narsarsuaq museum, detailing through exhibit and publication the nature and role of the US presence there (Figs. 4.7, 4.8, 4.9). Kangerlussuaq (Bluie West Eight) operated as a backup runway and as a search-and-rescue, radio and weather reporting station. It also now houses a museum dedicated to its US military heritage. The town of Kangerlussuaq was founded as an American air base in 1941, but the area around the town has been used as hunting grounds for over 4,000 years. Many places nearby and deep into the hinterland hold remains of the region's Inuit history. With a resident population of just 500, Kangerlussuaq was being promoted during the Covid-19 crisis as a staycation destination (Chapter 12).

Both bases found new leases of life as civilian airports, acting as important visitor gateways with their relatively long runways (Section 8.2).

(*Continued*)

Box 4.6. Continued.

4. A third air base – Bluie East Two in East Greenland – has bequeathed a different heritage. Established in 1942 on Ikateq island with a 1,500m runway, Bluie East Two also operated mostly as an alternative airfield, as well as a base for meteorology, navigation and search-and-rescue. The US military abandoned the site in 1947, leaving behind more than 10,000 rusting aviation fuel barrels, collapsed iron frames of an asbestos-laden hangar and rusting trucks, all slowly disintegrating into the small watercourses that criss-cross the land, leaching into the otherwise pure waters of the Sermiligaaq Fjord. Now treks come – such as a 12-day 'Unplugged Wilderness' East Greenland trek – for visitors to marvel at the paradox of a toxic dump set in an otherwise pristine Arctic wilderness.

In January 2018, the Danish government announced that it would contribute DKK180mn (€23mn/£21mn/US$31mn) over six years to begin cleaning up some of the 30 abandoned military bases on Greenland, although experts warned that the amount was unlikely to be sufficient (Lang, 2017; Hyldal, 2018). Nonetheless, Bluie East Two has become a macabre 'last chance' tourism attraction (Germany, 2018).

Any involvement of the USA in the clean-up process remains an outstanding issue: Article 11 of the US–Denmark defence agreement, noted above, states that equipment and facilities may be disposed of in Greenland by the US government after consultation with the Danish authorities.

5. Other countries have taken an interest in decommissioned military bases. In 2016, the Chinese General Nice mining company offered to buy an abandoned naval base at Grønnedal (Kangilinnguit) – put up for sale by the Danish government which had declared it to be no longer of strategic importance – that was constructed by the US military in 1942 and known as Bluie West Seven. Reflecting 'Western' apprehension at China's Arctic role (Section 2.6.3), Denmark declined the Chinese offer (Lang, 2017; Lulu, 2017), discovering that the base retained strategic significance, and would be reopened as a logistical location for storing fuel and materiel and for training personnel (Breum, 2016). A Danish informed source was reported to say: 'The government does not want to sell the naval base to China, as we have a long, close partnership with the US and a defence agreement for Greenland with the US from April 1951 to think about' (Matzen, 2017).

Fig. 4.7. Entrance to the Narsarsuaq Museum, adorned with a Stars and Stripes flag. Author's own photograph.

(Continued)

Box 4.6. Continued.

Fig. 4.8. 1996 plaque at the former US Bluie West One air base dedicated to 'a joint effort' between the USA and Denmark, 1941–1958. Author's own photograph.

Fig. 4.9. Graffito in Sisimiut city centre. Author's own photograph.

In 2016 it became evident that climate change could lead to the leakage of nuclear waste from another former US base, Camp Century (Box 4.6). Subsequent debate between Greenland, Denmark and the USA over who was responsible for paying for clean-up operations, coupled with the transfer of the service contract for the Thule air base from a Greenlandic state-owned enterprise (SOE) to a subsidiary of a US corporation in 2014, generated concern over both the apparent lack of benefits that Greenland derived from the US presence and the Danish government's perceived inability to stand up for the interests of its autonomous territory (Colgan *et al.*, 2016; Grydehøj, 2020a: 228).

4.5 'Modernisation'

During World War II, the export of vital cryolite was re-directed to North America, shipped from Ivittuut in southern Greenland to Philadelphia PA: for the first time since the eighteenth century, Greenland's trade was not with Denmark. Most communication between Greenland and Denmark ceased at this time.

The war demonstrated that Greenland could survive without direct Danish support, and this is claimed to play an important part in Greenlandic national consciousness today (Grydehøj, Kelman and Su, 2020: 293).

With the stationing of American troops on Greenlandic soil, the island's cultural isolation, which the Danish government had reinforced, was breached. Greenlanders were exposed to large quantities of American products (Emmerson, 2010: 306), and American influence in Greenland remained strong throughout the subsequent Cold War period (e.g. Gad, 2017b). Echoing past practice, in 1946 US Secretary of State James F. Byrnes reportedly suggested that the US should purchase Greenland outright for \$100mn (Birkhold, 2019).

Denmark thus emerged from World War II with its North Atlantic realm under threat. Iceland, after a referendum, gained full independence in 1944, while Denmark was still occupied. The Faroe Islands, occupied by the Allies, attained greater self-governance in 1948. Denmark needed to stall any Greenlandic moves in this direction as a matter of national pride: Greenland was a critical diplomatic and geopolitical asset. However, in 1946 the territory was visited by a delegation from the Danish press who sent back reports of poverty, inhumane living conditions and a tuberculosis epidemic ravaging the country.

Subsequently, Copenhagen substantially increased investment in Greenland and doubled salaries for government employees there. Unlike other European colonies, Greenland had never represented a net economic benefit to Denmark apart from a few notably good years for the export of cryolite after 1935 (Emmerson, 2010: 306).

After almost 300 years, Greenland lost its colonial status in 1953 with a constitutional referendum confirming its full integration into Denmark – changing its status from a colony to a Danish county (*amt*), with representation in parliament – whereby Greenlanders became Danish citizens. Economic and political dependence was near total: Denmark signified Greenland's role internationally as merely a 'non-governing territory' (Rud, 2017).

Subsequently Denmark embarked on a vast and speedy modernisation programme. Danish efforts to industrialise cod fishing required concentrated populations rather than dispersed hunters. This resulted in some forced population relocations, rapid urbanisation and a restructured labour market. Government investment in Greenland increased five-fold between 1948 and 1958. Danes migrated to the island to run businesses, and their numbers increased from 5.25 per cent of the population in 1950 to nearly 20 per cent by 1975. Growing numbers of Greenlanders were on the payroll of the Danish state, but new laws dictated that civil servants born in Greenland could be paid only 85 per cent of the salary of those born in Denmark (Petersen, 1986, 1995; Sørensen, 1995; Birkhold, 2019). Expanded urban settlements emerged from among the 'colony towns' through the introduction of mass housing and neighbourhood construction in the form of apartment blocks and open spaces that, at the time, represented the cutting edge in Nordic urban design (Grydehøj, 2014, 2020a).

The modernisation process – bringing housing and living standards into line with Danish levels – appeared to generate more problems than it solved. A fundamental tension lay between the apparently conflicting objectives of social justice and economic self-sufficiency. Improved social justice meant increasing Danish state intervention and Greenland's economic dependency.

Improvement in health caused a rapid decrease in mortality and increase in birth rates, leading to a substantial growth in demand for teachers, doctors, nurses and various skilled workers throughout public and private sectors, and particularly in construction. This required more workers from Denmark, with higher salaries, thereby widening the economic gap and exacerbating inequalities that the modernisation process had sought to eradicate. Subsistence hunters and women found themselves becoming wage earners. In 1964, the introduction of legislation that distinguished between birthplaces created further economic division between Greenlanders and Danes in Greenland (Petersen, 1995). In response, the Inuit political party was created.

It was clear that the modernisation policy had been pursued too hastily and without listening to local voices. Subsequent social problems, which continue to inform the stereotypes that many Danes appear to hold about Greenlanders, can be attributed to this process (Lidegaard, 1991; Sørensen, 1998).

A main obstacle to achieving equality both with Danes and within Greenland was the territory's infrastructure shortcomings and the great distances between outwardly isolated inhabited areas which were costly to maintain. In the early 1960s this issue spawned a centralisation policy, commonly known as G-60, whose aim was to eliminate inequalities by closing many small settlements and relocating entire populations to larger settlements where there would be fish processing plants and urban facilities.

Flora (2019: 21) has viewed such forced relocations as an opportunistic way for Danes and other southerners to become wealthy, providing a means to gain ownership and control over the land and the people who lived there. The impacts of G-60 varied in different parts of Greenland. A few residents refused to move, and in some cases the authorities relented and agreed to retain their settlement and its infrastructure – school, church, trade station, supply shop, postal service, fuel station, telegraphic communication lines and sea travel routes. But this was rare, and Flora (2019: 22) has questioned how and why was it that so few Greenlanders ever challenged or refused to do what was asked of them.

The most notorious example of the population concentration policy is often said to be the town of Qullissat (on Disko Island/Qeqertarsuaq), which was closed in 1972, leading to the forced, dispersed resettlement of its population. Qullissat was not, however, a hunting or fishing settlement but a coal mining town that had been founded as recently as 1924, and had grown to become one of Greenland's largest towns by the 1960s (Andersen, Jensen and Hvenegård-Lassen, 2016). Yet its notoriety reflects a wider complaint regarding the limits of Danish paternalism and the historical tendency for Greenland's 'development' to occur on Danish terms (Grydehøj, 2020a: 225).

The centralisation policy was not only spatial, but cultural, educational and oriented southward towards Denmark. Danish became the official language (Flora, 2019: 16–19, 170).

Much of the Arctic, including Siberia, has experienced a similar history of forced re-settlements and swift modernisation followed by a string of social problems, including suicide (e.g. Sampath, 1990; O'Neill *et al.*, 1994; Mancini Billson and Mancini, 2007). Suicide rates in Greenland, like other Arctic regions, grew rapidly during the 1960s and 1970s, and continued to do so well into the 1990s (Thisted, 1992; Lynge, 1994; Leineweber *et al.*, 2001; Hicks, 2007; Flora, 2011; Bjerregard and Larsen, 2015).[3]

4.6 Sovereignty?

Demands for autonomy increased during the 1960s and 1970s. There were two strands to this movement: a Danish-educated 'campus' anti-colonialism on the one hand, and a pan-Inuit movement on the other (Emmerson, 2010: 306–307). It expressed itself in a number of ways.

In the early 1970s, Greenlanders were opposed to EEC entry, largely to protect their offshore fish resources (Section 5.2). But, as a constitutionally integrated part of Denmark, Greenland unwillingly joined the Community in 1973. This experience generated demands for greater autonomy, and eventually, in 1979, Greenland was granted limited self-government in the form of Home Rule. An exit from the EEC followed in 1985 (Section 5.2).

As devolution of political responsibility spread across the northern hemisphere, in 2009, following a national referendum, the Self-Government Act (*Lov om Grønands Selvystre*, 2009), provided Greenland with political autonomy except in matters of defence and foreign affairs, with the Danish monarch remaining Greenland's head of state. The Act raised the possibility of eventual independence for Greenland, but this could only come about as a matter of negotiation.

In the Arctic, nation states and the Inuit Circumpolar Council (ICC) disagree over the interpretation of 'sovereignty': is it a fundamentally binary concept or a contested one in flux (Jacobsen, 2019)? As an incipient state with 80 per cent of its population of Inuit origin, Greenland appears to be uneasily positioned: on the one hand, as a self-governing component of the transnational Inuit community, while, on the other hand, aspiring to become a (nation?) state with full formal sovereignty (see Chapter 12 for the position of tourism mediating between these 'logics'). In striving for this latter objective, greater foreign policy sovereignty is an important factor in the enhancement of Greenland's international status and in its ability to attract investment. This has been emphasised with what appears to have been a reinvigoration of US interest in Greenland.

Chastened by increasing Chinese interest in Greenland and in particular by the prospect of Chinese involvement in (re-)developing two international airports (Section 8.3), the US embassy in Copenhagen published a 'Statement of Intent on Defense Investments in Greenland', laying out

> ... the principles for investment in Greenland to enhance U.S. military operational flexibility and situational awareness in order to address the changing security environment in the Arctic. ... the U.S. Department of Defense intends to pursue potential strategic investments vigorously, including investment that may serve dual military and civilian purposes. For example the U.S. Department of Defense intends to analyze and, when appropriate, strategically invest in projects related to the airport infrastructure in Greenland.
>
> (Quoted in Bislev, Gad and Zeuthen, 2018: 2)

A subsequent tripartite accord between the US, Greenland and Denmark was agreed in October 2020, following a $12.1mn (€9.8mn/£9.1mn) aid package for Greenland and the opening of a US consulate in Nuuk consequent upon the intentions quoted above. The agreement included:

- provision of improved civilian benefits from the Thule air base, including the ability of Air Greenland to use it for bad weather diversions, greater commercial air access for nearby communities, and access to acute and emergency care for local populations;
- the maintenance contract for Thule air base, which historically was held by a Danish–Greenlandic company until 2014, would revert to Greenlandic control when it was up for renewal in 2024 (representing around 10 per cent of Greenland's tax revenues);[4]
- the US and Greenland would more closely co-operate in trade and investment, energy and mining, education, tourism and nature management:
  - Washington was assigning a senior development advisor to Greenland to assess and direct future assistance;
  - the US was to promote investments in the mining sector, to strengthen technical training and expertise, and to improve the sector's overall security and resilience. A subsequent three-year $3mn project 'Leveraging Decades of Arctic and Mine Training Experience to Assist Greenland' would be based in Sisimiut (Bykova, 2021);
  - the US would co-operate with universities to build programmes focused on land and fisheries management, hospitality and tourism; in part this would be managed through the US National Park Service, co-operating in the development of sustainable tourism, eco-tourism, parks and cruise ship management (Humpert, 2020).

A renewed geopolitical interest in the Arctic has been welcomed by successive Greenland governments in providing an opportunity for forging new international relations while loosening dependence upon Denmark. At the same time, however, the Arctic has also become one of Denmark's top-five foreign policy priority areas, resulting in something of a reverse dependency relationship, given that it is only Greenland's geographic location and membership of the Danish Realm that legitimises Denmark's Arctic state status (Jacobsen, 2019).

Of growing interest, therefore, has been the way in which Greenland has been able to take advantage of international Arctic interest and Greenland's geostrategic location to articulate, act upon, and appear to hold, a more sovereign position internationally (Jacobsen, 2019). Ostensibly, the 2020 agreement with the U.S. might be seen as a prime example of this in that it: '... has a significant symbolic value for Greenland in the way that it enhances the Self-Government's role in the triangular relationship while strengthening its autonomous appearance on the international scene' (Jacobsen quoted in Humpert, 2020).

4.7 In the next chapter

Moving on from an historical perspective, Chapter 5 focuses on the hesitant evolution of Greenland's recent economic development, its interrelationship with aspirations for political independence, and the philosophy of economic 'self-sustainability' that has emerged in the face of growing global climate crisis pressures.

Notes

[1] 56,081 in 2020 (https://bankstat.gl/).

[2] Greenlandic school grammars are titled ABD as there is no letter 'C' in the alphabet.

[3] In 2014 there were 45 suicides representing 10.2 per cent of all deaths in that year. In 2015 the number fell to 32 and 7.4 per cent of deaths (Statistics Greenland, 2020: 10).

[4] Representatives from Vectrus (vectrus.com) – the company currently responsible for the operation of Thule Air Base – visited Ilisimatusarfik (Greenland University) in February 2021. The University was working on establishing collaboration with Vectrus to enable students to undertake internships at the base (news@uni.gl, 27 February 2021).

5

Greenland's 'Self-sustainability'

'... tensions [...] arise when the imaginary of climate change projected by science comes into conflict with subjective, situated, and normative imaginations and agendas of human actors who engage with nature and day-to-day politics'

(Bjørst, 2019: 123)

5.1 Introduction

This chapter briefly assesses Greenland's economic development, and how it has become intertwined with both political aspirations and climate change. Evolving the concept of economic 'self-sustainability', politicians have, on the one hand, adopted for Greenland the role of a less developed Indigenous society to avoid the responsibilities of international climate agreements, while on the other hand pursuing economic development as the basis for moving towards establishing a politically independent state.

5.2 The legacy of fishing and hunting

Fishing and hunting have been the sources of social and economic survival in Greenland's harsh environment since migratory groups first settled there. In recent years the fishing industry has flourished, in part due to retreating sea ice.

By the 1920s, the traditional foundations of Greenland's economy were beginning to disintegrate. Notably, the number of seals caught declined from eight per head of population in 1910 to below five by 1935. Backed by government development policy and a warm period of Arctic climate, fishing became increasingly important. Greenlanders owned 157 wooden boats in 1900; by the late 1930s there were 1,500. Thus a hunting-based economy was transformed into a fisheries-based one, and the relationship between Greenlanders and their environment was fundamentally transformed. Hunters were becoming industrial employees, and the construction of fish processing plants required a concentration of the island's population in fewer places (Emmerson, 2010: 305), a contributory factor to the G-60 concentration policy of the 1960s (Section 4.5).

Greenlanders had been hostile to Faroese fishing rights permitted in Greenlandic waters in the 1920s and 1930s, and they were equally opposed to Denmark's moves towards EEC membership, being fearful of wider European access to fishing grounds. Thus Greenland voted against accession in the 1972 referendum. Outvoted by 'mainland' Danes, however, Greenland acceded to the EEC with its 'parent' state in 1973.[1] This experience eventually led to Greenland gaining limited self-government in the form of Home Rule in 1979 (Section 4.6).

The same year saw a 'nationalisation' of the halibut industry. Greenland took over ownership of the Royal Greenlandic Trade company (KGH) and renamed it *Kalaallit Niuerfiat* (KNI). This was

DOI: 10.1079/9781789246728.0005

an important development for local communities in the years following the traumatic G-60 policy. KNI was now able to bypass the old monopoly with Denmark, and export halibut to such markets as the UK and Japan (Flora, 2019: 22–23).

Apart from Algeria back in 1962, Greenland became the first territory to withdraw from the EEC, in 1985, following continued unease over fishing rights (e.g. Rebhan, 2016). Some commentators would argue that such (de-)integration decisions were more closely related to identity concerns than with economic rationality (Gad, 2014),[2] arguments that could equally apply to a similar withdrawal undertaken much further south three decades later (Hall, 2020).

After prolonged multilateral negotiations, this large island, representing more than half of the EEC's total area, and as part of the territory of a committed member state, was thus exempted from membership. Greenland was transferred into the category of Overseas Countries and Territories (OCTs) that had a constitutional link with a member state (Gad, 2017a). The differing statuses of Greenland, the Faroe Island and Denmark itself in relation to the EEC/EC/EU have provided Denmark with space for diplomatic manoeuvring, that Gad (2016, 2020; also Jacobsen, 2019) refers to as 'postcolonial sovereignty games'. Greenland's first agreement with the EC within the OCT framework was, unsurprisingly, over fisheries (in 1985) and was intended to ensure that the territory received compensation for keeping its waters open to European fishing fleets.

Subsequently, the government has tried to help the local fishing industry by forcing Royal Arctic, Greenland's domestic shipping monopoly, to keep freight rates artificially low. After disappearing in the 1990s, cod are returning to the fishing grounds off western Greenland. Shrimp are now caught off Upernavik, further north than earlier.

Hunting quotas regulate the harvest of selected maritime species, and while seal, arctic cod and halibut are abundant; other animals, such as narwhals and belugas continue to remain subject to government regulations. All whale meat and *mattak* (whale skin with blubber, a local delicacy) is sold locally. Whale quotas are set by the Greenland government (*Naalakkersuisut*) annually and are subsequently distributed to local districts, where the municipal authorities decide on the quotas allocated under commercial and leisure hunting licences. Other species which local residents hunt include birds, walrus, seals, polar bears, musk ox and reindeer.

Because of the retreating ice, ice fishers are now increasingly required to travel with their sleds and dogs in boats. However, the growing number of boats, which can now be used for a longer period in the year because there is less ice, also bears witness to fishing prosperity. The price of halibut has tripled in a decade. The annual catch in Ilulissat, for example, is worth some DKK500mn (€65mn/£60mn/$80mn) (Hook, 2019), and the busy harbour here is flanked by a halibut-processing facility and a shrimp-processing plant (Fig. 5.1).

Only a few hundred kilos of fish can be loaded onto a sled. But a small local boat can manage two tonnes: local entrepreneurs thus welcome global warming. This position is reinforced by a perceived reduction in domestic heating costs, and the fact that supply ships can access harbours for a longer season to restock grocery shelves.

5.3 Self-government

In 1999 the adjacent part of the Northwest Territories of Canada became the new Inuit territory of Nunavut ('Our land' in Inuktitut) with devolved powers. Ten years later, Greenland was granted self-government (Section 4.6), with the Danish annual subsidy guaranteed[3] (Emmerson, 2010: 308–310). The Self-Government Act also signified the possibility of eventual independence.

Exploitation of Greenland's mineral wealth could provide an economic basis for independence and the creation of the first truly Arctic state, with a majority Inuit-origin population. The prospect of income from natural resource exploitation, including tourism, raised the optimistic aspiration that Greenland could become economically self-sufficient within a generation. Hesitant onshore mining could be eclipsed by offshore oil and gas. Onshore (Chapters 7 and 9) and 'offshore' (Chapter 11) tourism would also play their part.

Significantly, part of the Self-Government Act stipulates that as Greenland's income from mineral wealth rises, the subsidy falls by half that amount (with a threshold of DKK75mn received by the Greenlandic government from mineral extraction before the rule kicks in). Despite faltering industrial projects, for many

Fig. 5.1. Ilulissat harbour: in the middle distance the shrimp processing plant is on the left and the halibut processing facility on the right. The large number of small boats in the foreground emphasises the everyday role of the sea for both fishing and transport. Author's own photograph.

Greenlanders independence depends almost mechanically on rising resource income. On the other hand, Greenland's geostrategic position as Denmark's international bargaining chip justifies subsidising Greenland as a security guarantee, perhaps justifying it as defence expenditure.[4]

Could an (independent) Greenland be in a position to give up Denmark's 'protection', and with it the friendship of NATO, the EU and all that this implies (Emmerson, 2010: 291–293), with the annual block grant from Copenhagen and transfers from the EU making up around 42 per cent of Greenland's government revenue (Statistics Greenland, 2020: 34)?

5.4 Mining the 'resource frontier'

Greenland's natural resources have been perceived by outsiders as a source of (potential) wealth since at least the sixteenth century (Section 4.3), but until the twentieth century mining was small scale and artisanal. The most significant was the cryolite mine at Ivittuut, opened in 1854 (Fig. 5.2), which provided a steady income for the Danish state. When the mine closed in 1987, 3.7mn t had been produced. In the twentieth century cryolite had been important as a key flux in the electrolytic processing of bauxite to produce aluminium.

Interest re-emerged in exploiting Greenland's mineral resources at the beginning of the twentieth century, although mining was only permitted under strict conditions. Thus a concession granted in 1902, which covered everything other than cryolite, stipulated that no local labour was to be used in the mines and there was to be no contact with the local population (Emmerson, 2010: 305). Such segregation could be interpreted as assisting the sustainability of the local culture and economy by protecting them from twentieth-century encroachments. Yet how far did the claimed Danish fear of an influx of capital and foreign workers undermining the traditional structure of Greenlandic society represent a colonial mind-set bent on preserving the 'noble savage'?

Under these conditions, a small number of mines opened: for copper (1905), graphite (1915) and coal (1924). But these were hazardous, short-term enterprises, and only the cryolite

Fig. 5.2. Location map of mineral exploration and exploitation. Author's own figure, modified and redrawn from Têtu and Lasserre, 2017: 6.

mine appeared sustainable. Indeed, in the 1920s demand for cryolite boomed following technological improvements in aluminium smelting, with a consequent increase in demand. The mineral became a major source of Greenland's revenue, and by the late 1930s 500,000t was being exported annually, mostly to Denmark (until the country was occupied by Nazi Germany in 1940), and then to the US. By 1942 exports had reached 85,000t (Emmerson, 2010: 305–306).

By the time Greenland's Bureau of Mineral and Petroleum (BMP) took over control of the minerals sector in 1998, there was declining international mining interest, resulting from poor commodity prices compared to high production costs.

In 2008 the Danish–Greenlandic Self-Rule Commission concluded a series of negotiations on mineral rights, ownership of sub-soil resources, and the administration of revenues from mining and hydrocarbon development. It emphasised that minerals in Greenland's sub-soil belonged to Greenland and that the country had a right to their extraction (Dodds and Nuttall, 2016). Consequently, *Naalakkersuisut*, Greenland's Self-Rule Government, encouraged an expansion of extractive industries to achieve a strengthened economic basis for Greenland's future development. To improve local skill levels a mining school was opened in Sisimiut.

The Mineral Resources Act 2009 regulates all extractive activities, and places an emphasis on securing local benefits through employment opportunities (Hansen, Larsen and Noble, 2018: 386). This marks a significant shift from the 1990s and early 2000s when the agenda set by the Arctic Environmental Protection Strategy (AEPS) and Arctic Council prioritised issues relating to conservation, pollution, climate change and sustainable development (Chapter 2). The circumpolar agenda had become more business focused.

The relationship between the perceived political desirability of mineral extraction and a malleable conception of sustainable development was articulated at the eighth Arctic Council Ministerial Meeting's Kiruna Declaration of May 2013 recognising that 'Arctic economic endeavors are integral to sustainable development for peoples and communities in the region ...' (Arctic Economic Council, 2014).

The US Geological Survey's publishing, in successive years (2007/8), optimistic estimates for hydrocarbon resources around Greenland's seas largely triggered the hyperbole of Greenland as a 'resource frontier' (Chapter 2). Estimates of 31.4bn barrels of oil equivalent off north-east and south Greenland (USGS, 2007), complemented by estimates of up to 7.3bn barrels of oil, and further 52 trillion cubic feet of natural gas in the West Greenland–east Canada basin (USGS, 2008) were widely reported as equating to 30 per cent of the world's 'undiscovered' natural gas deposits and 13 per cent of its crude oil (Klare, 2013). Uncertainty surrounding these figures was substantial (Emmerson, 2010: 314).

Nonetheless, Greenland's 'Oil and Mineral Strategy 2014–2018' (Kirkegaard and Kielsen, 2014) estimated that an income of over DKK30bn (c. €4bn/£3.6bn/$4.8bn) could be expected over a period of 15 years from minerals and DKK435bn (c€56.5bn/£52bn/$70bn) over 40 years from oil and gas. Geopolitical declarations of a twenty-first century version of the 'Great Game' (Crawford, 2013), and of superpowers jockeying for position (Rosenthal, 2012), firmly fixed attention on Greenland's apparent fulcrum position in global resource frontier rhetoric.

At the same time, environmental groups such as Greenpeace had been warning and demonstrating that the Arctic needed to be protected from further industrial exploitation (Dodds and Nuttall, 2016). An international perception that climate change, coupled to a hitherto geopolitically stable region, was assisting access and exploration, now co-existed with reactions against the consumption of fossil fuels.

A Danish royal order of 1963 had established the territorial sea of Greenland extending out to three nautical miles. Denmark ratified the United Nations (1994) Convention on the Law of the Sea (UNCLOS) in 2004, and is consequently responsible for the marine areas in the exclusive economic zone (EEZ) around Greenland. In accordance with the Act on Greenland Self-Governance, Greenland exercises authority over the mineral resources but has not taken over responsibility for the protection of the environment in the marine areas beyond three nautical miles. In the Danish version of the 2013 European Parliament and Council Directive on the safety of offshore oil and gas operations, the sea around Greenland was not included, suggesting that EU environmental law did not cover the EEZ (Pelaudeix, 2017: 315–316). This has implications for offshore working conditions in these areas (Box 5.1).

Exploratory work for oil has taken place in areas that are close to hunting and fishing communities and off the north-east coast, in waters adjacent to the world's largest national park (Fig. 4.1). Dodds and Nuttall (2016: 135) talk of certain landscapes and waters becoming 'zones of sacrifice', set aside for extensive development and intensive resource exploitation, within which 'the complexities of human-environment relations are not just ignored but erased'.

While the notion of deriving economic strength from industrial activities supports arguments moving towards independence, such activities also represent the beginning of new forms of dependency relationships as multinational corporations increasingly exert influence over Greenlandic politics and business (Nuttall, 2012, 2013; Grydehøj, 2020b) ... and concepts of sustainable development.

Agreements have been entered into with an array of international mining and hydrocarbon companies. The territory has potentially vast deposits of gold, iron ore, zinc and REEs/metals, with estimates of between 9 and 25 per cent of the world's deposits of the latter.[5] Notably, ilmenite sands here (mined at Dundas, close to the US Thule air base: Fig. 5.2), from which titanium is taken, are some of the world's largest and most pure deposits. The EU considers Greenland to be a major European future supply source, although ilmenite is currently mostly shipped out to Quebec for refining, providing some 30 per cent of the world's titanium (Milne, 2019).

Chinese companies have been active in mineral extraction in Greenland, including the General Nice Group, developing iron ore at Isukasia, while others have been involved in gold exploration in Ittoqortoormit on the east coast (Fig. 5.2) (Têtu and Lasserre, 2017; Dubois and Gagaridis, 2018). Isukasia was the first large-scale mining project granted in Greenland. It is located in a traditional reindeer hunting area, deep into the fjord system 150km from Nuuk (Foley, 2018).

Initial plans here were to transport the ore as slurry, through a 105km pipeline to a purpose-built west coast port in Taseraarssuk Bay, regarded as ice-free year-round. But issues of access – road, pipeline, harbour – and the threat of local social and environmental impacts, brought substantial opposition within the country on the grounds that the project would disturb sensitive ecosystems and reindeer habitat, as well as excluding hunters and fishers. Further, a government Large Scale Project Act had created a framework for companies to bring in foreign workers exempt from the country's labour standards for the construction of large scale mining projects: such plans for Isukasia generated additional opposition.

Bilateral relations with China started in 2012 when Beijing's minister of land and resources visited Greenland and proposed to create a cartel of REE-producing countries (Mered, 2018). Representatives of ten large Chinese companies visited Greenland in 2013 to seek investment opportunities, and the Greenland minister of industry and mining visited China later

Box 5.1. The EU dimension

EU Arctic policy has been developing since 2008, alongside a growing awareness of the global impacts of climate change on international security. Brussels' sincerity in encouraging sustainable development has been questioned in view of its explicit energy and mineral resource interests in the Arctic (Solana and Ferrero-Waldner, 2008). The Union remains dependent on oil and gas imports from Norway and Russia, and has needed to diversify its imports of rare earth elements (REEs) (Pelaudeix, 2015).

An Observer status application was made to the Arctic Council in 2008 which has been repeatedly deferred, both because of Canadian opposition in response to the EU's ban on seal products, and a block by Russia seeking to limit the EU's influence in response to the sanctions imposed following Russia's annexation of Crimea.

An early interest specifically in Greenland was reflected in a European Commission Communication in 2008 arguing that additional efforts were required to position the EU as a more important partner for managing Greenland's fragile environment and 'the challenges confronting its population' (Pelaudeix, 2017: 311). The low level of European companies' representation in Greenland's raw materials development is a source of continuing concern for the EU.

After China placed restrictions on its export of raw materials and REEs in 2009, EU officials expressed concern at the EU dependence on REE imports and the prospects of Chinese investment in Greenland's resources and infrastructure. Subsequently, a 2011 European Parliament resolution on sustainable policy in the High North re-emphasised Greenland's resource importance (Pelaudeix, 2015). Indeed, in June 2012 the EU sought a deal with Greenland: to share exploitation rights to REE metal ores in return for technological and environmental mining know-how. A letter of intent was signed by both parties, but concern was expressed over the heavy-handedness of the partnership agreement within which such deals fall (Pelaudeix, 2017).

that year. In 2014, China Nonferrous Metal Industry's Foreign Engineering and Construction Company (NFC), prominant in REE processing, signed a memorandum of understanding with the Australian company Greenland Minerals and Energy regarding joint development of the Kuannersuit/Kvanefjeld REE-uranium project (Box 5.2) (Dodds and Nuttall, 2016; Dubois and Gagaridis, 2018).

Towards the end of 2020, as part of its renewed interest in wanting to invest in Greenland, the US Export-Import Bank (EXIM) indicated that it was prepared to loan $216mn to Australia-based Ironbark Zinc to develop the Citronen zinc and lead mine in Greenland's far north-east (Fig. 5.2). The loan would be in support of 'US content'. This arose shortly after – in his last weeks in office – the then President Trump issued an executive order declaring America's reliance on foreign suppliers for 31 of 35 critical minerals a 'national emergency'. The order called on federal agencies both to find ways to increase domestic production and to ensure that imports could not be disrupted by China. Zinc was not on the list of critical metals, but is considered strategic in its use for galvanising steel and other metals. The US imports more than 80 per cent of its zinc requirements.

Ironbark Zinc, which is partnered by SOE China NFC, was granted a 30-year licence in 2016 to operate at Citronen. Production was due to begin in 2020. But from a 2017 peak, the world price of zinc fell. As the price began to recover, Ironbark Zinc was to produce a revised feasibility study for 2021. The site is said to have sufficient deposits for at least 14 years' mining operations (McGwin, 2020a).

Yet questions remain over how the company intends to transport 300,000t of ore from the mine during the six weeks that Citronen Fjord, located at 83 degrees north, is accessible to large ice-classed bulk carrier ships.

The line between the economic and political are increasingly blurred in the context of China's ever-expanding 'Belt and Road Initiative' (BRI), which is simultaneously a top-down central government strategy, a collection of bottom-up local strategies for harnessing

Box 5.2. A resource frontier: uranium mining

Exploitation in Greenland had been ongoing for some time, such that by 1980 4,200t of material from the Kuannersuit/Kvanefjeld plateau, 8km from Narsaq, was being shipped to Denmark for the extraction of uranium. But in the 1980s such activities were prohibited.

There are REEs in Kuannersuit that are physically bound to the uranium, and it is claimed to be impossible to extract these elements without also extracting uranium: both a constraint and source of added value.

In 2013, Greenland's parliament passed by one vote a law to reverse the previous uranium zero-tolerance policy, and in 2015 Prime Minister Kielsen announced that the government would support recommencing uranium/REE mining.

Subsequently, Chinese firm NFC has been a minority investor in the Kuannersuit mine (Têtu and Lasserre, 2017), while Shenghe Resources has been partnering Greenland Minerals and Energy to develop the project (Gronholt-Pedersen, 2018). Although the regulation of extraction and trade in uranium materials is governed within the EU by the Euratom Treaty, and thus Denmark is required to abide by it, it does not apply to Greenland following its withdrawal from the EC: the OCTs association agreement excludes Greenland from any obligation regarding uranium.

However, Denmark retains competence and control capacity in the trade of uranium by virtue of the material's security element, and in January 2016 the governments of Greenland and Denmark signed a set of agreements on co-operation regarding future mining and export of uranium from Greenland.

Protests and resistance have reflected mistrust about the real intentions of stakeholders. Andersson, Zeuthen and Kalvig (2018) argue that Chinese involvement in REE projects is primarily motivated by interest in the strategic resource itself, and that the country has a well-developed and clearly defined national strategy for REE. By contrast, other Chinese involvements in Greenland, the authors claim, are more determined by China's foreign policy priorities of securing raw materials while gaining greater access to the Arctic region.

Sources: Danish Ecological Council, 2013; Nuttall, 2013; Bjørst, 2016

Chinese investment and a framework by which Chinese companies (in some cases SOEs) can expand their activities internationally. With the emergence of a Chinese 'Polar Silk Road' strategy some observers have argued that it may no longer be enough to say that Chinese interest in Greenland is only concerned with business and scientific research, given that these are themselves of strategic significance. But then much the same could be said for US interest(s) in Greenland.

5.5 Hot Greenland?

In contrast to the ice-'cool' imagery of Greenland (addressed in Chapter 9), climate warming has been perceived as helping to roll back the frontiers of natural resource exploration and exploitation. As a result, according to Ren (2012, 2016), in the 'hot' vision of Greenland, roles have shifted: local empowerment, global interest and foreign investment now steer the development agenda, while representations of sustainability and the need to address climate change have become, at best, muffled (Ren and Bjørst, 2018).

This vision of social and economic transformation enables a 'new' Greenlandic society to claim its right to act, to industrialise and to enter and engage with a global economic and political agenda, as 'the ice' makes way for exploring and accessing Greenland's natural (and cultural) resources. In this reality '... climate is a multiple object but it does not necessarily connect to everyday life; not even for the people who are said to be at risk' (Bjørst, 2019: 132, citing Bjørst, 2012). Sustainability has become a political concept in that it defines and shapes competing trajectories of the present into the future. Arctic stakeholders, not least in and of Greenland, agree that development needs to be 'sustainable', but there has been little outward agreement concerning exactly what the object and objective of such sustainability actually is. Originally in conservationist discourse, 'the environment' was the referent object, but as sustainability discourses have expanded, the concept has been linked to an increasing number of object(ive)s such as society, economy, culture and identity (Gad, Jacobsen and Strandsbjerg, 2019).

Although Greenland, as part of Denmark, ratified the Kyoto Protocol in 2001 almost without discussion (Bjørst, 2019: 123), Nuuk's own positioning in terms of climate change, sustainability and economic development was emerging. The 'hot' nature of Greenland's development trajectory was becoming intertwined with the country's pathway away from climate concern and its narrative of sustainable development as a socio-economic goal (Ren and Bjørst, 2018). But increasingly enmeshed in big power geopolitics, Greenland has also found itself in a state of 'hesitant transition', with economic and political implications for the future.

5.6 Greenland's pathway away from climate concern to a conception of sustainable development

The Greenland political establishment has tried to maintain a two-pronged climate strategy as a way of coping with its commitments. On the one hand the government has encouraged a global reduction of CO_2 emissions, while, on the other, it has declared a wish not to be constrained by the same global standards for its domestic national development agenda (Bjørst, 2008, 2019; Bjørst and Ren, 2015).

It is relatively new in Greenlandic politics to talk about the climate as global (as opposed to local: see Sejersen, 2015, 2019), and of climate change (Bjørst, 2012):

> As the world becomes more aware of the importance of the Arctic, the concept of Arctic risk is becoming globalised, advancing technocratic discourses and solutions that suit metropolitan (rather than Arctic) interests ... The peoples of the Arctic are not responsible for having produced global environmental threats, yet metropolitan society calls upon us to forego economic development and extractive industries, take symbolic action against global challenges and stand at the vanguard of a new era of 'conspicuous sustainability' (Grydehøj and Kelman, 2016).
>
> (Markussen, 2017: 305, 307)

Greenland's first climate strategy – for the reduction of CO_2 emissions to meet the Kyoto goal for 2012 – was introduced only in the autumn of 2009 (Grønlands Selvstyre, 2009), some eight years after Greenland had ratified the Kyoto Protocol with Denmark (Bjørst and Ren, 2015: 95). At this point Greenland was failing to meet its Protocol obligations to reduce emissions by 8 per cent, and its commitment was only fulfilled through an agreement with Denmark for the purchase of CO_2 credits (Bjørst and Ren, 2015: 95).

For the second Kyoto period (2013–2018), Greenland was exempted from international reduction commitments through a territorial exclusion (ClimateGreenland.com 2014; Bjørst and Ren, 2015: 95). At COP16 in Mexico City, the Greenland premier (Kuupik Kleist) successfully advocated special provision for 'developing countries' (which included Greenland) and for Indigenous peoples to embrace rights that enabled development. This critical point is expressed by Bjørst (2019: 128):

> By reintroducing the embedded inequalities in the UN system (Greenland not being recognized as an independent state), the set-up for CO_2 reduction (differentiated between developed and developing countries), and the importance of indigenous peoples' rights, he [Kleist] hinted that what was to be sustained was indigenous peoples' right to (sustainable) development. This argument is still in line with the advocacy for a 'common but differentiated responsibility' but is now positioned as an indigenous peoples issue. This position came to be endorsed by Danish politicians ...

Greenland's obligation to be part of a binding climate deal was thereby circumvented, and the country emerged presenting itself as 'more Indigenous and less national' (Dahl, 2013). While consonant with the philosophy of the ICC, for example, this was in sharp contrast to the Westphalian state-centric approach Greenland was (and is) taking in its aspiration for political independence (Gerhardt, 2018), yet an independence that would, it was hoped, be funded by the large-scale industrial development and mining projects that were to generate increasing emissions:

> ...we will double our CO_2 emissions – but not developing Greenland because the rest of the world has been polluting is not an option ... I do not feel guilty for developing new industries. Of course, we will do it as environmentally as possible.
>
> (Henrik Leth, Chairman of Business Greenland, 2015, quoted in Bjørst, 2019: 121)

The mid-2010s prominence of the UN Sustainable Development Goals (SDGs) (United Nations, 2015) saw the Arctic sustainability narrative in Denmark and Greenland come to reflect an explicit business agenda (Ministry of Foreign Affairs, Denmark, 2017b), while the Greenland government's oil and mineral strategy had already talked about 'sustainable mineral resource development' (Government of Greenland, 2014: 90; cited in Bjørst, 2019: 125).

On the other hand, the Greenland government pointed to the fact that it had invested 1 per cent of GDP annually in renewable energy throughout the previous decade (ClimateGreenland.com, 2014; Bjørst and Ren, 2015). But for successive Greenland governments, building up new industries and gaining increased political and economic independence from Denmark was of prime importance, relegating climate concern to an, at best, secondary position. 'Sustainability' for Greenland now firmly meant 'economic self-sustainability' (Coalition Agreement, 2014: 3; Bjørst, 2019: 129). This also framed tourism development plans (Government of Greenland, 2012) and conferred new roles on tourism as generating income for supporting independence (Bjørst and Ren, 2015).

For its part, climate change was not considered a looming catastrophe but simply an element of the quotidian that had to be contended with, even by those said to be at risk (Bjørst, 2012, 2019). Thus Bjørst (2019: 130) could conclude that 'Greenlandic climate policy can be best described as a policy that never really existed'.

As climate agendas disappeared from Greenlandic politics, the concepts of sustainability and sustainable development seemed to revolve largely around the development of extractive industries as the road to 'sustainable development' *in* and *for* Greenland (Bjørst, 2016). The referent object for this interpretation of 'sustainability' was the national budget, with a focus on creating local jobs, ensuring capacity building and a better living standard: climate concern was not seen as relevant (Bjørst, 2019: 131), except in positive terms as facilitating greater accessibility to resources.

In this narrative, therefore, the sustaining of Greenland's road to political independence is through economic development and the potential, brought about by climate change, to support ideals of 'self-sufficiency'. Ice retreat and increased run-off from glaciers have contributed to the widely accepted wisdom that 'Greenland's rivers and lakes have never been as large', thereby boosting hydro-power potential. There is a reported belief that Greenland could also become self-sufficient in sheep farming and vegetable production with a warming climate (Emmerson, 2010: 316–317).

Overall, in her critique of Greenland government and business sector narratives, Bjørst (2019) points out that while Greenland and the Arctic environment are subject to profound change, and that there is a need to move away from 'business as usual' to substantially reduce greenhouse gas emissions, politicians and business communities still seem to think that this does not apply to Greenland. While 'sustainability' and 'sustainable development' are repeatedly invoked, climate change is virtually absent in official discourse.

5.7 Not so much 'hot' as 'hesitant transition'?

Returning to Ren's (2012, 2016) suggestion of Greenland's imagery being transformed by climate change from cold to 'hot', she qualified her musing by concluding that in fact nature has a tendency of getting in the way of hot visions:

- as environmentalists challenge the building of facilities for mining extraction in vulnerable areas;
- as wind and weather conditions preclude the (profitable) building of a long-haul airport at central tourism sites;
- as drilling for oil turns out to be too difficult due to the geological composition of the bedrock;
- as global climate advocates urge Greenland to support an agreement on lowering carbon emissions (Ren, 2012: 155).

Indeed, over the past decade, although the country's air transport accessibility is changing, raising far from supportable expectations of a tourist 'boom' (Chapters 7 and 8), industrial development has not met expectations, and active mining projects remain limited, mostly confined to smaller and mid-sized companies. While increasingly subject to geopolitical manoeuvrings, industrial development has been '... slowed by a lack of infrastructure and heavy red tape' (Gronholt-Pedersen, 2018), as well as increasingly variable climatic conditions and uncertain world commodity prices.

For the short-term at least, notwithstanding tourism aspirations, Greenland remains dependent upon Denmark, with the block grant continuing to constitute around a quarter of Greenland's GDP. And while all nationally elected politicians are Greenlandic, the bureaucracy, education system and business community remain highly dependent on imported Danish skilled labour (Chapter 6), which, in Grydehøj's (2020a: 221) opinion, results in a prevalence of Danish policy solutions to Greenlandic problems. Equally, Denmark needs Greenland in order to be able to sustain its role as an Arctic power and as a player within NATO.

Given the underwhelming mineral resource development performance that Greenland has enjoyed thus far, perhaps the concept of 'hesitant transition' is appropriate for the country's uncertain journey towards independence. Although rarely defined, 'hesitant transition' appears to have been initially employed to broadly characterise a slower than anticipated pace and depth of political and economic transition (in post-communist Central and Eastern Europe) (O'Neill, 1996). Later it was employed more effectively, conceptually, to qualify aspects of 'lagging' socio-economic 'transition', and to contrast with the immediate and substantial impacts of 'shock therapy' policies (Gentile and Marcińczak, 2014).

Subsequently, the term has been applied in such sectors as business modelling (Feng *et al.*, 2016), the adoption of renewable energy (Adve, 2019), education (Bloomfield, Harreveld and Fisher, 2020) and sustainable product marketing (Coburn, 2019). The similarly interpreted term 'ambivalent transition' has also been employed in relation to socio-economic change in Central and Eastern Europe (Creed, 2010) as well as in the less apposite context of paediatrics (Lundqvist, Nilstun and Dykes, 2002).

In relation to Greenland's economic development processes, 'hesitant transition' can perhaps be applied in three senses. First, it can be observed relating to the hesitant negotiations with multinational oil and mineral exploitation corporations resulting in uncertainty and hesitance in capital investment decisions. Second, application of the notion of 'hesitant transition' recognises the different trajectories of different sectors – a 'variety of paths of socio-spatial development' (Marcińczak *et al.*, 2014: 1399). This also reflects the concept of 'multiple transformations', which characterises different spheres of socio-economic endeavour being transformed at different paces (Sýkora and Bouzarovski, 2012). Sub-sectors within onshore and offshore development, as well as in tourism, for example, will experience impacts within different time frames, with varying speeds and in relation to different aspects of political decisions. Third, the term 'hesitant transition' may be extended to the community dimensions of socio-economic change in the reported unwillingness of some Greenlanders to abandon their traditional pursuits for the sake of taking up regular full-time industrial employment (Grydehøj, 2020a).

By contrast, the impacts of Covid-19, particularly on the tourism and travel sectors, might be interpreted as (unplanned and unforeseen) 'shock therapy', in the way they have perforce stimulated and encouraged domestic tourism in Greenland alongside a re-assessment of international tourism marketing (Chapters 10 and 12):

> Infectious disease cannot be separated from geopolitics, due to its inseparability from relations of space, power, and resources ... [and the Covid-19] situation has also highlighted Kalaallit Nunaat's capacity to stand by itself, effectively designing and implementing policy in tense and rapidly developing situations ... the Covid-19 pandemic may come to play a similar role in Kalaallit Nunaat culture, helping to clarify and rebalance what, precisely, Kalaallit Nunaat is capable of doing for itself, what must come from Denmark, and what can come from elsewhere in the world.
>
> (Grydhøj, Kelman and Su, 2020: 295, 293)

Further, it remains to be seen at the time of writing what level of hesitancy, not least among potential investment companies, the Danish moratorium on North Sea oil and gas activities (Box 5.3) will have on Greenlandic economic development attitudes.

It has been estimated that Greenland would need 20 large-scale mining projects in order to be able to finance its political independence, a situation which is far from today's reality (Taagholt and Brooks, 2016: 362). Further, the Committee for Greenlandic Mineral Resources (2014: 23) argued that

> Extracting sufficient mineral resources to Greenland's independence within 20 to 30 years would require such extensive foreign investment and massive inflow of foreign labour that there is a real risk that the current Greenlandic population would become a minority in Greenland.

Moreover, since a significant portion of the Greenlandic workforce is under-qualified and under-educated, independence through resource exploitation does not seem likely in the short term (Taagholt and Brooks, 2016). Nonetheless, while Greenland continues to view its resource potential – both for tourism and for mineral extraction – as critical to achieve a greater economic independence from Denmark and thereby lever political independence, a potential paradox is created with the stated determination to maintain its traditional lifestyle and culture. As Dubois and Gagaridis (2018, emphasis added) starkly put it: 'Greenland cannot achieve *economic* independence without foreign workers.'

With Greenland considering exploiting its rare earths reserves, China is impossible to circumvent, as it holds a near monopoly on REE processing after extraction was given up by the

Box 5.3. Danish moratorium on North Sea oil and gas activities

In December 2018 Copenhagen announced that all Denmark's offshore hydrocarbon activities – with 55 platforms operating across 20 oil and gas fields – would cease by 2050, and that current and future licensing rounds had been cancelled with immediate effect.

Although small by global standards, Denmark's production of 103,000 barrels equivalent per day in 2019 rendered the country the largest oil and natural gas producer in the EU, representing an annual income of DKK13bn (€1.7bn/£1.5bn/$2.1bn) (Ambrose, 2020; *Herald*, 2020).

Following France's announcement in 2017 to phase out its hydrocarbon exploration and production by 2040, the Danish move was seen as a further incentive for countries such as Norway and the UK to follow in reducing emissions and meeting their 2015 Paris COP 21 commitments (Helm and McKie, 2020).

West (Zeuthen, 2017). While large numbers of foreigners may arrive in Greenland either by air or by cruise ship and be welcomed as tourists (Covid-19 notwithstanding), their presence is relatively short-lived and reasonably concentrated seasonally (despite efforts to extend the tourist seasons). By contrast, scenarios of large numbers of low-cost Chinese labour staying in Greenland for perhaps two years or more could represent a threat to Greenlandic identity, at least (Taagholt and Brooks, 2016; Têtu and Lasserre, 2017).

In addition to the security and cultural dimensions, Copenhagen has expressed concern that for China, Greenland's mineral resources, not least uranium, possess a potential dual use (Rasmussen and Merkelsen, 2017). Significant Chinese access to these may complicate Copenhagen's relations with Washington, which considers Beijing as its main strategic competitor, and, partly as a result, is taking a much closer interest in investing in Greenland's (economic and strategic) development (Sections 4.7 and 8.3).

And what of the role of China in a sustainable post-Covid Arctic?

> Some Chinese policy analysts seem to regard the economic setback of Western powers caused by coronavirus as an opportunity to advance China's interests and BRI operations at a regional level. ... [Under these circumstances] there seems to be little prospects [sic] that the BRI, or its offspring Polar Silk Road, would constitute a driver of sustainable development or low-carbon coronavirus recovery in the Arctic or beyond.
>
> (Kopra and Nojonen, 2020: 14)

And while

> the Polar Silk Road is not likely to be prioritized in any way but will continue to be a sub-initiative of the broader BRI ... there might be a risk that China will take further advantage of the global economic predicament and attempt to gain stronger control of the two major sectors in the Arctic region: infrastructure building and natural resources.
>
> (Kopra and Nojonen, 2020: 21)

5.8 In the next chapter

From this brief economic development context-setting, Chapter 6 shifts the focus onto the nature of Greenlandic culture and identity, and their role in the development of tourism and tourism-related processes.

Notes

[1] As part of their 1948 Home Rule arrangement, the Faroe Islands were later able to remain outside of the EEC.

[2] '... Greenexit was a poor precedent for anything' (Gad, 2017a: 92).

[3] A sum of more than DKK3.4bn (€442mn/£408mn/$578mn) – around €7,880/£7,275/$10,305 for every Greenland resident.

[4] During the Cold War the Soviet Union sought consular representation in Greenland (and the Faroe Islands), which Denmark denied. The appointment of an honorary consul was finally approved in 2020, and it followed the opening of a US consulate in Nuuk earlier that year. There are 11 honorary consuls in Greenland representing 14 countries (Bye, 2020).

[5] REEs are a set of 17 chemical elements which have become important in alternative energy technologies, lifestyle and industry electronics, and the defence industry.

6

Greenlandic Identity and Culture within Development Processes

'The way of the Inuit is to struggle with nature ...' (Karl Kristian Kruse in Gavron and Katznelson, 2012)

'... with Greenland opening up to the rest of the world, along with tourists, research scientists, mineral prospectors, environmentalists, foreign politicians and investors, the efforts of journalists, artists, photographers, television crews and documentary makers have connected Greenland to the world in new ways. Social media, the Internet and information technology have likewise made the world beyond seem much closer than in the past and have exerted a profound influence on local life'

(Flora, 2019: 20)

6.1 Greenlandic identity

This chapter aims to unravel the strands of Greenlandic identity and culture and to then examine the elements of these that can contribute to (sustainable) tourism development and sustain the road to political independence. This is later complemented by a critical evaluation of the 'ethnic' and 'modern' logics contributing to Greenland's future, through a tourism lens (Chapter 12).

Greenlanders have for a long time engaged in discussion about what Greenlandic identity might be, how it is manifested and expressed. While some 80 per cent of Greenland's population may be regarded, ethnically, as Inuit, some 88 per cent of the total population were born in Greenland and thus may regard themselves as true Greenlanders (Bogoyavlenskiy and Siggner, 2004: 28). But although most Greenlanders are Inuit, they do not usually refer to themselves as such, as, for example, Inuit in the Canadian territory of Nunavut do. The word *Inuit* is the plural of *Inuk* and means 'humans' or simply 'people' in Greenland as well as in Nunavut. In Nunavut, however, the term *Inuit* also implies ethnic identity (Flora, 2019: 20). By contrast, the term 'Eskimo' – meaning 'eater of raw meat' – is now considered both pejorative and archaic.[1]

The name for Greenland, *Kalaallit Nunaat*, means 'the land of the people'. Greenlanders, in turn, usually refer to themselves as *Kalaallit* (plural of *Kalaaleq*) which refers to both a national identity and an ethnic identity, commonly in opposition to *Qallunaat* (plural of *Qallunnaaq*): a pan-Inuit designation for 'whites' or southerners. In Greenland, the term relates directly to Denmark (*Qallunaat Nunaat*) and to Danes, distinguishing them from other Europeans (Flora, 2019: 170). One of the most significant ways in which the question of identity expresses itself in public discourse is by casting Danish-ness and Greenlandic-ness as natural opposites (Flora, 2019: 82).

In an invited contribution, Kirsten Thisted summarises the contemporary significance of these issues for Greenlandic identity.

DOI: 10.1079/9781789246728.0006

INVITED CONTRIBUTION: The Renegotiation of Greenlandic Identity
Kirsten Thisted

Greenlandic identity can be seen as an ongoing process of *cultural translation*: a term used by theorists within the fields of cultural and postcolonial studies (Stuart Hall, Homi Bhabha, Robert J.C. Young and others), to characterise a translocation of cultural forms into new contexts and regimes of power. Not only knowledge, ideas or worldviews but also people and their identities are translated: as when the Europeans translated the Inuit into 'wild' and 'primitive' people or 'heathens', and the Greenlanders decided to translate themselves first into Christians, and later into modern Greenlanders with Danish citizenship.

Characteristic of the Greenlandic process is that one translation has hardly been completed before a new translation has been under way. While the political strategies of the Greenlandic elite during the first half of the twentieth century had been based on the argument of *similarity* between Danes and Greenlanders, in opposition to the colonial rule based on the assumption of *difference*, during the 1960s the arguments were reversed. While the Danish state now spoke of the principled equality of all Danish citizens, Greenlanders were striving for self-government, based on the argument that they constituted a distinct people, according to international law. Hence, in Greenlandic discourse, Danes were positioned as the post-colonial, powerful and oppressive 'Other', by which Greenlanders measured their 'Greenlandicness' (while in Danish discourse, Greenlanders remained in the position of the primitive 'Other' by which Danes measured their own modernity). Initially, negotiations led to the Home Rule Act, implemented in 1979. Later, with the Act on Greenland Self-Government, implemented in 2009, Greenlanders were recognised as a people pursuant in international law with the right of self-determination.

Essential to the Greenlandic debate on identity is the fact that the nation as such perceives itself as a mix of Inuit and European heritage. Outside Greenland, it is generally believed that the Greenlandic name for Greenland means 'Land of the People'. However, the Greenlandic word for human being or person is *inuk* (plural: *inuit*), and the name of Greenland is not *Inuit Nunaat*, but *Kalaallit Nunaat*. Kalaallit is the West Greenlandic term for modern-day Greenlanders, who trace their ancestry to two sources: the Inuit to the west and the Scandinavians to the east. The Greenlanders are recognised as an Indigenous people, and they consider themselves one people, even though different Inuit languages are spoken in West Greenland, East Greenland and Thule (the North). The Act on Greenland Self-Government is not based on ethnicity: 'The Greenlandic people' mentioned in the Act means people living in Greenland. Thus, the government is a civic government, not an ethnic government – even though the fact that the vast majority of the Greenlandic population is also part of the Indigenous population is reflected in the make-up of the government's cabinet.

The Greenlandic government coalition agreement of 2016–2018 opened with the words: 'Greenland is irreversibly on its way to independence.' A Minister for Independence was appointed in 2016, adding substance to the statement. Governments have changed rapidly in recent years, and the Minister of Independence has concurrently been in charge of other areas. The words quoted above were not included in any of the subsequent coalition agreements, but a committee was appointed to draft a constitution with the purpose of state formation.

The desire to formulate an independent Greenlandic constitution arises from a strong wish that Greenlanders become the executive subject of their own law. Thus, a forthcoming constitution might well begin with a phrase about 'we, the people of Greenland', replacing the current royal 'we' (representing the Danish state), which stands as the formal donor of both the Home Rule and the Self-Government Acts. As part of this process, Greenlanders need to identify who they are and what kind of society they want to develop. One of the big questions is what role Greenlanders' status as an Indigenous people should play, and what impact it should have on the decisions that are made regarding the development of Greenlandic society. This is, not least, an important issue in the management and use of Greenland's living resources and subsoil. The question is whether the status of Greenlanders as an Indigenous people will alleviate or possibly further complicate the work, due to the collective ownership of land and resources being a constitutive factor in Greenlanders' perception of Inuit heritage.

Indigeneity played an important role in the 1970s and 1980s, and has gained renewed relevance in recent years, where some Greenlanders literally take on Inuit identity by getting so-called Inuit tattoos. However, there are divided opinions about how big a role indigeneity should play in modern Greenlandic identity and for a future Greenlandic constitution, although the pride of being descended from Inuit remains unaffected.

The debate about Greenlandic identity is thus complicated. But a historical insight into the debates about Greenlandic identity and changing demands for recognition can lead to a better understanding of contemporary debates such as those surrounding the Hans Egede statue in Nuuk (Box 4.3, Fig. 4.6) or continued use of the word 'Eskimo' in Danish ice cream marketing (AFP/The Local, 2020; Ren, Chimirri and Abildgaard, 2020; Ritzau, 2020).

6.2 Language, identity and development

> The scalar mismatch ... between Denmark (population: 5.73 million) and Kalaallit Nunaat (population: 55,847) is replicated in many other large state–micropolity relationships around the world, which are similarly characterised by economic, political and cultural power imbalances as well as a presumption that the large or metropolitan state holds a monopoly on professional expertise. The discourse of economic and governmental weakness and failure that often surrounds micropolities is built upon these territories typically being described and evaluated through the lens of the large state (Baldacchino, 1993, p. 33). This discourse is inseparable from the neocolonial processes by which metropolitan powers maintain and entrench their influence over polities at the periphery.
>
> (Grydehøj, 2018a: 70)

In 2009, Greenlandic became the only official language in Greenland. In the following year, however, legislation stipulated that provision be made for both Danish and English in education and in public life.

As noted above, this official Greenlandic language, *Kalaallisut*, is based on the West Greenland form of the Inuit tongue. The dialects/languages spoken in East Greenland (*Tunumiit oraasiat*) and north-west Greenland (*Inuktun*) are significantly different to *Kalaallisut*. A large minority of Greenland's population speak none of these, notably Danish migrants and the descendants and families of Scandinavian settlers.

On the one hand, the Danish language has been considered by some as an almost indispensable tool for the development and modernisation of Greenland, while on the other, Danish has been identified as a threat to Greenlandic. According to Gad (2019), the controversies about language reflect divergences over the desired character of the Greenlandic nation – whether ethnic or civic (a debate reflected on in Chapter 12) – and how elements of tradition and modernisation should be prioritised in the ultimate realisation of Greenlandic independence. During intense discussions in *Inatsisartut*, Greenland's parliament, one suggestion was to adopt English as the first foreign language as a means of escaping the dilemmas arising from Greenland's 'double relationship' with the Danish language (Gad, 2019).

Such dilemmas were aired in November 2020 when Greenland's Premier, Kim Kielsen, writing in the political newspaper *Sermitsiaq*, suggested that all new public employees who do not speak Greenlandic should take language courses and pass examinations at a certain pace or risk losing their jobs.

This suggestion followed a separate proposition from *Siumut*, Kielsen's party, to reserve the right to vote in Greenland for people with two or more years' residence. Currently, Danes who have lived in Greenland for just six months may vote in local elections. Although the *Siumut* idea was soon withdrawn because it appeared to violate basic rights, it nonetheless established a precedent.

The two suggestions indicated a desire on the part of the governing elite for a faster pace of de-colonisation. In Greenland, as elsewhere, the national language is widely recognised as basic to the nation's pride and identity, but any suggestion that it should be used as an instrument of exclusion is inherently divisive (Breum, 2020).

To some critics, Kielsen's suggestion appeared to emphasise ethnicity based on out-dated, pejorative perceptions of Denmark, and that the language requirement would prevent badly needed skilled foreigners from seeking employment in Greenland's public sector. Greenland's population is not growing, and these political signals could also repel the many young Greenlanders studying abroad, whose required skills and experience are needed back home for capacity-building and strengthening the country's resilience. Currently much effort is expended in encouraging them to return after their overseas education. Kielsen's plan could also further ostracise Greenlanders who are Danish monoglots or who speak Danish better than Greenlandic. Members of this sizeable group have previously complained of a sense of exclusion and discrimination (Breum, 2020).

Kielsen's proposals were a response to the expected increase in the number of foreigners in Greenland involved in the airports development projects (Chapter 8) and the prospect of further mining operations. While he suggested that free language courses should be available online for all those preparing to move to Greenland, it was the suggestion that language training and proven progress should be *obligatory* for all future public servants that particularly provoked controversy.

Yet language teachers are scarce, and greater efforts are needed to educate the necessary number of skilled language trainers and scholars, and to safeguard the continued development of the Greenlandic language. For example, many foreign terms have yet to be fully integrated into Greenlandic.

Support for Kielsen's suggestion regarded it as logical and long overdue. Not least because although Greenlandic is the official language, all public communication is still translated *from* Danish, the language of colonisation, which is still widely used in *Inatsisartut*, in government departments, municipal administrations, institutions of higher learning, the health sector – where most doctors are still Danish – and in publicly-owned companies. A crucial element here is the lack of sufficiently educated, Greenlandic-speaking administrators, economists, lawyers and other professionals, requiring a constant in-migration of personnel – mostly from Denmark – a process which is both expensive and inefficient, as many only stay for one or two years. This has been particularly crucial in the health care sector in recent times. And the consequent need to translate all public documents places a significant burden on the administrative machine, especially since around half of all people living in Greenland do not speak Danish, or speak it only rudimentarily.

For these reasons, the question of language has been central to public discourse on identity, power and discrimination in Greenland for generations. Kielsen therefore promoted the logic that his plan would help save much needed resources (Breum, 2020).

The two political parties that support close links with Denmark and rely significantly on votes from resident Danes (*Demokraatit* and *Samarbejdspartiet/Suleqatigiissitsisut*) naturally oppose any downgrading of the role of the Danish language. Nationalist parties with support away from Nuuk (*Siumut*, *Partii Naleraq* and *Nunatta Qitornai*) are less attached to Danish, while *Inuit Ataqatigiit*, the main opposition socialist party, has been equivocal.

Issues of language, identity and even location therefore render Greenlandic politics less amenable to interpretation in the simple binary terms of Greenlandic–Danish, left–right or centre–periphery divisions (Grydehøj, 2020a: 227), and distinguish Greenland on the Arctic stage.

6.3 Cultural heritage

Greenland has been inhabited discontinuously for 4,500 years (Grønow, 2000). As noted in Chapter 4, the earliest known arrivals originated from Central Asia and journeyed to Greenland via Siberia, Alaska and Canada. The last major wave of migrants was of the Thule culture from Canada around AD 1200 (Gulløv, 2000), forming the basis of the current Inuit people, who depended mostly on hunting and fishing. Hunting tools became highly developed during this period and larger boats, kayaks and dog sleds provided this culture with greater mobility than earlier peoples. While most of the prehistoric remains to be found in Greenland originate from this Thule culture, adaptations of a number of their cultural artefacts remain in use today.[2] The tenth-century in-migration of Norse people who settled around the fjords of southern Greenland, farming, hunting and fishing (Arneborg, 2000), has left many archaeological remains, but no definitive answer as to why the settlements were abandoned in the fourteenth century (Chapter 4).

Many of the cultural inheritances, including story-telling (Fig. 6.1), as well as present-day culture, including commemorative artefacts (Fig. 6.2), are now part of the tourism product as well as contributing to national imagery. Tourism is helping to preserve and even revive some cultural aspects in Greenlandic society that are undergoing climatic and economic change. Examples would include sustaining traditional dog sledding (rather than employing snowmobiles). An ancient craft, dog sledding can be traced back to at least the Palaeo-Inuit Saqqaqq people who lived in Greenland until around 800 BC. Evidence of this comes in the form of a dog's jawbone discovered on the island of Nipisat, just south of the town of Sisimiut, where a Saqqaqq settlement once existed.

> The dog sled is still a fundamental part of a hunter's way of life. And many hunters earn a welcome supplement to their income by taking tourists for rides – a phenomenon which more than any other links dog sledding with pictures of Greenland.
>
> (Jensen, 2007: 68)

Sled dogs are traditionally an essential part of winter ice fishing, when hunters travel long distances on sleds across the frozen ocean

Fig. 6.1. A 'Norse' story-teller at the partly reconstructed settlement of Brattahlíð (Qassiarsuk). Author's own photograph.

Fig. 6.2. Visitors passing the Leif Eriksson Hostel, overlooked by a hill-top statue of Leif's father, Erik the Red, founder of the adjacent Norse settlement, Brattahlíð (Qassiarsuk). Author's own photograph.

(Axelsson, 2020). But as the sea ice retreats, that way of life is also in retreat and needing to adapt, with numbers of Greenlandic sled dogs declining steeply in recent years (Fig. 6.3, Table 6.1); and old knowledge and know-how is in danger of being lost for future generations, although use for tourism purposes can assist the retention of traditional skills. Special permits are required to move dogs from the north and east Greenlandic sled areas, in which the dogs are primarily used for fishing, hunting and tourism, to the west and south, and equally in the reverse direction. This is enforced to prevent mixed breeding (Statistics Greenland, 2020: 26).

Today, the dog sled is used mainly for hunting, social trips, tourism and for transport to and from the sea ice to fish for Greenland halibut. Hunting trips are mostly pursuing seals, which are caught in large nets. In spring they are shot as they sun themselves on the ice. In North and East Greenland long dog sled trips have been undertaken to hunt for polar bears (Jensen, 2007: 68).

The dog sled has been part of almost all Inuit cultures, 'as a technology [it] has been a crucial factor in the survival of the Inuit' (Sisimiut Museum display, August 2019), and over time has taken many physical forms and regional variations, achieving a high technological standard that renders it functionally efficient over several types of snow and ice conditions. For example, the dog sled of eastern Greenland is relatively light with wide runners, and is unique in having a strong metal foot brake to cope with hilly terrain and heavy snowfall. The long sleds of the northern Thule region originally had the largest teams of dogs for hauling heavy loads over sea ice, but subsequent modifications to the eastern sled has resulted in dog teams of similar size in both the east and west of the country (Jensen, 2007). 'These sleds are a physical testimony to a living culture in which humans have adapted to natural circumstances, survived and thrived' (Sisimiut Museum display, August 2019).

As indicated in Table 6.1 there was an overall halving in the number of sled dogs nationally between 1990 and 2018. While some volatility in numbers between years is evident in the data, local factors would appear to account for

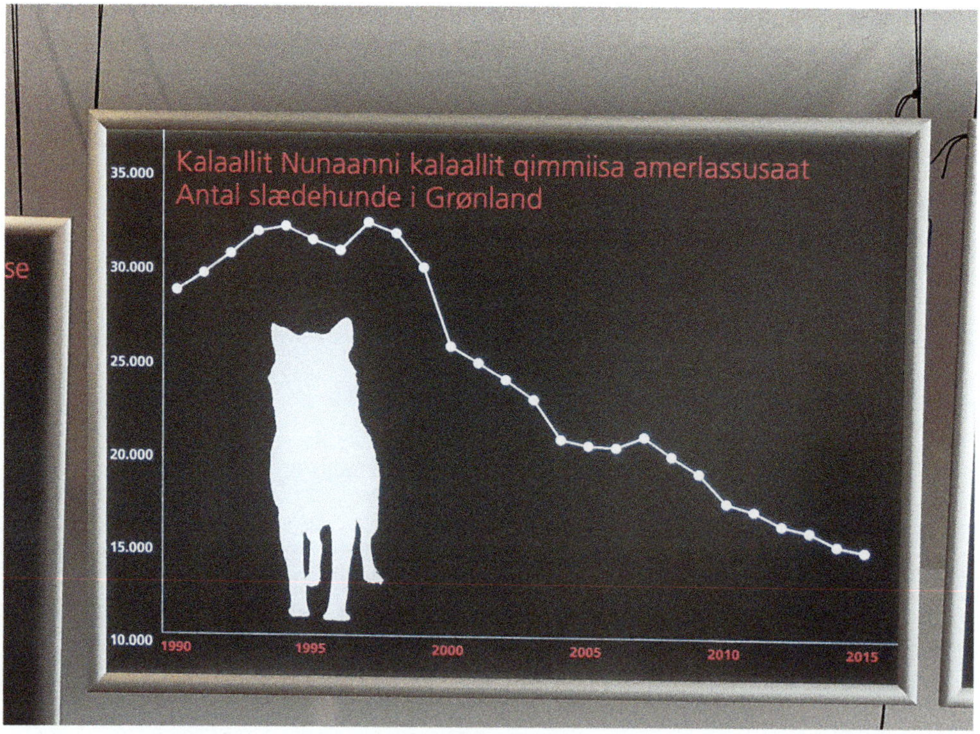

Fig. 6.3. Graph showing the national decline in sled dog numbers, 1990–2015, as exhibited in Ilulissat Museum, 3 August 2019. Author's own photograph.

Table 6.1. Greenland's sled dogs: numbers and distribution, 1990–2018. Author's own table derived from https://bank.stat.gl/TU/TUX01.px/TUE01; Statistics Greenland, 2018: 20, 2020: 20, derived from http://bank.stat.gl/ENEMO2DI; and author's additional calculations.

Location	1990	1995	2000	2005	2010	2015	2018	1990–2018 % reduction	Registered snowmobiles 2017	Registered snowmobiles 2019
Total Greenland	28,449	31,932	25,024	18,755	17,971	14,607	14,130	50.3	1402	2467
Aasiaat	1,850	2,220	1,437	840	551	680	656	64.5	32	82
Ilulissat	5,605	8,262	4,408	4,024	3,228	3,058	2,933	47.7	159	226
Ittoqqortoomiit	1,000	870	680	430	341	202	189	81.1	42	48
Kangaatsiaq	1,190	1,742	1,131	824	473	494	550	53.8	0	18
Qaanaaq	1,800	1,636	1,590	1,398	1,112	1,036	1,053	41.5	3	11
Qasigiannguit	1,300	1,601	1,867	1,264	964	1,102	756	41.8	40	72
Qeqertarssuaq	1,400	1,498	1,084	750	669	460	327	76.6	6	155
Sisimiut	1,735	1,872	1,777	1,485	1,647	946	656	62.2	572	665
Tasiilaq	2,066	1,787	1,585	1,156	1,318	1,136	1,012	51.0	25	34
Upernavik	4,561	4,867	3,958	2,584	3,846	2,680	2,954	35.2	139	215
Uummannaq	5,942	5,577	5,512	4,000	3,822	2,813	2,633	55.7	12	353

variations in overall rates of decline between locations. By contrast, there has been a rapid growth in the use of snowmobiles, with a 76 per cent increase in their number just from 2017 to 2019.

Although the location furthest north, Qaanaaq (Fig. 4.1), had a below average reduction in the number of sled dogs, the next most northerly location on the west coast, Upernavik, experienced the lowest rate of decline at 35 per cent, while Ittoqqortoomiit at a similar latitude but on the east coast exhibited the greatest reduction at 81 per cent; Tasiilaq, further south on the east coast, returned a decline at about the national average. In the vicinity of Disko Bay on the west coast, and the area of major tourism activity, the three locations revealed marked variations: Ilulissat experienced a 47.7 per cent decline, just under the national average; Aasiaat showed a reduction of 64.5 per cent; and Qeqertarssuaq, located on Disko Island, had a drop in numbers of more than three-quarters over the period, by far the greatest proportional reduction at any location.

6.4 Indigenous tourism?

In such a destination as Greenland, where Indigenous life is necessarily adapted to the natural environment, and interactions between nature and culture become complex (Smed, 2017), tourism may act to impose an exogenous categorisation and false separation of the two. This can act to constrain an understanding of the profound and pervasive significance of nature in shaping Arctic communities (Olwig and Lowenthal, 2006; Hansson, 2012).

Thus, despite, or perhaps because of, the complexity of the nature–culture relationship located at the core of Indigenous Greenland (and the 'southernicity' of tourism marketing, concepts and labels?), this relationship has become more of an implicit factor in tourism promotion, not being addressed overtly, perhaps because it is difficult to promote and sell to 'sophisticated' southern consumers (Smed, 2017: 144), but also perhaps because of the nuanced nature of Greenlandicness.

Apparent demand for cultural tourism has not been high in Greenland, and until very recently the ice has been the main visitor focus (Chapter 9). But has this demand been stimulated through marketing imagery or has the imagery followed perceived demand? After all, the prominent nature-related activities – dog-sled rides, sailing and fishing trips, hikes, even experiencing the Northern Lights (Box 1.5) – all engage with cultural elements, however implicit or mundane. Nonetheless, recent attempts by Visit Greenland to 'repopulate' Greenland's landscape imagery through branding campaigns have been aimed at raising interest and awareness of Greenlandic culture (Ren, Chimirri and Abildgaard, 2020: 15–16) (Chapter 12).

Visit Greenland had previously promoted the 'Indigenous nation with an Indigenous people', but this characterisation of 'a feature of the Greenlandic community that has been carried into modern living' appeared to imply a passive rather than an active role for culture (Smed, 2017: 140). In their analysis of evolving interconnections within the Greenlandic tourism development discourse, Bjørst and Ren (2015) emphasise how apparently conflicting or paradoxical visions of natural and cultural environments, rather than being mutually exclusive, actually entangle with each other as well as with climate change and sustainability narratives, political discourses and social realities. They employ the term 'collaborative friction' (Tsing, 2005) to identify such entanglements and their contribution to visions of a developing Greenland (see also Chapter 12).

If culture is therefore critically significant in terms of being intertwined with nature, how important is it as an agent of development and change, not least on the tourism development stage? The dual characteristics of 'Inuit' being not strictly Indigenous to Greenland, while Inuit descendants constitute the majority of a hybridising population and a self-rule nation, render the notion of 'Indigenous tourism' at best contested (if not irrelevant) in the Greenland context. Indeed, compared to other Arctic tourism regions (Section 1.6), '(I)ndigenous culture plays a different and much less prominent role in promoting, developing and articulating tourism in Greenland' (Ren, Chimirri and Abildgaard, 2020: 8).

One accepted narrative is, of course, that active involvement as decision-making stakeholders in development processes such as those that are tourism-related can be a way of empowering 'Indigenous' populations and an opportunity to highlight history, heritage and culture (Section 3.6), perhaps offering an alternative voice to post-colonial narratives. But, as Müller and Viken (2017b: 287) argue, is this any different from other forms of cultural tourism embracing local stakeholders? Further, the imagery of exoticism and small scale can render 'indigeneity' an exclusive 'product' encouraging commodification.

Certainly, in Greenland, 'Indigenous tourism' – as patronising and potentially misleading an exogenous imposition as that term may be – appears to be an elusive concept. If it exists, is it the commodification of Inuit dress, reproduction of *tupilak* models (Box 6.1) and visits to long-abandoned Inuit settlements? Or is it Greenlanders managing local land- and sea-based tour companies, administering and undertaking tourism planning, and teaching and researching tourism development and hospitality?

Arguably, the modernity of a hybrid majority Kalaallit people should obviate any need to employ such terms as Indigenous. And while more than 35,000 Greenlandic artefacts have been repatriated to the National Museum in Nuuk from Copenhagen over the past 40 years (Deboick, 2021) to reinforce the material store of cultural heritage, contemporary culture in the form of music, art, story-telling and literature (Section 1.4) is increasingly contributing to Greenland's trajectory towards independence (Thisted, 2015).

This debate is developed in Chapter 12.

6.5 Cultural sustainability through tourism?

> It is important to broaden the understanding of the value of tourism for the community and for tourists to feel welcome and received by the local community. In sustainable development there is room for both tourists and the local population.
>
> (Visit Greenland, 2020e: 25)

With the (pre-Covid) tourism sector witnessing slow but steady growth in Greenland, the largest tourism actors, such as Visit Greenland, had been taking a more explicit role in orienting tourism in a 'sustainable' direction: in 2018 it joined CSR Greenland,[3] and, at a local scale, created a 'how to *kaffemik*' guide concerning appropriate etiquette for tourists to follow when participating in visits to people's homes during special occasions such as birthdays, to share coffee and other aspects of family hospitality.

In recent years, decision-makers in tourism (Chapter 7), as in other fields of social and economic development, have sought to emulate models from other countries. Iceland has often been viewed as offering such a model of development, and tourism stakeholders are aware of the 'Iceland Pledge' whereby (potential) tourists are invited to commit themselves to responsible tourism. There would certainly appear to be merit in developing a similar vehicle for articulating mutual expressions of responsibility in Greenland (Ren, Chimirri and Abildgaard, 2020: 14–15). Such mutuality in terms of respect and responsibility applies equally well to the relationship between residents and their environment.

The natural environment in Greenland comprises habitats closely adapted to high, low and sub-Arctic conditions that are more sensitive compared to those in more temperate areas because of their slow pace of regeneration and short growing season. Although Greenland has a low overall population density, because of this fragility, the use of natural resources has been a critical element in Greenlandic society and is increasingly impacting upon sensitive areas (Due and Ingerslev, 2000): a less than fully collaborative friction.

The idealisation of traditional Inuit culture holding a deep respect for nature and employing only the resources needed for sustainable living has been significantly modified in recent decades, with instances of over-hunting and wasteful use of natural resources (Hansen, 2001). Man-made visual pollution has also been an area of contention: both local administrative and tourism authorities have sought to reduce it. A programme initiated in Ilulissat (Nordic Council of Ministers, 2001) has encouraged residents to clean up their localities, emphasising health and associated benefits for the community, for the wider environment and for existing and potential tourism development. As part of this approach, the head of Visit Greenland actually called on fellow Greenlanders to 'clean up their act' in response to visitor complaints about rubbish and waste in ports, towns and other settlements (Section 8.1), and to be more aware of using non-renewable materials (Ren, Chimirri and Abildgaard, 2020: 26-27).

Inuit descendants in Greenland comprise the majority of the population, and, as *Kallaalit* (Section 6.1), no Indigenous *minority* issues exist as in some other Arctic regions. However, as the main minority, Danes and Danish migrant labour in Greenland, until recently, have received such privileges in Greenlandic society as higher salaries and priority access to accommodation (Chapter 4). As a consequence of this and the continued, troubling presence and influence of colonial history, Ren, Chimirri and Abildgaard (2020: 21–22) suggest that distrust and latent racism are not unusual. In tourism this has been reflected, for example, in young Danish guides coming to Greenland for the summer and 'usurping' potential roles and opportunities for locally-trained people (Ioannides, 2019). Although such potential tensions arise with the seasonal influx of a 'foreign' workforce,[4] it is reported that, thus far, local stakeholders have tended to regard them as low level.

Concerns at the local level focus on how to integrate tourism into building and strengthening the community, in terms of both culture and economy. Encouraging participation in attracting visitors to experience domestic activities, such as through *kaffemik*, highlights a 'traditional' motivation in wishing to support family and the local community (Wennecke, Jacobsen and Ren, 2019). It can act as one foundation stone upon which to build partnerships for developing local products and services based on cultural heritage and contemporary experience. For example, souvenir-producing artists usually meet before and after the summer season to discuss successes and areas for improvement (Ren, Chimirri and Abildgaard, 2020: 21). Johnston and Viken (1997: 981) further reported that in Ilulissat, dog-sled tours are arranged by a tourism agency which hands over responsibility to the dog-sledders' association. The association then chooses which sledder should take on the job and sends a representative along on the trip, thereby ensuring that local people control the quality of the product and gain most of the rewards.

Currently no certification schemes exist in Greenlandic tourism in which local entrepreneurs and communities could participate, although some hotels are involved in environmental labelling schemes and many are signed up to TripAdvisor.

In terms of visitors' cultural codes of conduct, as also noted in Chapter 11, guidelines have been developed in Sisimiut by the local museum, Arctic Business Circle and the AECO (Association of Arctic Expedition Cruise Operators) as the first of several 'good tourist rules' particularly aimed at cruise tourists to emphasise respecting local people, their cultural and natural environment. There is recognition of a growing requirement for such codes across the country. Experience from a fact-finding visit to Svalbard led Sermersooq Business Council (Nuuk) to recognise the need for developing working guidelines for positive sustainable cultural interaction. While some stakeholders subsequently felt guidelines could be limiting and negative, acting to 'scare tourists away', simple infrastructural improvements such as signage would be a small positive step forward (Ren, Chimirri and Abildgaard, 2020: 22).

In their detailed interviews and focus group meetings with community representatives and local tourism stakeholders, Ren, Chimirri and Abildgaard (2020: 23) found that tourism in Greenland was not viewed as harming culture, but as having a positive effect, encouraging traditions such as drum and folk dancing. Alongside *kaffemik*, small-scale open dance and storytelling sessions at museums and cultural houses can be of a character that intertwines the public and private elements of such encounters.

Some controversies over the representation of traditional cultural artefacts have arisen from time to time, such as Ilulissat's Hotel Arctic developing igloo-inspired pavilions on a prominent headland (Johnston and Viken, 1997) (Fig. 6.4). Interrogation of the ethical nature of products being sold as artefacts has raised little controversy, although selling *tupilak* figures as souvenirs can be considered by some to be inappropriate as they were originally 'a voodoo thing' (Box 6.1, Fig. 6.5). More mundane but important quotidian artefacts such as the blubberbag (Box 6.2, Fig 6.6) have thus far escaped commodification but are well represented in Greenland's museums.

Local jewellery, handicraft and handicraft workshops would benefit from being more prominently promoted in conjunction with outdoor traditional activities such as fishing and dog sledding. 'Everyday life' could become a tourism product, involving, for example, short language courses, or local/self-caught food tastings, in

Fig. 6.4. Ilulissat's Hotel Arctic and its igloo-like pavilions on the headland entrance to the town's harbour, through lifting fog. Author's own photograph.

Box 6.1. The commodification of *tupilak*

A *tupilak* is a carved figure whose intended use was to cast a deadly spell on an enemy. Originally it was composed of materials from the natural world – animal, avian or human remains. The shaman, knowledgeable about witchcraft, gathered these elements together, chanted magic spells over the tied up bundle which then supposedly drew energy from the shaman's genitalia. The *tupilak* was then put into the sea and sent off to kill an enemy.

No original *tupilaks* remain as they were composed of perishable materials and not intended to be seen by others. When Europeans came to Greenland, *tupilak* figures were carved to satisfy the foreigners' curiosity about this custom. The more grotesque and terrifying the figure, the easier it was to sell and production became substantial, particularly in East Greenland with its rich carving tradition.

From the 1950s to the 1970s large numbers of these figures were produced in a mostly uniform style out of whalebone. Now they are usually made from reindeer horn or narwhal tusk.

Sources: Jensen, 2007: 5–8; Kreutzmann and Engell, 2020

Fig. 6.5. *Tupilak* figures (Nanortalik Museum). Author's own photograph.

Box 6.2. Making and using a blubberbag

Seal blubber is an essential part of the Greenlandic food tradition and is eaten in almost all kinds of meals.

The blubberbag is made of hooded seal skin. The skin is scraped roughly and sewn together. Sinew from the seal's throat is used as thread after being plunged into salty water and dried. The skin is sewn twice to tighten it. A wooden peg is fixed around each hole where a limb had been, and lashed tight with rope. The pegs are used for handles when carrying the bag.

Finally, the blubberbag is filled with air, like a balloon, and dried in the fresh air. When dry, it is filled with blubber for storage.

At home, the bag is placed within stones built up around it to avoid being damaged by sunlight. The blubber inside ferments and turns to oil and can be preserved for up to two years. When needed, blubber and oil are taken up through the hole in the top.

Source: Qaqatorq Museum, 2019

people's homes and/or after a foraging/fishing/hunting trip (Ren, Chimirri and Abildgaard, 2020: 25). When interviewed by Ren, Chimirri and Abildgaard (2020) local stakeholders displayed much pragmatism and a confidence in the future – also identified by Johnston and Viken (1997) more than 20 years earlier – while coping with the often difficult conditions of making a living in a tourism sector that remains, in most areas of the country, modest in size and highly seasonal: 'We were bad in our marketing for quite some years. Before, there were these pictures of Greenlanders in traditional costumes on a dog sled with a harpoon. We are now more aware that Greenland is so much more than that' (stakeholder view, quoted in Ren, Chimirri and Abildgaard, 2020: 26). Indeed, support for Visit Greenland was evident and its branding of a 'pioneering nation' appeared to meet with stakeholder approval.

While political debate on tourism development until recently revolved around raising numbers (Chapter 7), an ongoing discussion in the Greenlandic tourism sector is now taking place

Fig. 6.6. A blubberbag (Qaqortoq Museum). Author's own photograph.

regarding a more reflexive and smart approach to 'whom to invite in' and for mutually respectful and rewarding engagement (Ren and Chimirri, 2018; Ren, Chimirri and Abildgaard, 2020: 26). Such considerations can only increase in importance as Greenland moves towards independence while simultaneously developing the infrastructure for a potentially significant increase in visitor numbers, perhaps reducing the high costs to reach and stay in Greenland. Carrying capacities in fragile cultural and natural environments have thus far been barely tested. But as Johnston and Viken (1997: 981) warned almost a quarter of a century ago: 'The danger for Greenland in this context is that the level of tourism that is too high to enable protection of the environment and culture will not be recognized until it has been surpassed.' Today, Greenland is surely at a critical stage in its economic, cultural, political and environmental 'development'. Can it become a model for sustainable (cultural) tourism in an era of dynamic environmental and (geo)political change?

6.6 Greenlanders towards independence

Domestically, several issues of cultural significance arise in any likely achievement of political independence (Emmerson, 2010; Kruse, 2017; Grydehøj, 2020b).

- The unspoken question of whether the self-rule government would continue to financially support remote settlements which would not otherwise be economically viable.
- Fisheries would continue to be Greenland's most important economic sector, but value-added processing would need to be further developed both in relation to marine stocks and to mineral resources, complemented by creating more industry training schemes to produce a skilled workforce with the capacity to support processing sectors. This would require new sources of investment and systematic improvement of the framework conditions for strengthening business competitiveness. This is likely to further encourage migration to the major towns (see below).
- Increased growth and employment in the private sector – in an economy otherwise heavily reliant on the public sector – has been viewed by the Greenland government as critical in securing the foundation for future welfare and prosperity. Growth and employment are aimed to be boosted in several sectors, rendering Greenland less vulnerable to changes in any single sector, such as in volatile export prices or fish catch volumes ... or a global pandemic placing the international tourism sector on hold.
- Wage inflation caused by inward investment could crowd out other sectors of the economy, whereby Greenland's dependence on Denmark could simply be substituted by a dependency relationship with multinational corporations that could result in an homogenisation of culture.
- Similarly, would cultural traditions be lost in the rush for self-sufficiency, not least, paradoxically, through an influx of foreign workers?

These issues would be in addition to the continuing social problems such as alcoholism (Fig. 6.7), which have been used at times by outsiders to stereotype Greenlanders. But the suicide rate, especially among young men, continues to be a major public health concern.

Fig. 6.7. A graphic cautionary poster outside a supermarket in Qaqortoq. Author's own photograph.

Compared to its overall population size, Greenland's pattern of internal migration is substantial, at around 40 per cent equivalent of the total population per year. A stepped migration process, from outlying settlements to small towns and from small to larger towns, albeit with some reverse migration (Table 6.2), has been proceeding for half a century, such that the proportion of the population living in towns now approaches 90 per cent (Table 6.3) and almost half of the country's population lives in the five largest towns: Nuuk, Sisimiut, Ilulissat, Aasiaat and Qaqortoq (Fig. 4.1) (Statistics Greenland, 2018: 7–8). A high level of intra-urban migration represents about 28 per cent of all town dwellers moving residence each year within their town, while a further 10 per cent are undertaking inter-urban migration.

According to Statistics Greenland (2020: 37), average life expectancy is 73.0 for women and 68.3 years for men, far below the level in Denmark, resulting from a high rate of infant mortality and of suicides, especially for men aged 15–35 years. Towns differ from settlements in fertility rate. Women in settlements give birth to more children than those in towns, although the settlement birth rate has been decreasing in recent years. Greenlandic women give birth to their first child at an average age three to five years earlier than women in Nordic countries. Greenland has also had an annual net out-migration of around 1 per cent for over 30 years and net emigration for over 40 years. As a consequence the country's population has been declining, such that, given current trends, by 2040 it will be below 50,000 (Statistics Greenland, 2020: 7).

The labour force includes 74 per cent of 18–64-year-olds living in Greenland the full year; more than half of the employed population has only nine years of school education. The unemployment rate oscillated during 2012–2015 between 8 per cent in the third quarter of the year to 12 per cent in the first quarter. It is higher in the settlements than in towns, and for younger adults (Christensen *et al.*, 2020: 4). A key aim of the Greenland government is to raise the general

Table 6.2. Greenland: numbers and patterns of internal migration, 2015–2019. Author's own table derived from Statistics Greenland, 2020: 20, derived from http://bank.stat.gl/BEEBAF2B; http://bank.stat.gl/BEXSAT; and author's additional calculations.

Nature of migration	2011	2013	2015	2017	2019
Within the same settlement	1,316	1,219	1,232	993	910
Between settlements	370	332	363	417	301
From settlement to town	1,392	1,272	1,306	1,317	1,224
Between towns	4,507	4,672	4,875	4,919	5,133
Within the same town	13,200	14,000	13,791	14,158	13,671
From town to settlement	1,124	1,115	1,242	1,082	1,112
Others	32	56	22	16	9
Total population	56,615	56,370	55,984	55,860	55,992
Percentage of the total population moving residence within the year	39	40	41	41	40

Table 6.3. Greenland: growth of urban population, 1978–2020. Author's own table derived from Statistics Greenland, 2018: 7, 2020: 7, derived from http://bank.stat.gl/BEEST3; and author's additional calculations.

Year	Total population	Towns' population	Population 'outside towns'	% of total population living in towns
2020	56,081	48,985	7,096	87.3
2018	55,877	48,492	7,385	86.8
2008	56,458	47,099	9,359	83.4
1998	56,072	45,485	10,587	81.1
1988	54,524	43,297	11,227	79.4
1978	49,147	36,796	12,351	74.9

level of education and training to combat high levels of unemployment among persons under the age of 30 who have not completed secondary education, are unskilled or lack work experience (Kruse, 2017). Low accessibility and peripherality (Chapter 8), as well as the legacy of colonialism, have influenced employment opportunities, out-migration, development and economic performance. A critical objective for future tourism development is to extend the seasons and render employment in the sector less seasonally pronounced (Section 7.5.2).

Although geographical and socio-economic distances are great in Greenland, Flora (2019: x) argues that each place has something that makes it distinctive, and we should be cautious not to depict these places as remote or the people who live in them as isolated. While some Greenlanders may stay in a place for generations, others frequently move, as noted above, and can belong to many different places at the same time. Seasonal and semi-nomadic movement is an old and recurrent theme in research on Inuit people throughout the Arctic (e.g. Boas, [1888] 1964; Mauss, [1906] 1979; Rasmussen, 1925) and clearly retains contemporary agency (Flora, 2019: 89–91).

The relationship between the cultural elements of such ethnicity and their political expressions has been conceptualised by Gad (2016: 56) in relation to *Kalaallit* identity in three basic positions underlying the apparent consensus in favour of independence:

- self-governance as a precondition for [economic] self-support;
- self-support as a precondition for self-governance; and
- a more complex construction of 'self-supportedness' as a joint project shared by Greenland and Denmark where self-government would be a gradual affair.

Public opinon polls on independence have been far from overwhelmingly in favour (Skydsbjerg and Turnowsky, 2016; Turnowsky, 2017). In a more recent poll, 11 per cent of the sample expressed a preference for independence no matter what, and 12 per cent wanted independence even if it meant a small drop in living standards. On the other hand, 44 per cent indicated that they would only support independence if living standards were not going to 'drop', while 27 per cent preferred a strengthened autonomous position for Greenland rather than outright independence (Grydehøj, 2020a: 222–223).

There is a general understanding that, because Greenland controls the timing of the decision to become independent, the territory is being given the opportunity to lay the economic, political and social groundwork for a *successful* independence (Grydehøj, 2020a, 2020b). But Grydehøj (2020a: 225) considers there is a growing divide in opinion in Greenland which brings spatially expressed socio-economic divisions to the fore, between the residents of Nuuk and the other towns and settlements. Because Nuuk acts as the apex of internal migration processes, it sustains a continuous demand for new housing development and 'metropolitan' lifestyles. The capital also continues to be associated with Danish-ness (Grydehøj, 2020a: 226), generating further core–periphery identity dichotomies (Box 6.3), incorporating the paradox – so important in the Greenland context – of the state acting both as a redistributor and as a centraliser.

6.7 In the next chapter

In Chapter 7 the focus moves decisively onto the organisation, development and significance of tourism to, in and for Greenland.

Box 6.3. Centre and periphery in envisioning Greenland's future

Centre–periphery divides in visions of Greenland's future may be expressed in tensions between the ideal of centralisation as a means of achieving economies of scale versus decentralisation as a way of supporting the livelihoods of peripheral communities.

This has been exemplified in the debate over a fishing industry law that contains internal contradictions: the need to support industry-wide competitiveness versus a desire to redistribute fishing quotas to a greater number of smaller actors for whom fishing is a central element of their cultural identity. And while state-owned enterprises (SOEs) can represent a key mechanism through which the state can redistribute resources and employment, they may also be perceived to be a core element of the central state and thus centralising and anti-competitive.

Source: Grydehøj, 2020a: 227

Notes

[1] Copenhagen University's *Institut for Eskimologi* has been exceptional (see Thuesen, 2016).

[2] For example, familiar English language nouns such as 'kayak' and 'anorak' are derived from Inuit usage, the artefacts themselves being derived from Inuit innovation.

[3] A UN-recognised NGO framework for corporate social responsibility.

[4] Hook (2019) reported a summer labour shortage in Ilulissat, requiring 'some cafés and supermarkets' to recruit workers from the Philippines. In 2020 there were1,246 non-Danish citizens resident in Greenland – 2.2 per cent of the total population – the largest non-Nordic groups being Filipinos and Thai (Statistics Greenland, 2020: 7).

7

Tourism Supporting Greenland's Aspirations

'... tourism has been introduced as a central pillar to build a strong economy... perceived as an economic tool to secure the wish of many Greenlanders, namely full independence from Denmark' (Ren *et al.*, 2021: 4)

7.1 Introduction

In this chapter Greenland's tourism development and administration are reviewed before examining the actual scale of the sector in supporting 'self-sustainability'. Such critical issues as the responsibility of tourism and seasonality are raised that are central to current and future management of Greenlandic tourism.

The chapter underpins and informs subsequent chapters that relate tourism processes to other aspects of Greenland's development path. In the next chapter these are issues of peripherality; in Chapter 9 tourism's engagement with climate change focuses on Greenland's central characteristic, ice. Chapter 10 considers branding, imagery and the supposed soft power of tourism, and draws on the discussion of adventure tourism in the previous chapter to examine aspects of cultural discordance in its marketing. Chapter 11 evaluates the further dimension of Greenlandic tourism: cruise ship tourism, its nature, impacts and role in imposing dependency. The final chapter in Part 2 is the invited contribution from Carina Ren and Mette Simonsen Abildgaard: 'Greenlandic Independence and Tourism Futures – Exploring *Modern* and *Ethnic* Logic' which helps to draw a number of the previous strands together by interweaving debates concerning the future of Greenland as a likely independent state with the future of tourism and the role of the Greenlandic people in that future.

7.2 Tourism evolution in Greenland

Greenland's first 'excursionists' are claimed to have arrived in 1869 when artist William Bradford and polar explorer Dr Isaac Hayes brought a party of sightseers and artists on the steamship *Panther*. In 1902 Mylius-Erichsen unsuccessfully sought permits to bring a ship with a hundred tourists twice each summer from England. In the 1930s, ships from the US and France carried tourists along Greenland's forbidding coast.

After World War II the Danish government closed Greenland to civilian outsiders and restricted military personnel to their bases. Nonetheless, in 1953 charter flights were introduced to Kulusuk in the east and Narsarsuaq in the south, making use of former US military airfields (Chapter 4). But it was only in 1959 that Copenhagen sanctioned regular air links with Europe, and in 1960, 500 tourists were recorded. Until the 1980s, however, Greenland hosted only a few thousand tourists a year; high costs and lack

DOI: 10.1079/9781789246728.0007

of infrastructure were major constraints along with low levels of tourism investment, highly fragmented businesses and the absence of an effective marketing programme (Christensen, 1992; Johnston and Viken, 1997; Kaae, 2002, 2006; Snyder, 2007). Although a 1973 tourism white paper had aimed at 35,000 by 1980, visitor numbers had only reached 10,000 by 1981 (Grønlands Landsstyre, 1998; Ren, Chimirri and Abildgaard, 2020). Not for the last time were optimistic expectations left unrealised.

Copenhagen's earlier decisions to constrain visitor growth had helped to cultivate Greenland's image as an enticing, if inaccessible, destination. With Greenlanders voting for self-government (Home Rule) in 1979, the subsequent autonomous Greenlandic parliament (*Inatsisartut*) was able to express its support for tourism development (Smith, 1982).

In 1990 the Home Rule government began to intensify its focus on tourism, viewing it as a future staple sector of the economy. To this end, a 'Tourism Plan 1991–2005' was developed, and in 1992 'Greenland Tourism' was established (Box 7.1). Later becoming 'Visit Greenland', throughout the 2000s this body worked to establish tourism as Greenland's second most important sector, aiming for it to be in harmony with nature and culture and rooted in the people of Greenland (Greenland Tourism, 2001a; Hynne, 2002; Ren, Chimirri and Abildgaard, 2020). The relatively recent incorporation of Greenland into the global tourism system therefore reflects the impress of politics intertwined with the territory's hitherto perceived peripherality and isolation.

Growing attention from political and other public actors generated several reports, policy documents and strategies conducted during the 2010s to explore Greenland's tourism potential (Udenrigsministeriet/Udenrigsdirektoratet, 2011; Instituttet for Fremtidsforskning, 2013; Rambøll Group, 2014). Although they all acknowledged development of a strong tourism sector as one pathway towards growth (complementing fishing, mining, farming and hunting), they also recognised a number of persisting constraints (Rantala *et al.*, 2019) (Chapter 3): short season, lack of infrastructure, limited capacity, low customer service standards, low tourism growth rate, short length of stay, low per capita tourist spending, low priority/lack of concrete government initiatives, lack of online information and shortcomings in internal and external co-ordination, not least in marketing (Rambøll Group, 2014: 40–52; Ren, Chimirri and Abildgaard, 2020: 8).

Although a number of these impediments continue, in recent years the combined effects of tourism strategies, research and development projects, funding opportunities and networks have helped to improve service levels, enhance entrepreneurship, develop products and improve digital and physical infrastructures. Ren, Chimirri and Abildgaard (2020: 8) cite as significant in this respect the 2018 conference 'Towards More Tourism'[1] hosted by Visit Greenland and Air Greenland and attended by 137 tourism stakeholders, mostly representing micro- and small businesses. The authors argue that the emphasis of Greenlandic tourism development had shifted to actively enhance the role of local and regional actors, rather than being colonised by large foreign investors (Ren, Bjørst and Dredge, 2016).

7.3 Tourism administration

Greenland is confronted with a challenging tourism environment coupled to the evolving complexity of political and public discourse surrounding the role of tourism planning and development (Box 7.2). One element of this is the dominance of a few major players – administratively, such as the Government of Greenland (*Naalakkersuisut*) and Visit Greenland, and operationally, such as Air Greenland and large foreign tour operators and cruise lines. By contrast, numerous small-scale businesses and entrepreneurs are located across the country and operate mostly locally. With varying degrees of access to resources and disparate locations, both groups have tended to focus on different issues within tourism and have pursued diverse development plans (Chimirri, 2020: 25–26).

Naalakkersuisut, for example, facilitates tourism growth by developing areas, cities and towns through spatial and urban planning. Laws and guidelines aim to regulate and manage the arrival of tourists while enabling tourism actors to pursue their businesses with fishing and hunting licences and cruise-ship passenger fees (Government of Greenland, 2015). Greenland's destination marketing organisation (DMO), Visit Greenland, focuses on the marketing and global promotion of the country as a tourist destination.

Box 7.1. Greenland: tourism evolution

1973
Tourism in Greenland white paper published.

1975
Tourism conference. Emphasis on need for ownership and control of development by Greenlanders themselves in order to strengthen cohesion among the people and to improve understanding of Greenlandic culture. An ethos re-emphasised in the 2001 *Greenland Tourism* report (below), and the 2018 conference 'Towards More Tourism'.

1979
First tourism strategy: Home Rule and Danish governments in co-operation with community leaders.

'Arctic wilderness' seen as a primary attraction, appealing to all likely market segments. Key elements identified: cruise ship and air transport mass markets, cultural events, wildlife viewing, kayaking, sport fishing, mountaineering, mountain biking, heli-skiing, adventure racing and dog sledding. Specific locations identified: towns and Inuit settlements 'receptive to tourism', Ilulissat Icefjord, protected areas, and the North East Greenland National Park (the world's largest) (Fig. 4.1).

1979–early 1990s
Characterised by initial attempts to produce and market tourism opportunities after relaxation of the closed door policy. Minimal infrastructure: four ex-US military airstrips (Section 4.4).

1982 Introduction of regular scheduled airline from North America. Promotional slogans employed such as 'New Adventure in Tourism'. Coastal ferry services, intended to carry cargo and local passengers between settlements, were increasingly used to transport tourists.

1984 Policy emphasis towards employment and income for local communities (re-)stated.

1991
Greenland Tourism and Business Council created. Objectives: to encourage forms of tourism that acknowledged Greenland's unique features and to maintain the country's cultural values.

Greenland Tourism (later Visit Greenland), government-owned, to concentrate on five key areas: consultancy, marketing, education, documentation and customer services. Nuuk central office for product and structure development, consultancy, analysis, documentation, internal marketing and international co-operation. Greenland Tourism Information Office in Copenhagen for external marketing, distribution and information flow to the media, travel agencies and private individuals.

Greenland became the second 'country' in the world to have a fully digitised service network.

Human resource development programmes and community participation programmes were implemented to support the tourism development plan.

'Turismeplan for Grønland' 1991–2005, master plan. Home Rule government formally identified tourism as one of three key areas for commercial development to reduce reliance on fishing.

Key elements included:

- increase the annual numbers of incoming visitors to 35,000 by 2005;
- generate DKK500mn (€65mn/£60mn/$80mn) annual income by 2005, at a level of DKK15,000 (€1,950/£1,800/$2,400) income per tourist;
- develop and conduct tourism in an environmentally and culturally responsible manner;
- develop local tourism competence to ensure the highest possible financial return for local people, seeking to create 2,200–2,500 full-time jobs, plus 1,000-1,500 in associated enterprises;
- four tourism development zones established: Kangerlussuaq (Sondre Stromfjord), Narsarsuaq (South Greenland), Kulusuk/Ammassalik (east coast) and Nuuk (capital city).

At the same time Kangerlussuaq, Nuuk and Ilulissat (west coast) and Kulusuk (east coast) were targeted for upgrades to their air and sea access, accommodation and infrastructure. These emerged as strategically critical tourist gateways, including for the development of international events (Box 9.4) such as the cross-country Ski Arctic Circle Race.

Continued

Box 7.1. Continued.

2000
Parliamentary deputy for tourism highlighted the importance of developing and marketing Greenlandic products, including handicrafts and food, with tourism integrated into the country's culture.

2001
Greenland Tourism report emphasised the need for tourism to be rooted in and supported by the local population.

2003–2005
Tourism Strategy Plan placed an emphasis on:

- sustainable tourism development – environmentally, economically and socio-culturally;
- capacity building; and
- alliances.

Reiterated 1991–2005 targets. It was later realised that expenditure and multiplier effects were too ambitious, and the desired annual tourist numbers were raised to 61,000 by 2005. The original target of 35,000 was met (Table 7.1).

2009
Achievement of Self-Rule. This was critical for tourism development as the government embarked on a series of measures to diversify the economy with an eye towards eventual full independence from Denmark. Tourism was identified as a key sector for attaining this goal.

2012
Greenland Tourism and Business Council functionally separated.

2014
Co-operation between the Danish Ministry for Industry, Business and Financial Affairs and the Greenland government resulted in the issuing of a number of tourism concessions attracting international investments in experiential tourism (deemed necessary following some years of stagnation in the sector).

2016–2020
Tourism Development Policy (Government of Greenland, 2015). A key instrument for shaping the sector's future. The Greenlandic government's approach to developing tourism policy was seen to be top-down, reflecting a 'government' rather than a 'governance' approach. Barriers to the formation of policy networks identified included a lack of a shared image for the future, lack of trust among actors, insufficient time and a spatial fragmentation hindering interactions, with shortcomings in the institutional enabling of information and knowledge-sharing.

2016–2019
National Tourism Strategy (Visit Greenland, 2016b). Aimed to increase arrivals by 5 per cent per annum and to enhance international marketing for establishing Greenland as a quality adventure and cruise tourism destination. Invested in a business to business platform for local–international co-operation.

2020–2023
Towards More Tourism: Visit Greenland's marketing and market development strategy, in preparation for …

… 2023
Three new airports planned to be operational at Nuuk, Ilulissat (international) and Qaqortoq (regional) (Chapter 8).

Sources: Home Rule Authority, 1992; Greenland Tourism, 1995, 2001a, 2001b, 2003; Johnston and Viken, 1997; Lycke, 1998; Timothy and Olsen, 2001; Kaae, 2002, 2006, 2007; Snyder, 2007; Government of Greenland, 2015; Visit Greenland, 2016b, 2020e; Ioannides, 2019; Petridou, Olausson and Ioannides, 2019; Lumholt, 2020; Ren, Chimirri and Abildgaard, 2020

Box 7.2. Greenland tourism administration

Governance and legislation
Tourism is the responsibility of the Ministry of Industry and Energy within the Greenland Self-rule government. It is also regulated by Danish rules and legislation, for example regarding transportation and aviation.

Visit Greenland
The national tourist board is a government-owned agency responsible for tourism development. The six-member board of directors is appointed by the government. Its vision is to create responsible growth with its partners through focus, shared knowledge and idea development. It acts as an advisor to the government on matters related to tourism. Its stated roles are to:

- serve as the national tourism resource centre of Greenland;
- work to brand and promote Greenland as a tourist destination; and
- work to promote and develop the tourism industry and tourism activities in Greenland.

Visit Greenland's five key focus areas of action are:

- market development;
- visibility;
- season expansion;
- accessibility and framework conditions; and
- statistics and documentation.

Each area is worked through a project-based approach by collaborating with stakeholders regionally, nationally and internationally, an approach reflecting the company's size and need to optimise human resources.

Sources: Visit Greenland, 2016b: 20; Árnadóttir, 2019: 38–39

As ground-level stakeholders (Box 7.3), small-scale tourism entrepreneurs are confronted with such quotidian challenges as changing regulatory frameworks, unreliable and high-maintenance technical infrastructure, complex geographical logistics and strong seasonality (Ren and Chimirri, 2017; Chimirri, 2020: 26).

While the last 15 years or so have seen a growing focus on developing strategies for entrepreneurship education in Nordic countries, Greenland has developed only limited experience (Ren and Chimirri, 2018). By 2017, the country had no national strategy or goals for entrepreneurship education, no ministry involvement nor any national definition of the term (Reffstrup and Christiansen, 2017). However, tourism actors at the local level have taken greater interest in entrepreneurship and have applied this to new opportunities created through tourism development (Ren and Chimirri, 2018).

Entrepreneurial tourism businesses in Greenland are typically small, often seasonal initiatives, offering services or products such as equipment, souvenirs, food products or excursions, and are often characterised by collaboration within communities, employing co-operative entrepreneurship to innovate. In this way, emerging local tourism initiatives have been able to embrace and interlink education and training, entrepreneurship and community development, and to take advantage of new opportunities collaboratively (Ren and Chimirri, 2017, 2018). Nonetheless, local tourism practitioners, while acknowledging the significance of collaboration, may underestimate the complexity of business networks and the various forms collaboration can take (Chimirri, 2020: 27).

While, superficially, it might appear that there is untapped potential for development, community-level collaboration is characterised by a series of informal trade-offs (Mulford and Rogers, 1982), and of multiple, interrelated mutual activities (Chimirri, 2020: 34).

7.4 Tourism development and strategies

While tourism may be just one of a number of agencies that have been recruited to contribute

Box 7.3. Greenland tourism stakeholders

Government of Greenland
Ministry of Industry and Energy
Visit Greenland
Greenland Holding – Greenland Business – Greenland Venture

Governance
Municipal governance – destination management offices
Government of Denmark – regulatory framework

Industry organisations
Greenland Business Association (GBA)
Greenland Employers Organisation (NUSUKA)

Labour market associations
The Association of Fishers and Hunters in Greenland (KNAPK)
Labour Union Greenland (SIK)
Employers Association (AK)

Education and training
Ilisimatusarfik (University of Greenland)
Niuernermik Ilinniarfik (business school)
Greenland Food Industry College (INUILI)
Campus Kujalleq (tourism studies)

Research and statistics
Statistics Greenland (*Naatsorsueqqissaartarfik*)
Visit Greenland

Support and business development
Visit Greenland
Greenland Holding – Greenland Business – Greenland Venture
Municipalities' business councils.

Source: Árnadóttir, 2019: 40

to Greenland's economic 'self-sufficiency', so, in its turn, tourism may be impacted upon by both deliberate policy and happenstance from across a wide range of political, economic, cultural, environmental and conflict realms (Hall, 2020).

Although visitor numbers to Greenland have been relatively low compared to other regions of the Arctic, recent heightened interest from researchers and practitioners – not least in intensified tourism marketing – has closely followed increased international attention being paid to the Arctic (Chapters 1–3), not least because of the direct and indirect consequences of climate change. Consequently, political and societal interest has consolidated as tourism has emerged as one of Greenland's projected three major economic pillars (alongside fishing and mining), and as a potentially critical element in the path towards independence (Ren, Chimirri and Abildgaard, 2020).

Tourism may be viewed in source countries as representing values of freedom and democracy, if only for some (and the constraints of Covid-19 further highlighted the inequalities of access to multiple forms and dimensions of mobility). Tourism may be employed as a symbol of escape from the mundanity of the quotidian. Yet tourism is deeply embedded in politics (and, symbiotically, politics in tourism) (Hall, 2017a), and can be concerned with the quotidian, albeit usually of 'Others'. Tourism contributes significantly to global ordering (e.g. Franklin, 2004; Tribe, 2008), and embraces both the symbolism and reality of 'neoliberal hegemony'.

(International) tourism is an implicitly geopolitical activity in its reliance upon working relationships between blocs', countries' and regions' administrations and businesses (e.g. Hall, 2004, 2017b). And complementing tourism-related mobility is the need to facilitate cross-border flows of the capital, commodities, labour and skills required to sustain the (international) tourism process.

The spatial consequences of international relations at supra-national and national levels can impact substantially on activity at sub-national levels, both at tourism destinations and at international borders. But local tourism-related responses to such impacts can vary considerably, are often place- and context-related (perhaps suggesting path dependencies – see Chapter 11), and thereby articulate variable local conditions. Hazbun (2004), among others, has referred to this as re-territorialisation: the increased relevance of location and the particular characteristics of place. Such relevance would appear to be particularly significant when addressing the nature and impact of tourism development in Greenland's dispersed and relatively isolated destinations.

Yet, '... tourism spaces primarily are articulated to serve the interests of non-locals (i.e. tourists, foreign investors, organisations)' (Saarinen and Rogerson, 2014: 25) (Box. 7.4).

While the needs and aspirations of locals and non-locals may not be necessarily contradictory, critical issues can be raised by tourism

Box 7.4. Sustainable tourism supporting 'self-sustainability'?

Without wishing to rehearse long-standing debates, major shortcomings in the concept and practice of 'sustainable tourism'/the sustainability of tourism have included:

1. While tourism destinations may claim to pursue sustainable practices at a local level, as critiques of Greenland's sustainability stance argue (Sejersen, 2015, 2019; Bjørst, 2019), such declarations wilfully ignore the role of scale and the fact that visitors, in the case of Greenland, are required to travel long distances in transport modes that, up to now, have been significant emitters of greenhouse gases, thereby contributing to the climate crisis.
2. The notion of sustainable/the sustainability of tourism is too often used as marketing greenwash, with 'fashionable' employment of 'green' and 'eco' labelling to both encourage the traveller to feel undeservedly good about themselves while allowing the promoter to charge a premium price for the privilege. The over-use and misappropriation of such terms has rendered them almost meaningless while encouraging sloppy thinking ... and travelling.
3. When tourism sector practitioners talk about sustainable tourism, it is often the sustaining of the tourism sector which is uppermost in their thoughts, rather than any positive environmental or cultural impacts tourism's activities may render. The pertinence of this perspective has been dramatically foregrounded by the global impacts of Covid-19 on the sector.

practices that reflect potential contestation and unequal power relationships within and between them. This can perhaps be exhibited most explicitly in the development of 'Indigenous tourism' by outsiders which can contribute to the 'Othering' and exoticising of an Indigenous people (Keskitalo and Schillar, 2017; Keskitalo *et al.*, 2019), a process that is being determinedly circumvented in Greenland.

Tourism policy per se may be a recent phenomenon in Greenland and may offer little more than a blunt instrument in the face of multiple exogenous agencies such as tour companies and cruise line operators. Yet tourism processes have been deeply influenced by policy decisions in areas which impinge upon the nature and practice of tourism, such as the freedom of movement of people, goods and ideas, regional development policy and biosecurity. As such, given Greenland's unique geographical conditions, its facility to manage freedom of movement and biosecurity were well exhibited in the territory's response to the Covid-19 threat, suspending, more effectively than many territories and countries, both the symbolism and reality of international tourism's neoliberal hegemony.

Although Greenland is engaged in the long-term process of shedding some of its bonds with Denmark, as a small country – at least in terms of population numbers – and one that inherently incurs high costs, it will need to sustain and extend its means of international collaboration, in tourism as in other areas of 'development'. Along with Iceland and the Faroe Islands, Greenland thus far co-operates in areas of tourism development through the North Atlantic Tourism Association (NATA).[2] NATA's (2020) vision was to:

- promote the West Nordic countries as a tourist destination;
- encourage co-ordinated marketing of the region;
- offer grants and travel support for projects that express the vision;
- help improve the quality of tourist services in the region;
- build partnerships between the participating countries; and
- promote knowledge-sharing, product development and innovation.

The work of NATA is extolled as an example of successful collaboration and viewed as a model which could be extended to other Nordic countries (Árnadóttir, 2019: 16–17).

In its evaluation of participating countries' tourism structural strengths and weaknesses, NATA'S (2015: 8) strategy document noted for Greenland the strengths of: political focus/strategy; UNESCO Heritage/'environmental issues'; increased focus on an established system to collect statistics; and increased co-operation with Icelandic tourism operators. Major challenges were seen to be the familiar issues of: lack of critical mass; perceived as expensive; insufficient range of attractions; seasonality; absence of a quality system; a shortage of infrastructure investments; and education (human resources). Certainly, for some time (e.g. Johnston and Viken, 1997), there has been an acknowledged need for trained local tourism workers who can communicate in several languages and for the development of a tourism service tradition in Greenland.

Since 2015 the situation has moved on in terms of visitor numbers gradually increasing (Table 7.1; Section 7.5), improvements in education and training at all levels (Box 7.3), including the training of local operators, the coming upgrade/construction of three airports, provision of more visitor accommodation and the adoption of quality systems. Plans are also being formulated to extend visitor activity across the long seasonal shoulder periods, to enable tourism to become a full-year employer (Árnadóttir, 2019; Lumholt, 2020).

Visit Greenland (2016b: 29) particularly values NATA's networking events where adventure operators meet, share knowledge and experiences and discuss how to improve quality and safety. In a generic sense Visit Greenland regards tourism to the country as either (responsible) adventure tourism (Chapters 9, 10) or cruise tourism (Chapter 11). This binary understanding underpins much of Visit Greenland's approach in terms of vision and mission statement, as set out in its 2016 strategy document (Visit Greenland, 2016b). Its ten-year vision states:

> By 2025, the tourism industry of Greenland is one of the leading industries of the country, and Greenland is an internationally known and sought after adventure and cruise destination. The Greenlandic tourism industry is known for its innovative profile, unique Arctic experiences and a tourism development focused on sustainability, responsibility and a local foundation.
>
> (Visit Greenland, 2016b: 23)

In its mission statement, Visit Greenland argues that it:

- implements prioritised and targeted international marketing to adventure and cruise segments in co-operation with partners within and outside of Greenland;
- also works continuously to improve the framework conditions for the tourism industry, including the development of how to receive tourists and develop new experience products; and
- continues to adjust and further develop the 'Pioneering Nation' brand (see Chapter 10) in order to build an internationally recognisable and positive Greenlandic image that strengthens the marketing of Greenlandic experience products and Greenland itself.

An emphasis upon responsible tourism is interpreted as prioritising:

- the involvement of local people in tourism development to generate economic benefits for local communities;
- a positive contribution to the conservation of natural and cultural heritage;
- minimising negative environmental and social impacts;
- the offering of experiences that create more meaningful relationships with locals, thus providing a greater understanding of local cultural as well as social and environmental challenges; and
- the creation of areas of commercial co-operation between local tourism actors and foreign travel agencies that target the tourism segments relevant to Greenland (Visit Greenland, 2016b: 23).

But, for communities to benefit from such prioritisation, serious questions need to be asked (Higgins-Debiolles, 2020: 615): 'who defines the boundaries and operations of tourism; who benefits from these forms of tourism; how is power exercised; and how is power limited and controlled'.

7.5 Tourism scale and impact

7.5.1 Foreign guests and the statistical basis

Figures of 'around 100,000' are often quoted in relation to (pre-Covid) visitor numbers to Greenland. There are perhaps three ways of sourcing data contributing to such numbers. As Table 7.1, column 2 indicates, registered overnight stays have been recorded since at least 1994. And while such figures indicate (registered) accommodation occupancy, and thus income for the hospitality sector, they do not provide a picture of the actual visitor numbers involved. This has been to some extent rectified since mid-2000

Table 7.1. Enumeration of travellers in Greenland, 1994–2019. Author's own table derived from https://bank.stat.gl/pxwb/en/Greenland/Greenland_TU_TU30/TUXHOT.px/ and https://bank.stat.gl/pxwb/en/Greenland/Greenland_TU_TU_20/TUXPAX.px/; author's additional calculations.

Year	Total registered overnight stays	Total registered guests	Total average nights per guest	Foreign (including Danish) registered overnight stays	Foreign (including Danish) registered guests	Average nights per foreign (including Danish) guest	Annual growth in foreign overnights	Annual growth in foreign guests
1994	112,316*			81,517				
1995	137,121			72,810			–10.7	
1996	170,730			82,950			13.9	
1997	181,043			86,807			4.7	
1998	200,573			88,256			1.7	
1999	205,047			99,820			13.1	
2000	219,454	34,223+	–	107,457	20,415+	–	7.7	–
2001	191,299	62,648	3.1	94,502	30,532	3.2	–12.1	–
2002	179,491	62,603	2.9	84,781	30,529	3.6	–10.3	–0.01
2003	195,231	62,912	3.1	92,955	29,393	3.2	9.6	–3.7
2004	190,774	68,433	2.8	95,262	33,400	3.5	2.5	13.6
2005	215,916	78,551	2.7	109,204	38,801	3.6	14.6	16.2
2006	225,712	76,919	2.9	122,026	39,291	3.2	11.7	1.3
2007	234,604	83,031	2.8	114,634	40,818	3.6	–6.1	3.7
2008	236,727	84,211	2.8	121,624	41,811	3.4	6.1	2.4
2009	224,801	75,389	3.0	103,931	34,534	3.3	–14.5	–17.4
2010	216,141	72,368	3.0	96,719	33,403	3.5	–6.9	–3.3
2011	213,656	73,801	2.9	98,334	34,758	3.5	1.7	4.1
2012	209,896	79,881	2.6	99,439	37,137	3.7	1.1	6.8
2013	214,012	81,236	2.6	98,288	36,966	3.8	–1.2	–0.5
2014	209,560	79,873	2.6	95,342	36,596	3.8	–3.0	–0.1
2015	218,527	87,225	2.5	96,545	38,291	4.0	1.3	4.6
2016	240,631	96,222	2.5	101,330	38,721	3.8	5.0	0.1
2017	262,793	107,618	2.4	118,433	46,661	3.9	16.9	20.5
2018	259,282	101,300	2.6	119,414	44,132	3.7	0.8	–5.4
2019	264,711	105,680	2.5	132,748	51,573	3.9	11.2	16.9

* Indicates this figure includes a significantly large number of 'unknown' nationality (39,736); + indicates these data began being collected mid-year.

with individual registered guests being enumerated (Table 7.1, column 3), thus also permitting an indication of the average length of stay per registered visitor (column 4), which has been between 2.4 and 3.1 nights throughout the 19 pre-Covid years, latterly exhibiting a downward trend.

But these published figures include Greenlanders staying in registered accommodation. And in 18 of the 19 years 2001–2019, Greenlanders outnumbered foreigners as guests in registered accommodation (Table 7.1, column 3 minus column 6). When their numbers are subtracted, the foreign guest average length of stay (column 7) is shown to be longer, at between 3.2 and 4.0 nights. However, the year-to-year volatility in numbers of foreign registered overnights (column 8) and in the growth of numbers of foreign guests (column 9) will need some attention in a post-Covid recovery.

The numbers of foreign guests in registered accommodation (column 6) are consistently below foreign international air passenger (departure) figures (Tables 7.2, 7.3), suggesting that the difference may be made up of foreign guests staying in non-registered accommodation and those who may arrive by cruise liner but depart Greenland by air or vice versa.

A number of trends can be detected from Table 7.2. First, significant increases can be seen in numbers from the major markets over the five-year pre-Covid period, albeit with some

Table 7.2. Greenland: international air passenger departures by country of origin, 2015–2019. Author's own table derived from https://bank.stat.gl/pxwb/en/Greenland/Greenland_TU_TU20/TUXFLY.px/; and author's additional calculations.

	Year					% change	
Country of origin	2015	2016	2017	2018	2019	2018–19	2015–19
Denmark	20,427	26,865	25,672	26,778	27,270	1.8	33.5
Germany	3,292	4,508	3,963	4,925	4,459	−9.5	35.4
USA	2,649	3,668	3,054	3,960	3,465	−12.5	30.8
France	1,542	2,179	2,029	2,709	2,750	1.5	78.3
UK	1,053	1,597	1,242	1,479	1,637	10.7	55.5
Canada	1,169	2,590	1,211	1,342	1,514	12.8	29.5
Switzerland*	n/a	n/a	n/a	1,256	1,375	9.5	n/a
Norway	890	1,225	1,033	1,090	1,054	−3.3	18.4
China	892	921	1,251	938	995	6.1	11.5
Iceland	1,130	1,237	947	1,013	980	−3.3	−13.3
Sweden	908	971	1,088	907	855	−5.7	−5.7
Spain	736	900	745	771	828	7.4	12.5
Italy	503	614	471	624	509	−18.4	0.1
Japan	477	491	403	241	362	8.7	−24.1
Taiwan	992	586	357	375	337	−10.1	−66.0
Russia	122	232	140	209	313	49.8	156.6
Other Europe	3,083	3,267	4,161	4,325	4,267	−1.3	38.4
Other world	1,726	2,644	2,135	2,620	2,886	10.2	67.2
Unknown	15,940	6,326	11,020	7,029	5,713	−18.7	−64.2
Total foreign	57,531	60,821	60,922	61,335	60,194	−1.9	4.6
Total non-Danish foreign	37,104	33,956	35,250	34,557	32,924	−4.7	−11.3
% not Danish	64.5	55.8	57.9	56.3	54.7		
Greenland	23,331	28,394	28,970	31,342	32,641	4.1	39.9
Total international air passengers (including Greenlanders)	80,862	89,215	89,892	92,677	92,835	1.7	14.8

Most of the above tabulated figures differ from those in Lumholt (2020: 5) where the 'unknown/not registered' numbers are redistributed proportionally among named countries: the 'Visit Greenland redistribution' (Table 7.3).

*Figures tabulated for Switzerland are only available from Lumholt (2020: 5), and the above caveat should therefore be borne in mind in relation to these.

Table 7.3. Greenland: foreign air passenger departures by country of origin, 2018 and 2019, according to the 'Visit Greenland redistribution'. Author's own table derived from Lumholt (for Visit Greenland), 2020: 5, after https://bank.stat.gl/tuefly; and author's additional calculations.

Country of origin	Year 2018	Year 2019	% change 2018–19
Denmark	28,976	29,029	0.2
Germany	5,329	4,747	−10.9
USA	4,285	3,689	−13.9
France	2,931	2,927	−0.1
UK	1,600	1,746	8.9
Canada	1,452	1,612	11.0
Switzerland	1,256	1,375	9.5
Norway	1,179	1,122	−4.9
China	1,015	1,059	4.4
Iceland	1,096	1,059	−4.8
Sweden	981	910	−7.2
Spain	834	881	8.5
Australia	904	690	−23.7
Italy	675	542	−19.7
Austria	501	506	1.0
Netherlands	676	413	−38.9
Japan	261	385	47.5
Taiwan	406	359	−11.5
Russia	226	333	47.3
Poland	309	296	−4.2
Indonesia	95	217	128.4
Hong Kong	188	198	5.3
Singapore	142	144	1.4
India	0	110	'100'
Malaysia	80	94	17.5
Republic of Korea	25	62	148.0
Other Europe	1,937	1,952	0.8
Other Asia	500	524	4.8
Other world	901	1,035	14.8
Total	58,763	57,996	−1.3
Total non-Danish	29,787	28,967	−2.8
% not Danish	50.7	50.0	

The Statistics Greenland 'not registered' category (7.6% of the total in 2018 and 6.1% in 2019) was 'distributed proportionally between the "country of residence" segments to get more representative data' by Visit Greenland. The figures in Table 7.3 therefore do not correspond with those in Table 7.2. Both sets should be treated with some caution. Growth rates for segments under 1,000 are omitted by Visit Greenland 'as small samples often results [*sic*] in disproportionate high growth rates'. The current author's additional calculations have completed column 4 if only to emphasise the precarious and uncertain nature of source markets with relatively small numbers.

serious downturns for 2019 (Germany, USA, Italy). By contrast, Taiwan and Japan reveal significant proportional decreases (compared to China's modest increase), although the absolute numbers of travellers involved are relatively small.

Second, when Danish visitors are removed from the equation (and their numbers could be underestimated given the large numbers of 'unknown' in these figures – see Table 7.3), foreign air visitors (i.e. those not on cruises) only amount to around the mid-30,000s each year and show a reduction of 11.3 per cent over the five-year period. If, and when, the political relationship with Denmark changes, what will be the impact on Danish numbers visiting – given that many visits are business- and/or administration-related rather than for 'pleasure' tourism – and how far might this expose the relatively low level of non-Danish foreign visitation? Prior to the pandemic, Visit Greenland was beginning to address this issue by focusing on quality (and implicitly visitor spending) (Ren, Chimirri and Abildgaard, 2020).

Will the three new/reconstructed airports, *if* they open in 2023 (Chapter 8), herald a post-Covid resurgence and growth, as expected/predicted in the publication *Towards More Tourism?* (Visit Greenland, 2020e). Will the two new international airports encourage longer-haul arrivals, despite growing environmental concerns over such mobility? As an explicit 'victim' of climate warming, should Greenland be encouraging such travel? Growing numbers of Greenlanders have been travelling internationally by air, increasing by almost 40 per cent over the five pre-Covid years, raising the growth rate in international air travel numbers. Will eventual independence stimulate this process further or will it encourage some of the diaspora to return home and reduce the need for Greenlanders' visiting friends and relatives international mobility?

Lumholt (2020) is able to furnish the reader with a longer list of specific source countries (Table 7.3) than the Statistics Greenland data bank publishes (in English). Both the overall number of airborne foreign visitors for 2018 and 2019 and the proportions of non-Danish foreign visitors are clearly lower than those in Table 7.2. The latter at least is presumably explained by the proportionate redistribution of 'not registered' numbers to Denmark, as it represents by far the largest foreign visitor source.

Table 7.4 again highlights the relatively small proportion of non-Danish foreign visitors even taking into consideration the exceptional circumstances of 2019 (noted below). And although the numbers are (still) relatively small, the significant proportional increase in visitors from China and Hong Kong combined is noteworthy (see Chapters 3 and 10).

From figures in Tables 7.1 and 7.4, Lumholt (2020: 3) highlights for 2019 a double-digit growth in the number of international (including Danish) overnight stays and guests (as well as a 2 per cent increase in the number of cruise passengers; see Chapter 11). On the other hand there was a 1.3 per cent decline in the number of international flight passengers (Table 7.3). Lumholt (2020) suggests that this apparent anomaly indicates that a greater proportion of visitors to Greenland in 2019 stayed in a hotel that fully registered their detailed presence, and that these were likely to be mainly business and civil service travellers staying in connection with issues concerning the airports development process[3] (Chapter 8) and the short-lived activity related to the then President Trump's offer to purchase Greenland (Chapter 4), factors that probably called for a more than usual number of meetings and conferences with participants from Denmark, whose overnights increased by 33.5 per cent. Again, their presence raises the question of distinguishing Danish business/administration visitors from 'pleasure' tourists.

With the bankruptcy of low cost Icelandic WOW Air in March 2019, accessibility to that country was reduced, and this could have had a knock-on effect for access to Greenland from Europe.

Normally Visit Greenland focuses on promotions to five core markets: Denmark, Germany, North America (US and Canada), France and the UK. Yet visitors from Germany (10.9 per cent down) and the USA (13.9 per cent down) exhibited a decline in overnight stays in 2019, while overall registered guest numbers from France (−13.2 per cent), the UK (−17.0 per cent) and Canada (−26.3 per cent) also fell significantly (Table 7.4).

In its 2016 strategy document, Visit Greenland (2016b: 18) mused that if the number of flight passenger tourists continued to grow by

Table 7.4. Greenland: foreign guests by country of origin in registered accommodation, 2018 and 2019, according to the 'Visit Greenland redistribution'. Author's own table derived from Lumholt (for Visit Greenland), 2020: 9, after https://bank.stat.gl/tuehot; and author's additional calculations.

	Year		% change
Country of origin	2018	2019	2018–19
Denmark	21,223	29,140	37.3
Germany	3,048	3,145	11.7
USA	2,770	2,689	−2.9
Iceland	1,157	1,553	34.3
France	1,730	1,503	−13.2
UK	1,366	1,134	−17.0
Spain	1,230	1,109	−9.9
Switzerland	800	830	3.8
Norway	803	708	−11.8
Canada	938	691	−26.3
China	442	644	45.7*
Sweden	657	640	−2.6
Hong Kong	170	628	269.4*
Australia	425	488	14.8
Netherlands	335	420	25.4
Italy	546	419	−23.3
Japan	318	301	−5.3
Austria	317	263	−17.0
Poland	100	137	37.0
Singapore	102	104	2.0
Russia	160	82	−48.8
Taiwan	63	74	17.5
Republic of Korea	39	69	76.9
Indonesia	168	42	−75.0
Malaysia	35	22	−37.1
Other Europe	1,726	1,011	−41.4
Other Asia	204	430	110.8
Other world	1,270	1,095	−13.8
Total	42,144	49,593	17.7
Total non-Danish	20,921	20,453	−2.2
% not Danish	49.6	41.2	

*Indicates taken in combination, the growth rate for China with Hong Kong is 107.8%.
As in Table 7.3, Statistics Greenland's 'not registered' category was 'distributed proportionally between the "country of residence" segments to get more representative data' by Visit Greenland. This does not explain why the total figures in Table 7.4 do not correspond with those in Table 7.1. As Lumholt/Visit Greenland (2020) states, both sets of data should be treated with 'appropriate reservations'. Highlighting biases arising from non-registration, no less than 71.5% of guests overnighting in East Greenland during 2019 were not registered for nationality.
Again, as in Table 7.3, growth rates for segments under 1,000 are omitted by Visit Greenland, but are included following the current author's additional calculations, in order to highlight planning difficulties when confronting the volatility of source markets with relatively small numbers.

Box 7.5. How many tourists?

As many developing countries before it, until recently Greenland had been unquestioningly inviting to international tourism. In pursuit of economic 'self-sustainability' (Chapter 5), proposals for continued tourism growth (Box 7.1) had not heeded calls questioning the volume growth model (e.g. Gössling, Scott and Hall, 2021). Thus Greenland's tourism numbers have appeared to remain important, if somewhat elusive. However, alongside the estimate that Greenland would need 20 large-scale mining projects in order to be able to finance its political independence (Taagholt and Brooks, 2016: 362) (Section 5.7), has anyone realistically attempted to appreciate what the equivalent minimum number of annual high spending tourist overnights might be?

In the task of pinning down the answer to the seemingly innocuous question, 'How many foreign tourists actually visit Greenland?' a number of considerations are required. Putting aside the question of whether we are dealing with 'pleasure tourists', 'business/administration visitors' or various combinations of the spectrum between, we need to consider the following, where the last pre-Covid year, 2019, is taken as the comparator:

1. Calculating the Statistics Greenland figures (TUXHOT) for numbers of foreign guests staying in registered accommodation (Table 7.1) in 2019, we arrive at a figure of 51,573.
2. Visit Greenland's method of proportionately redistributing 'unknown' nationality figures actually provides a lower figure for foreign guests staying in registered accommodation: 49,593.
3. Calculating the Statistics Greenland figures (TUXFLY) for numbers of foreign international air departures (Table 7.2), some of whom will have arrived by cruise ship, gives us a figure of 60,194. On the other hand, those visitors who arrive by air to join a cruise ship are not captured in this figure.
4. Visit Greenland's method of proportionately redistributing 'unknown' nationality figures arrives at a lower foreign international air departures figure (Table 7.3) of 57,996.
5. An average of the above four figures produces 54,839.
6. Then there are cruise liner tourists, most of whom do not usually stay overnight on land (Chapter 11). According to Statistics Greenland figures (Table 11.1), 2019 saw cruise liners arriving in Greenland with a total of 46,633 passengers (presumably such figures exclude ships' crews, at least some of whom may come ashore at any particular port of call). Some of these passengers are likely to leave Greenland by air and thus be included in the enumeration of (3) above.
7. When looking at actual disembarkations at individual ports of call (Table 11.2) a total figure of 122,314 is arrived at.
8. In addition to the potential issue of double counting in some of the above figures, the almost existential question is raised: what is 'one tourist'? Although tourism practitioners, DMOs, national and regional governments are apt to quote absolute and relative tourist numbers, often in isolation, the utility of enumerating tourists would appear to be limited unless such enumerations are directly related to, for example, spending power/income derived/employment generated, levels of social and environmental impacts, seasonal and spatial distributions, activities and behaviour: in short, measures of contribution, equivalence and carrying capacity.

These thoughts are nothing new, of course, but when dealing with such relatively small numbers they can be significant for strategic planning and management.

5 per cent annually, their number could reach 90,000 by 2030. 'The growth, however, may be even bigger because of the improved framework conditions and the increased accessibility' that are posited from 2023 onwards.

If tourism growth is premised on direct arrivals through the new airports for Nuuk and Ilulissat, supplemented by heavy pulses of visiting cruise tourists (and Iceland's considerable growth in tourism numbers is employed as a model), a substantial increase and upgrading of infrastructure – notably accommodation – is required, along with, potentially, formal constraining mechanisms, where there are currently none, such as for access to the Ilulissat Icefjord (Chapter 9). The paradox of Greenland being such a vast land yet its settlements and tourism attractions being so generally compact and relatively isolated from each other raises very serious questions concerning current and future carrying capacities, their computation and integrated management.

Further, Greenland currently has no direct civilian flight connections with North America, but only to Copenhagen and Reykjavik (Fig. 8.5). Iceland's substantial post-financial crash tourism growth has been partly the result of Icelandair's effective hub and spoke system connecting several North American cities to European destinations. Passengers on these routes are encouraged to break their journey and spend time in Reykjavik (Jóhannesson, Huijbens and Sharpley, 2010; Jóhannesson, 2016). If, following their completion, either or both Nuuk and Ilulissat airports attract an airline that is disposed to perform a similar role for Greenland, while this could be relatively lucrative it would also place significant additional pressures on the (tourism) infrastructure and environments of those two destinations.

At the time of writing, it remained to be seen what a post-Covid trajectory towards and beyond 2023 might look like.

7.5.2 Seasonality

Fig. 7.1 emphasises the seasonality of tourism activity in Greenland, in depicting the number of foreign overnight stays (including Danes) for 2018 and 2019. Both years see a significant summer peak in July–August (heightened by cruise tourists – Chapter 11), falling away in June and September/October, with a small secondary 'winter' season in spring (March, April, May) and low activity from November to February.

This 'challenge' has long been recognised by Visit Greenland. Looking abroad for appropriate models: 'In Iceland, tourists from Great Britain are particularly active in the periods January–April and October–December. Thus, the Great Britain market is an obvious target when focusing on increasing visits to Greenland in the shoulder season' (Visit Greenland, 2016b: 16). Tourism activity remains distributed very unevenly spatially (Section 8.1) as well as temporally. Only in Ilulissat have there been exhibited signs of what might be termed *overtourism* (Ioannides, 2019), reinforced by the irregular arrival of cruise tourists during summer months, with examples of social media blogs highlighting the difficulty of finding accommodation if not pre-booked. Otherwise numbers are relatively low, although again, smaller communities can be briefly swamped:

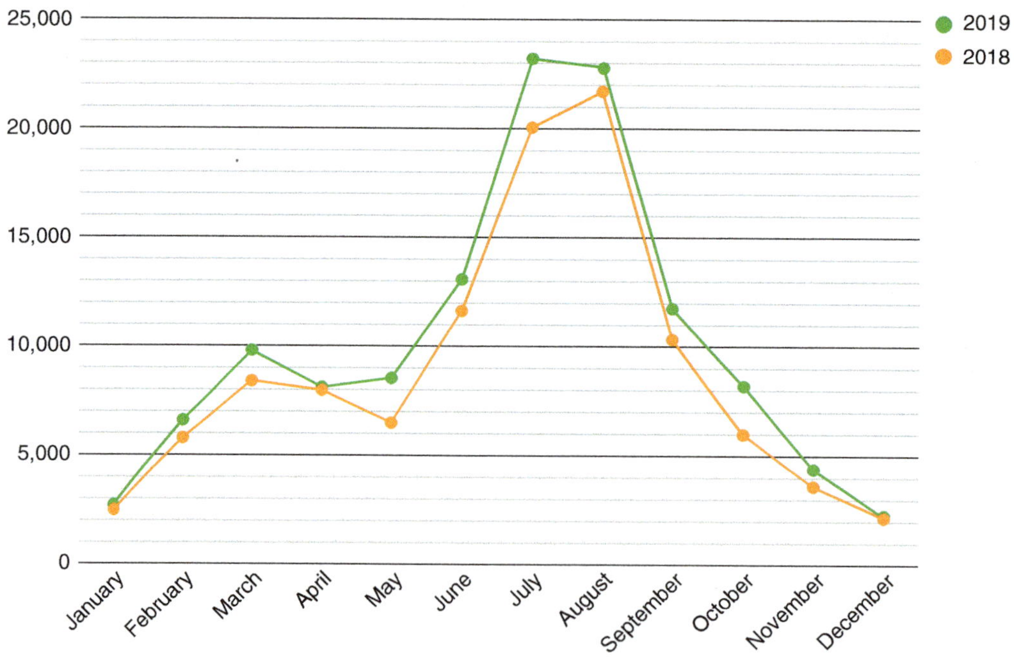

Fig. 7.1. Greenland: number of foreign visitor overnight stays in registered accommodation, by month, 2018 and 2019. Lumholt (for Visit Greenland), 2020: 8, after https://bank.stat.gl/tuehot. Figure courtesy of Statistics Greenland and Mads Lumholt.

> At the same time that Greenland does not have the overall desired volume of tourists throughout a year, several destinations experience capacity bottleneck problems with too many tourists for the local supply in the high season, in terms of availability of accommodations. Often times, cancelled group tours are a result. On the other hand, there is huge transportation and accommodations capacity available during the shoulder and low seasons.
>
> (Visit Greenland, 2016b: 36).

As Lumholt for Visit Greenland has more recently stated: 'In order to achieve a more sustainable tourism development it is necessary for the entire industry to work long-term and targeted to attract more tourists in the offseason' (Lumholt, 2020: 8).

Visit Greenland (2016b: 36-37) acknowledges that addressing the seasonality issue requires a regionally targeted approach, sensitive to local conditions in this enormous country, and has prioritised the following spatially-based strategies:

- in East Greenland, September needs to be included in the main season, and visitor volumes need to be increased in winter/spring;
- in South Greenland, it is desirable to both attract more tourists in the summer and to extend the season;
- in Nuuk, summer tends to be the low season, and there is need to attract more tourists in July and August;
- in the Arctic Circle region, season expansion in Kangerlussuaq should be focused on attracting more foreign tourists in the winter months as well as increasing visitor volumes in May (subsequently, this region has received increased attention through Covid-response staycation policies – see Chapter 12);
- in Sisimiut, where winter is the high season, there is a need to attract more tourists in June, July, and August; and
- in North Greenland, season expansion is focused on winter/spring and on extending the summer season.

Visit Greenland (2016b: 37) emphasises that because season expansion approaches need to be adjusted to fit each region's individual needs, it is especially important that regional governments create clear strategies on tourism development and season expansion as a part of the planning framework for their region, as well as contributing to national efforts.

7.5.3 Tourism's contribution

According to Árnadóttir (2019: 38), in 2017 the value of tourism for Greenland was around DKK500mn (€65mn/£60mn/$85mn), representing 11.7 per cent of the economy. Foreign tourists created 759 jobs through direct and indirect employment, although distinguishing the employment impacts of foreign and domestic visitors in a large hotel, for example, must be at best an inexact science.

Table 7.5 suggests that for the accommodation and food services sector at least, there has

Table 7.5. Greenland: accommodation and food services: businesses and employment, 2009–2019. Author's own table derived from https://bank.stat.gl/pxwb/en/Greenland/Greenland_ARXBFB3.px/; https://bank.stat.gl/pxwb/en/Greenland/Greenland_ESX1A.px/; and author's additional calculations.

Year	No. of businesses	No. of persons employed for at least one month in the year	Average monthly number of persons employed	Average monthly salary (DKK)
2009	105	1,859	762	14,506
2011	106	1,816	787	15,129
2013	116	1,828	798	16,769
2015	113	1,803	762	17,555
2017	101	1,913	807	18,448
2018	116	1,904	821	19,288
2019	116	1,849	800	20,422

been relative stability over the 2009–2019 period in terms of the number of businesses operating and people employed. In 2018 there was one accommodation and/or food service business for every 482 residents, and one for every 380 foreign overnight visitors.

The issue of accommodation has been something of an enigma, as even Visit Greenland publications complain about the inconsistent or non-registration of guests, particularly in the east. The extent of informal accommodation, including camping and the use of hikers' huts exacerbates incompleteness of data. Indeed, this sub-sector comprises a perhaps wider range of modes and standards than might be encountered elsewhere in the northern hemisphere: hotels, seamen's hostels, youth hostels, bed and breakfasts in towns, private accommodation in settlements, farm stays, camping, both at recognised provision and wild camping with no facilities, and hikers' huts[4] (Figs. 7.2–7.6).

Visit Greenland now helps to promote the upper range of accommodation options on its own websites and through links to the providers' sites. But the marked skewed spatial and seasonal nature of Greenland's overnight tourism activity has meant that drawing sensible conclusions from the nature of provision and its requirements is often difficult. One pattern is clear: relatively low annual occupancy rates (Table 7.6). Indeed, much accommodation provision may represent part-time/sideline/secondary and even hobbyist roles on the part of providers, such that official indications of 'occupancy' may be misleading in some cases.

The Disko Bay area (Ilulissat, Aasiaat, Qeqertarsuaq: Fig. 4.1) has the predominant part of hotel stays and cruise tourism, and is the best developed tourism market, catering for both summer and winter/spring activities. It also has the highest number of tourism operators.

7.6 Conclusion

Thus far, tourism in Greenland has not exhibited consequences of rapid growth and overexploitation, and numbers remain relatively small. Indeed, past efforts to attract more tourist arrivals have often fallen short despite

Fig. 7.2. Hotel Kap Farvel, Nanortalik. Author's own photograph

Fig. 7.3. Seamens' Hostel beyond a tourist group passing through the portals of Sisimiut open air museum. Author's own photograph.

Fig. 7.4. An Ilulissat bed and breakfast establishment. Author's own photograph.

Fig. 7.5. A campsite on Disko Island. Author's own photograph.

Fig. 7.6. 'Wild' camping among the cotton grass on the outskirts of Ilulissat, perhaps reflecting the shortage of formal accommodation available in the town in the 'high' season. Author's own photograph.

Table 7.6. Greenland: rooms available in paid accommodation, and % occupancy rates by region, 1996–2019. Author's own table derived from https://bank.stat.gl/tuekap; and author's additional calculations.

	1996	2000	2004	2008	2012	2016	2017	2018	2019
Total Greenland	367,642	456,918	431,463	511,505	491,956	458,026	480,308	484,980	487,630
	38.4	39.0	36.9	37.3	35.5	43.3	45.0	42.3	41.4
Ilulissat	38,186	41,064	52,083	81,450	86,650	85,070	88,281	95,204	95,582
	63.9	53.3	53.2	48.0	50.2	60.1	60.7	57.2	57.4
Sermersooq West (including Nuuk and Paamiut)	48,747	79,687	70,865	69,928	82,944	97,536	98,369	98,550	98,820
	57.7	53.5	49.0	63.2	50.7	50.5	54.6	48.3	47.7
Qeqqata (including Sisimiut, Kangerlussuaq)	161,462	159,393	122,901	149,755	123,611	103,086	101,575	101,710	109,034
	27.2	37.2	39.8	36.0	31.8	46.5	45.0	43.4	40.9
Kujalleq (including Qaqortoq and Nanortalik)	84,744	105,205	105,182	104,122	93,119	87,660	102,397	99,100	96,773
	37.1	30.3	23.7	27.3	25.6	25.2	27.8	25.9	25.9
Avannaata and Qeqertalik (including Aasiaat and Qeqertarsuaq/ Disko Island)	63,657	85,905	95,973	148,645	156,398	131,612	129,462	135,694	132,947
	51.9	46.1	43.6	38.2	38.8	51.4	56.1	52.1	52.9
Sermersooq East (including Tasiilaq and Ittoqqortoormiit)	9,032	26,730	36,542	39,055	35,884	38,132	48,507	49,926	50,050
	46.9	21.7	24.2	19.8	24.3	30.5	30.6	33.0	28.5

For communes' geographical extent see Fig. 4.1.

the investments committed. Yet both spatial and temporal concentration of activity pose problems. Areas along the coast not covered by ice sheets or mountains are relatively small with fragile ecosystems. A dispersed population of 56,000 living in 18 towns and 59 smaller settlements is vulnerable to tourism's socio-cultural impacts: both the opportunities and risks.

The current tourism organisational landscape continues to be relatively fragmented. Priority offerings remain focused on Greenland's nature through adventure tourism (Chapter 9), with Visit Greenland (2020d) emphasising the 'Big Arctic Five': dog sledding, whale watching, Northern Lights, ice/snow, and the nation's 'pioneering people'. These are complemented by sailing, hiking, fishing, hunting, kayaking, climbing, and photography tours, and cultural experiences such as Greenlandic food, traditional *kaffemik* coffee parties in private homes, guided tours to historic sites and visits to handicraft workshops (Chapter 6).

Projected development is at best ambiguous. On the one hand the new airports are looked to for substantially increased visitor numbers from 2023, while it has been expressed that a move toward higher quality activities rather than a dramatic increase in numbers may be more realistic. High-end products such as heli-skiing and the provision of exclusive accommodation have begun to exert an impact (Chimirri, 2020: 25), not necessarily a positive one.

Greenland's environmental and socio-cultural vulnerability to impacts from tourism is mediated, and to some extent both obscured and exacerbated, by impacts of climate change. Indeed, notions of 'once-in-a-lifetime' experiences and 'last chance' tourism (Chapter 3) are reflections of Greenland's front-line status in encountering the effects of rapid climate change (Chapter 9). Yet, as further discussed in Chapters 5 and 9, Greenland's official sustainable development and sustainable tourism discourses appear to provide little room for climate change narratives (Bjørst and Ren, 2015).

What emerges as the post-Covid landscape is likely to be challenging.

7.7 In the next chapter

In the next chapter, processes of changing accessibility and infrastructure provision are set within concepts of Greenland's peripheralities and within the increasing geopolitical competition for access to Greenland's infrastructure sector.

Notes

[1] A participatory exercise towards drafting the Visit Greenland market development strategy 2020–2023, *Towards More Tourism* (Visit Greenland, 2020e).

[2] With a secretariat based in Tórshavn, NATA was established in 2006 by combining three organisations: the West Nordic Tourist Board, SAMIK (a Greenland/Iceland tourism co-operative venture) and FITUR (an Iceland/Faroe Islands collaboration).

[3] The (re-)construction of two new international airports in Ilulissat and Nuuk and a regional airport in Qaqortoq was to have begun during the summer of 2020, for them to be operational in the autumn of 2023 (Lumholt, 2020: 4; see Chapter 8).

[4] Quoting from Statistics Greenland data, Kaae (2006: 102) noted that hotels had consistently provided less than 40 per cent of all accommodation. In 2002 this segment was complemented by 28 per cent private dwellings, youth hostels (13 per cent), 'camping' (7 per cent), seamen's hostels (4 per cent) and hiking huts or sheep farms (1 per cent).

8

Peripherality, Tourism and Geopolitical Dimensions of Accessibility in Greenland

'... peripherality in Greenland is a relative and complex phenomenon'
(Weaver and Lawton, 2017: 61)

'Covid-19 has highlighted both the fragility and the dangers of Kalaallit Nunaat's global connections, but it has also highlighted the strengths and opportunities connected to the territory's remote archipelagic geography'
(Grydehøj, Kelman and Su, 2020: 293)

8.1 Conceiving peripherality

As any map of Greenland will indicate (e.g. Figs. 4.1, 5.2, 8.5) the country's geographical periphery (both coastland and adjacent seas) actually represents its economic, social and political core. Conversely, the central ice sheet, covering four-fifths of the island, acts as its physical and psychological periphery. This paradox is heightened by the vast extent and often hostile nature of this central 'barrier–periphery', isolating components of the 'periphery–core' and severely constraining physical mobility and social interaction across Greenland. Further, with the abandonment of a number of settlements because of earlier centralisation policies (Chapter 4) and/or the exhaustion of mined mineral resources, some elements of this 'periphery-core' have reverted to a non-core (semi-?)peripheral role,[1] while others have embraced a potentially 'hotter' reconfigured core role as new centres of industrial activity, emphasising the relative and temporally dynamic nature of Greenlandic space and its core(s)' and peripheries' relationships. In short, within the Greenlandic space, there can be no simplistic core–periphery dichotomy.

These thoughts raise a number of issues in relation to concepts and perceptions of peripherality for the country's economic development – not least tourism – and its cultural and political cohesion. In Chapter 1, Box 1.4 emphasised a number of inter-related conceptual dimensions of this, including:

- the relative quality of peripherality: peripheral to what?
- peripherality can be temporally dynamic: what is regarded as peripheral at one time may not be at another, perhaps relating to the changed function of the space involved; and
- peripheral space is not homogeneous, but rather composed of 'nearer' and 'farther' peripheries.

In attempting to simplify an understanding of peripherality in relation to the spatial and temporal characteristics of tourism activity in Greenland, Weaver and Lawton (2017: 56, 61–65) constructed what they referred to as an inductive peripherality–tourist experience model, which, they claimed, could act as a framework for pursuing the sustainable management, development and marketing of tourism in Greenland. This model, a modified version of which is represented in Fig. 8.1,

DOI: 10.1079/9781789246728.0008

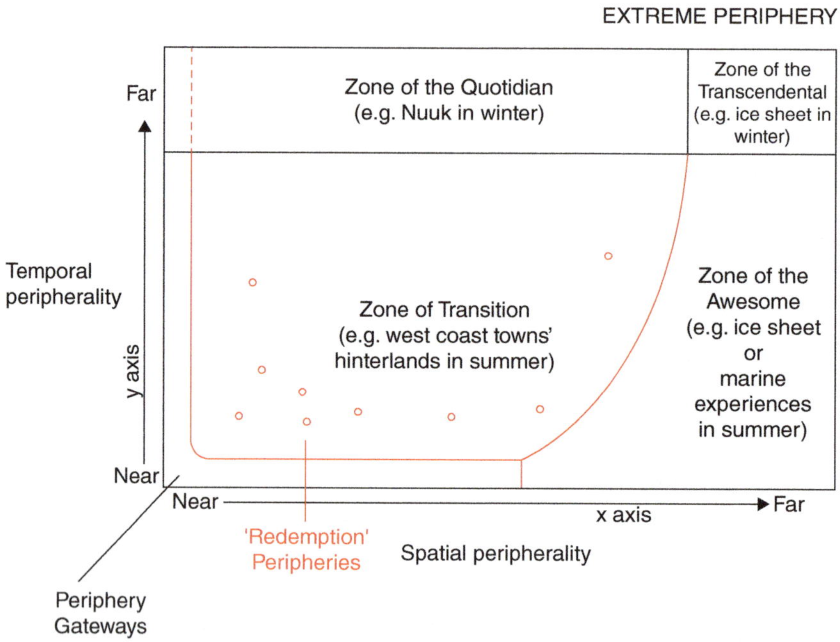

Fig. 8.1. Greenland: model of multiple peripherality. Author's own figure, modified and redrawn from Weaver and Lawton, 2017: 62

offers a matrix in which tourism activity is located in relation to its degree of spatial peripherality along the x-axis, arrayed against its temporal peripherality along the y-axis. As such, the model embraces the three important conceptual dimensions of peripherality noted above.

In Weaver and Lawton's (2017) model, peripherality is therefore represented as relative: in spatial terms this is set in relation to what they refer to as the gateway periphery – tourists' entry points into Greenland – while the referent for temporal peripherality would appear to be mid-summer, or thereabouts.

The model depicts, with various caveats, five major generalised spatial-temporal zones of activity and a sixth dimension of interaction between them, which can contribute to an understanding of the spatial and temporal dynamics of tourism experiences in Greenland's relative peripheries.

1. Periphery gateways (Weaver and Lawton's 'gateway periphery')

These are Greenland's coastal or near coastal/fjord tourist entry and potential dispersal points. The revised model extends these gateways into the temporal far periphery as they are employed by visitors all year round, albeit on a reduced scale in winter. As Weaver and Lawton (2017: 61) suggest, almost all inbound tourists have their initial and closing experience of the country within the gateway periphery centres[2] and may use these as the base for their stay.

2. Zone of transition (Weaver and Lawton's 'zone of the profane')

The example of Nuuk in summer is employed on the original model, but spatially this zone includes northern Sermersooq West (including Nuuk), a small area of Sermersooq East (Kulusuk and Tasiilaq), the whole of Qeqqata Commune (embracing Kangerlussuaq, Sisimiut and Maniitsoq), all Qeqertalik Commune (including Disko Island and Aasiaat) and the southern portion of Avannaata Commune (notably Ilulissat) (Fig. 4.1). Represented by the lower left cell on the model, this zone (incorporating the gateways) may account for 85 per cent or more of all tourism activity, according to Weaver and Lawton's

(2017: 61) original conception, although representing a relatively small area of the country.

It may be appropriate here to question Weaver and Lawton's terminology, as, firstly, this 'zone' – a nomenclature implying spatial contiguity – is, of course, comprised of a number of towns and settlements that are mutually isolated, being only connected, weather permitting, by boat or aircraft, over often considerable distances. Thus, in spatial terms at least, the 'zone' is comprised of many discontinuities: fragments clinging to the coast and fjords. The clear characteristic of this 'continental archipelago' was encapsulated in the Greenlandic government's rapid response to Covid-19, when Prime Minister Kim Kielsen, on curtailing all internal flights within the territory, pointed out that Greenland is unique in being composed of 74 isolated places (Lindstrøm, 2020; Grydehøj, Kelman and Su, 2020).

Second, although Weaver and Lawton (2017) employed the term 'profane' for this zone, reflecting a quotidian, perhaps anticlimactic aspect, in the current author's experience it represents, for the visitor, a zone of spatial and psychological 'transition'. Having made it through the gateway periphery of Greenland's portals, some personal adjustment will be required in respect of simply being here: of being an 'adventurer', a 'pioneer', before moving on, spatially and psychologically, if only temporarily, into the peripheries of awesomeness: (3) below.

The physical and social reality of the towns within this zone and time frame may be dissonant with the induced imagery of Greenland as the pristine Arctic wilderness conveyed by official marketing efforts: hence Weaver and Lawton's 'profane' nomenclature. Nondescript urban landscapes with apartment blocks scattered amongst discarded open spaces, industrial and warehousing facilities such as ugly oil storage tanks on prominent headlands adjacent to harbour mouths may act as striking landmarks for approach by sea or air. Litter and evidence of social problems may further present a contemporary landscape of 'utilitarian functionality and unvarnished social and cultural realities' (Weaver and Lawton, 2017: 62).

On the other hand, most visitors who make it to Greenland will be hardened travellers who are well familiar with life's nondescript, quotidian side, realities which can actually act to enhance the 'awesome' experiences to come by providing a stark contrast to them. But then again, it has been argued that tourism can act as a catalyst in clean-up efforts by local communities where hitherto low awareness of eyesores may be in sharp contrast to the perceptions of tourist expectations held by DMOs and local authorities, of people living in unfettered harmony with nature (Nordic Council of Ministers, 2001; Kaae, 2006: 108–109) (also Section 6.5).

Certainly, for some visitors, their initial negative visual impressions may influence the quality of their current and subsequent experiences, resulting in the construction of a (transient?) tourist psycho-spatial 'zone of disillusion' that can contribute to an undermining of branding exercises through, for example, the dissemination of negative social media commentaries by disillusioned visitors (Weaver and Lawton, 2017: 62).

Tourism stakeholders are clearly aware of this issue. This is perhaps one reason why, although located only two or three kilometres apart, local tour company minibuses whisk their clients away from central Ilulissat to the boardwalk leading to the Icefjord, permitting only fleeting en route glimpses of the town's unkempt urban fringe of industrial warehousing, rock piles, sled-dog kennels and tethered dogs (Fig. 8.2). In response to these issues, some two decades ago the authorities in Ilulissat, Greenland's main summer tourist destination, initiated a programme to remove objects of visual pollution (Nordic Council of Ministers, 2001). The question of visual aesthetics, can, of course, be both subjective and ethnocentric.

In the modified peripherality model (Fig.8.1), the role of contemporary isolated settlements, archaeological sites, waterfalls, beaches covered in icebergs, or human attempts to modify Greenland's biogeography, such as the arboretum outside Narsarsuaq[3] (Fig.8.3), can act as visitor destinations to redress the quotidian aspect of much of this zone of transition. For this reason, these minor peripheries are included in the model and referred to as 'redemption' peripheries, although their impact cannot compare to the attractions of the next periphery.

3. Zone of the awesome

In the original model, this zone, as represented in the bottom right sector of Fig. 8.1, is exemplified

Fig. 8.2. Zone of disillusion? Part of the Ilulissat urban fringe in summer, with tethered sled dogs, their 'kennels', unkempt open space, a large pile of rubble, some industry and apartment blocks. Author's own photograph.

Fig. 8.3. Within Narsarsuaq Arboretum. Author's own photograph.

by the ice sheet in summer. Here spatial extreme peripherality combined with temporal near peripherality can be experienced by individuals joining group tours, or self-contained cruise tour groups, visiting ice fields, icefjords and whale watching sites. The scope and accessibility of such settings is amplified by summer weather, length of daylight and water conditions. Weaver and Lawton's (2017) original model appeared to omit maritime experiences (apart from the allusion to cruise tour groups), to concentrate on landward experiences. Yet Greenland's vast multiplicity of small boats, and the employment of many of them for tourism purposes – viewing the Ilulissat Icefjord, getting close up to icebergs and glacier tongues, for whale and other maritime mammal watching, and combinations of these – should not be underestimated within the array of Greenlandic experiences. They also act to extend the spatial dimensions of this 'zone' into contiguous waters. Attracting considerably fewer numbers, one might add to this list of water-borne experiences more individualist pursuits within 'awesome' contexts such as kayaking and paddle boarding – both now recruited to Visit Greenland's menu of 'adventure tourism' activities (Por, 2020; Chapter 9).

But generally in this zone, experiences are likely to be mediated by travel within a group and probably being tied to the umbilical cord of a nearby light aircraft, fleet of snowmobiles or even cruise ship. Weaver and Lawton (2017) argue that crucially, for the prospects of market expansion, within this potentially threatening environment and unlike in the zone of the transcendental (5, below), participants may not be preoccupied with concerns of survival and physical danger, but they can concentrate on appreciating the 'awesome' attractions, and perhaps tick off another item on their 'bucket list' while posting far too many photographs on multiple social media platforms.

4. Zone of the quotidian (Weaver and Lawton's 'zone of the authentic')

This peripherality is exemplified by Nuuk in winter, represented in the top left sector of Fig. 8.1. Near periphery spatial settings here combine with extreme peripheral temporal periods. This is where the urban panorama of Nuuk and other gateway settlements transition from partial front- to back-stage, thereby possibly attracting tourists who seek 'authentic' cultural encounters. While cited by Weaver and Lawton (2017: 64) as exemplification of this zone, the 'Westernised' urban sprawl of Nuuk is perhaps not the most appropriate destination for any such 'authentic' encounters compared to smaller (and less accessible) settlements where traditional culture may be more evident. Winter weather and lack of daylight will inhibit outdoor activities, although the presence of settled snow – not to mention the possibilities of the Northern Lights (see Chapter 1, Box 1.5) – will bring some light and colour to the gloom.

Weaver and Lawton (2017: 64) cite evidence on social media of some tourists' inadvertent visits within this zone because of limited time availability or assumptions of lower prices and the presence of fewer other visitors.

Visits to this zone return us to questions about constructs such as authenticity, backstage and host/guest cultural contact, when the visitor is moving in unfamiliar spatial and temporal circumstances, and yet where 'authenticity' may be little more than an homogenised urban quotidian, albeit in its winter coat: hence the current author's renaming of this zone.

5. Zone of the transcendental

This zone is exemplified by the ice sheet in winter, and is located in the top right sector of the model. It represents extreme 'double peripherality' expressed in both spatial and temporal terms. In physical terms amounting to around 80 per cent of Greenland's land area, conceptually, this zone represents an extreme periphery experienced only by a deliberate few. Here severe spatial isolation and wilderness combine with the profound temporal isolation of winter. The extreme periphery experience can be a challenging non-mechanised trek – perhaps on skis or employing a dog sled – across the ice sheet during the snow-brightened dark of winter. This may be realised by just a few 'total-immersion' tourists and characterised by their peak emotional reactions and evaluations (Weaver and Lawton, 2017: 63). The element of exclusiveness is enhanced by the added requirement to obtain a government permit in order to access such experiences.

The severe constraints on accessibility and the high cost involved suggest that the growth prospects of tourism in such extreme peripherality are, almost by definition – and indeed of necessity – severely limited. The dramatic implications of climate warming could, however, generate a 'transcendentally soft' element of 'last chance' tourism pursued both by foolhardy 'bucket list' visitors and sincere environmentally aware travellers stimulated by the heightened experience of combined environmental and personal fragility.

6. Zone of interaction

In addressing the reality of tourists moving between zones, Weaver and Lawton (2017: 64) draw on Zurick's (1992) spatial model in which adventure tourism is represented as a hierarchy of peripheries and peripheral gateways. They acknowledge how the experience and perceptions of it change, as the tourist moves (spatially) from gateway to extreme periphery and back. An obvious example, alluded to above, is the location of Greenland's most visited attraction, the Ilulissat Icefjord World Heritage Site (Chapter 9), located just a few kilometres, and well within walking distance for most, from the centre of Ilulissat town. Aside from the larger human footprint (trails, noise, litter, other humans and the intervening urban fringe referred to in (2) above), how far does such proximity reduce the emotional impact of an extreme encounter (Weaver and Lawton, 2017: 64; Fig. 8.4)?

A second dimension of zone interaction is the temporal shifting of places from one zone to another. Weaver and Lawton (2017: 65) suggest the need for seasonally adjusted planning and management to take into account a temporary receding of the front-stage, shifting social dynamics and differences in visitor motivations, experiences and perceptions accompanying different seasonal settings, whether of a large urban area or the edges of the ice sheet. To some extent this is being addressed by Visit Greenland (2016b: 36–37) in acknowledging spatial variations in Greenland tourism's seasonality characteristics (Section 7.5.2).

A third factor, which Weaver and Lawton (2017) appeared to underplay (Weir, 2017), is the role of climate change, which Schrot, Christensen and Formayer (2019) have suggested will influence the nature and location of winter tourism in West Greenland as rising temperatures and

Fig. 8.4. Close encounters: visitors closing in on the Icefjord. Author's own photograph.

increased precipitation in the form of rain reduce the length and distribution of snow cover and thereby constrain (or occasionally enhance), the attractiveness and feasibility of winter sports activities (Section 9.6).

Earlier evaluations (e.g. Baldacchino, 2006; Kaae, 2006) anticipated that isolation, infrastructural limitations, high prices and political parochialism would conspire to characterise tourism in Greenland as small-scale and low-impact. Weaver and Lawton (2017: 65), however, recognise that patterns of periodically intensive tourism within constrained settlement front-stages, such as the mass convergence of cruise ship passengers, and the potentially negative experiences they may encounter (Chapter 11), would suggest such earlier assumptions to represent only one element of a larger and more complex picture of the spatial–temporal–volumetric sustainability of tourism processes as they impact upon natural and cultural landscapes in Greenland, and as the changing climate exerts its influence upon them.

8.2 Peripherality, accessibility and infrastructure

Turning from the relative peripheries that constitute Greenland, as independence beckons, Greenland's physically peripheral location internationally is also a major constraint – not least for tourism – that is difficult to mitigate. It can add considerable time and cost to imports, exports, inbound tourism and travel out.

In purely human travel terms, Greenland as a whole is highly peripheral to more central places in Europe and North America. It is difficult to visit a destination and undertake activities on a one-day trip because of the long distances and lack of direct connections (day trips from Iceland to Greenland's east coast being the exception). Many towns and all settlements are additionally highly peripheral in relation to the capital Nuuk (Christensen *et al.*, 2020: 4).

An island can be considered remote if the driving/ferry time to the closest inland city is more than four hours (Christensen *et al.*, 2020: 5). Greenland's settlements can be considered as 'urban islands' as they lack connecting roads or any other effective means of overland transport. Coastal sea transport is slow, weather-dependent and only operates in summer. Air services – the only realistic mode of domestic transport – are even more weather-dependent. Accordingly, in objective terms, Greenland and virtually all its cities, towns and settlements can be considered 'remote'.

As repeatedly emphasised, relatively high prices and difficulty of access have acted as a growth control mechanism, largely excluding conventional mass tourism. By contrast, an increase of cruise-ship tourism has been less constrained (Chapter 11).

Air transport offers the only feasible cross-country connections. Mountainous coastal terrain has limited the length of airport runways, only allowing for smaller planes. An exception is Kangerlussuaq, located inland, north of the Arctic Circle in Greenland's widest ice-free zone (200km). The area is considered desert, with an extreme but stable polar climate – ideal for an airport (Joyce and Burns, 2013). In a few cases, previous plans to expand infrastructure have collided with cultural, environmental and tourism interests. One example has been a plan to establish a road on the 160km corridor from Kangerlussuaq airport to the west coast at Sisimiut, but this corridor is a popular hiking and dog sledding area for tourists: the Arctic Circle Trail (Dillon, 2019). This Trail was specifically demarcated when, in the 1970s and 1980s, concerns were voiced about the possible impacts of hikers on both reindeer breeding and the fish population (Christensen, 1990; Johnston and Viken 1997). It was these concerns that led to the development of the hiking trail with cabins between Kangerlussuaq and Sisimiut, ostensibly to concentrate visitors and their impacts along a relatively narrow linear route, while actually encouraging hiking.

The pre-Covid air transport network is outlined in Box 8.1 and illustrated in Fig. 8.5.

8.3 The tangled web of airports, minerals and geopolitics

For more than 30 years, discussion proceeded concerning how to improve accessibility within and to Greenland in order to assist social development – helping to better integrate the dispersed

Box 8.1. The current air transport network

The present air transport network is built upon military decisions taken during World War II with the main international airport located at Kangerlussuaq (2,815m runway), having been taken over from the US Air Force as recently as 1991. This is located at the top of Søndrestrømfjord, 150km from the west coast and 10km from an edge of the ice sheet. The second former US base, at Narsarsuaq, has a 1,830m runway and is situated inland between two fjord arms, 50km from the sea in southern Greenland, and within sight of the inland ice sheet. These strategic locations were chosen based on considerations of lack of fog, distance away from populated areas and the immediate coast, but with access to the sea in order to transport construction materials but with minimal accessibility for German submarines.

Kangerlussuaq airport runway requires costly repairs because of subsidence from melting permafrost. Future options for it include a heliport or short runway for small fixed-wing aeroplanes, a reconstruction of the airport with a runway for regional connections (e.g. 1,500m), or with a long runway up to 3,500m for military purposes. Narsarsuaq airport will probably be replaced by a heliport. In the far north, Thule air base at Pittufik remains controlled by the US Air Force and has hitherto not been open to civil air traffic, but this was likely to change following a 2020 tripartite agreement (Section 4.7).

The capital Nuuk, with 18,000 inhabitants and located 320km to the south of Kangerlussuaq, only gained an airport (950m runway) in 1981. In the same year Ilulissat, the third biggest city with 4,500 inhabitants and situated 250km north of Kangerlussuaq, also saw completion of an airport (854m runway). Because of their short runways both Nuuk and Ilulissat airports have been served by smaller fixed-wing aircraft from Kangerlussuaq (about an hour's journey time) and Iceland (that to Ilulissat only as a summer connection).

In southern Greenland, Narsarsuaq airport is served by feeder traffic from Nuuk and Kangerlussuaq and has a summer connection to Iceland. Kangerlussuaq also has charter traffic to several international destinations supporting the growing cruise industry, and to the US Thule air base.

The main town in Southern Greenland, Qaqortoq (3,000 inhabitants) is currently only served by a heliport with a connection to Narsarsuaq.

Other air connections are shown in Fig. 8.5: with four of Greenland's five international airports located in the south-western quadrant. Kulusuk on the east coast is the exception, with around 15% of total air passengers.

As most attractions require travel beyond the international gateways, further travel is currently facilitated by a network of 12 airports, five heliports and 42 helistops in smaller settlements (Figs. 8.5–8.8). A parallel network of ports allows the cruise and ferry industry to access a number of settlements, although only during the summer months when there is the lower likelihood of adverse weather or blocking sea ice. Berthing for larger ships is limited.

A high dependency on a few air transport providers renders Greenland vulnerable to market changes. Air Greenland is government controlled with shareholding divided between the Home Rule government (37.5%), Scandinavian Airlines (37.5%) and the Danish government (25%). Diskoline is a relatively recent entrant to the market providing regional hub and spoke helicopter connections from Ilulissat and Narsarsuaq (Fig. 8.5).

Sources: Weaver and Lawton, 2017; Christensen *et al.*, 2020

settlements of this enormous land – and to stimulate economic development, not least international tourism: objectives viewed as necessary pre-requisites for attaining independence.

Much has been written concerning airports' economic effects on development in remote and peripheral regions (e.g. Tveter, 2017), including the encouragement of tourism (e.g. Khadaroo and Seetanah, 2007; Doerr *et al.*, 2020). However, airports serving remote areas can impose economic burdens, as small airports and routes servicing minor towns and settlements are usually unprofitable. Notably, there is a tendency for the number of jobs per passenger to increase when the size of airport decreases (e.g. Adler, Ülkü and Yazhemsky, 2013; Kazda, Hromádka and Mrekaj, 2017), and this is certainly the case in Greenland where even the smallest airports need to carry a high level of security, sustain risk avoidance measures and undertake winter infrastructure maintenance irrespective of passenger numbers.

The physical dependence upon air transport for internal and international connectivity

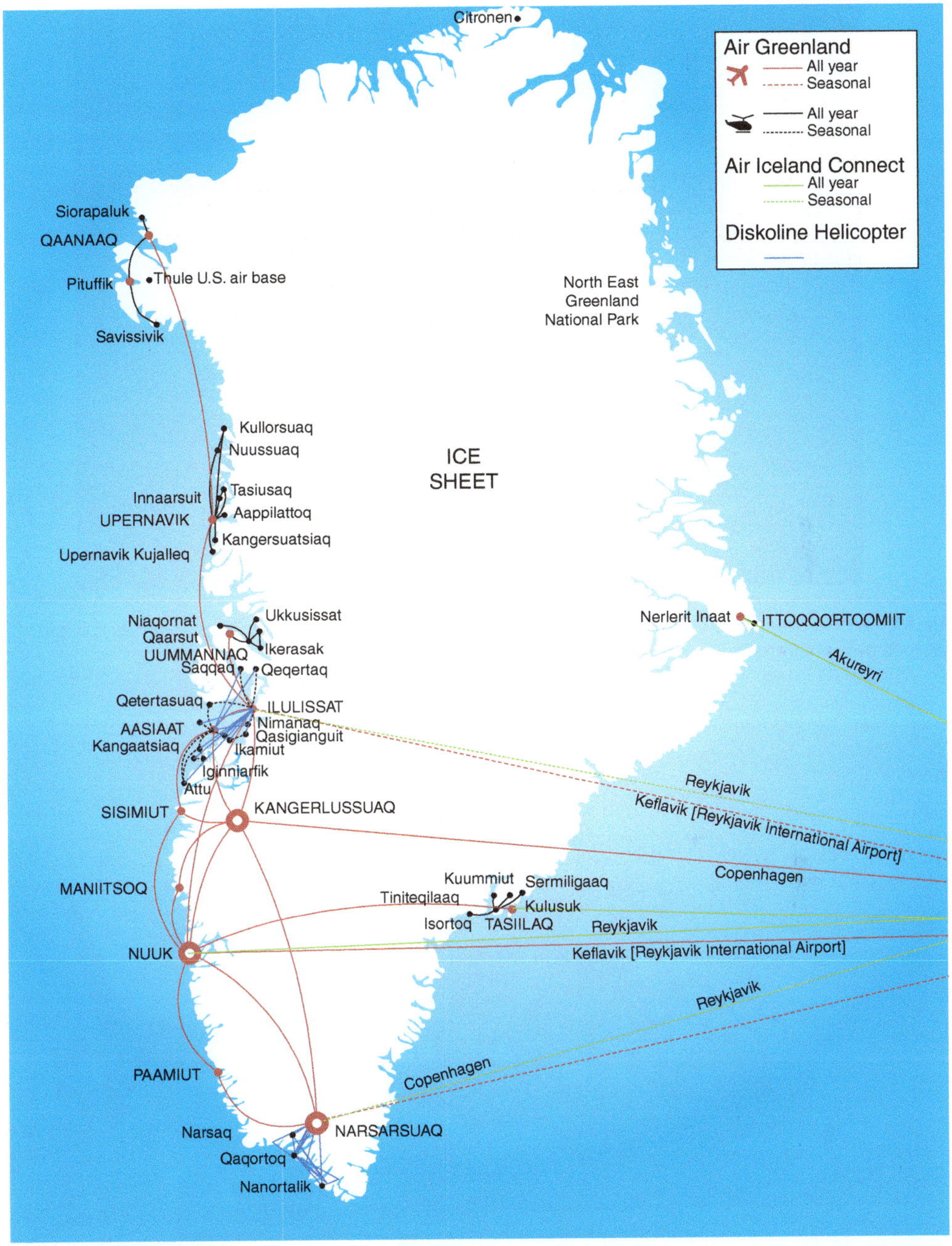

Fig. 8.5. Greenland's air connectivity. Author's own figure, modified and redrawn from an amalgamation of Air Greenland, 2020: 84, and Christensen, Nielsen and Knudsen, 2020: 3.

is one underlying factor in Greenland's equivocal approach to notions of sustainability (Chapter 5). And because of the increased recognition of air transport's central role in integrating Greenlandic society, finally in 2018, a new coalition government presented a bill to the *Inatsisartut* for a national airports upgrading programme. This was approved later in the year, and included extending the airport runways at Nuuk and Ilulissat to 2,200m, to render them as the main

Fig. 8.6. An Air Greenland DASH-8 catching the late evening summer sun as it approaches Ilulissat. Author's own photograph.

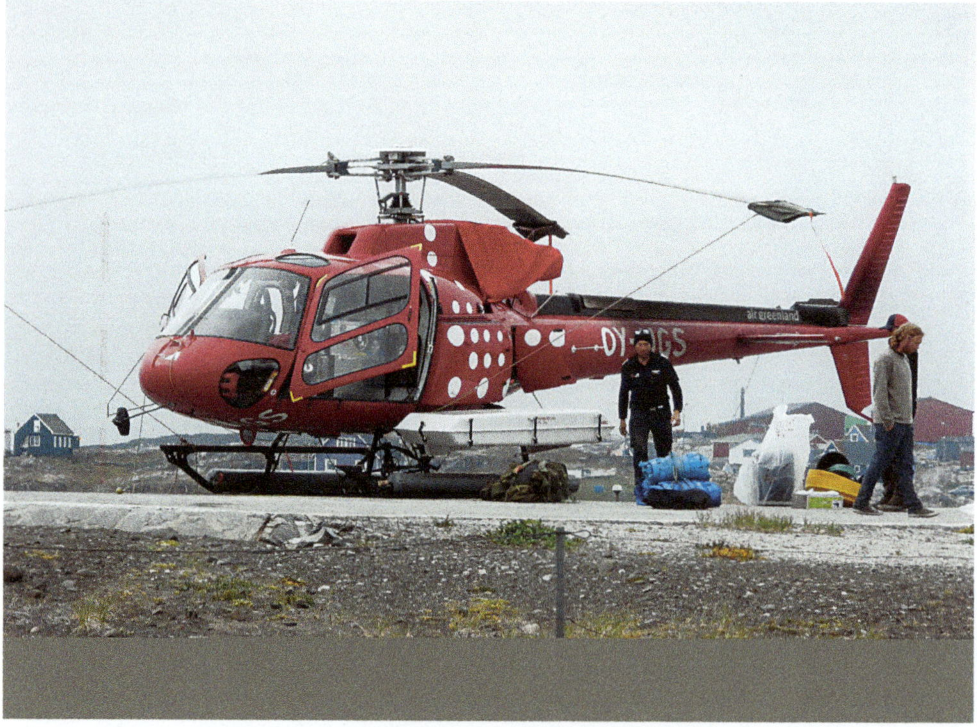

Fig. 8.7. The helistop at Qeqertarsuaq (Disko Island). Author's own photograph.

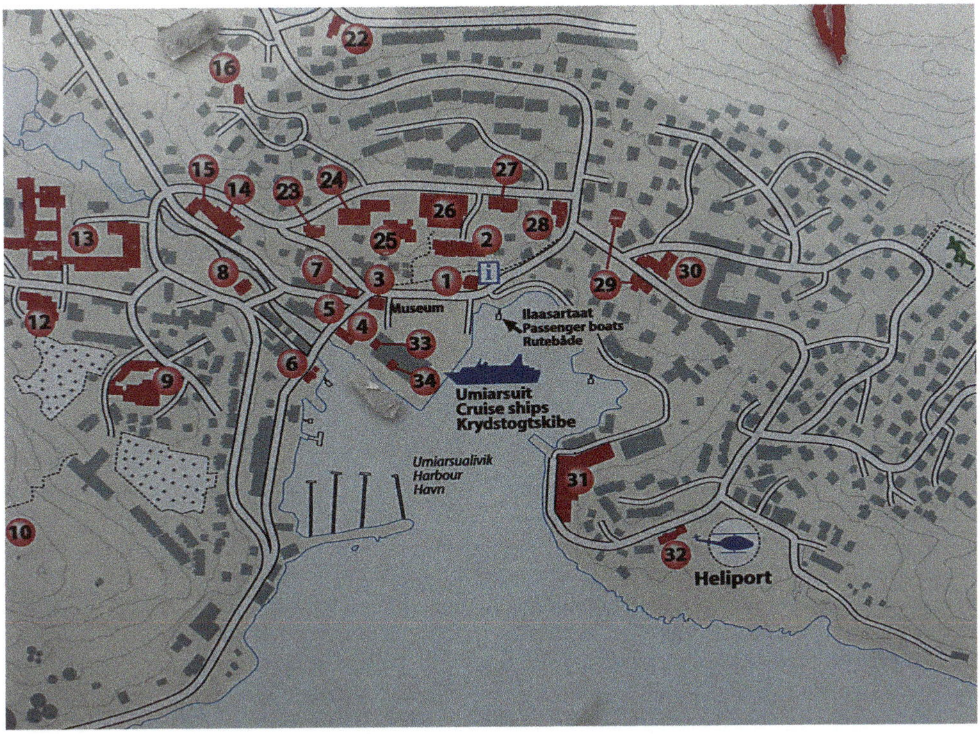

Fig. 8.8. The central portion of a Kommune Kujalleq/Destination South Greenland public information map in Qaqortoq, highlighting transport connection points. Author's own photograph.

gateways for international air traffic, and to build a new regional airport in the south, at Qaqortoq, with a runway of 1,499m. Notably, east coast facilities were not included in this upgrade.

The investment required for these projects corresponded to around 40 per cent of the country's annual GDP. The need to serve different parts of this large territory had rendered difficult any political decision seen to be favouring one part of the country without considering the needs of other regions. Such contested prioritisation only helped to delay decisions.

The domestic context for what was anticipated to be a step-change development in Greenland's air connectivity had comprised four evolutionary phases (Christensen *et al.*, 2020):

- During the first 20 years of Home Rule (1979–1999), a basic infrastructure with airports for fixed-wing aircraft adjacent to major cities and towns was established for internal connectivity and improved integration.
- From around 2000, the concern of the Home Rule authorities was to reduce the substantial subsidy to the traffic system in order to be able to finance development of the education sector. At the same time, however, an interest in air traffic as a way to increase tourism was also developing.
- From the beginning of self-government in 2009, the focus moved from internal integration through minor airports towards a wish to develop the bigger airports for the dual purpose of further improving both internal and external connectivity. Yet between 2000 and 2014, no major new decisions were taken.
- From 2015, the perspective widened to appreciate that air connectivity was a key factor in developing tourism and business and fundamental in underpinning arguments for economically 'self-sustainable' independence.

The airports development package became mired in geopolitics following China's growing interest in Greenland's potential mineral wealth and Trump's belligerence over perceived US interests.

Kalaallit Airports International, a state-owned company (two-thirds Greenland, one-third Denmark) was established to build, own and operate Greenland's airports projects (estimated cost: DKK3.6bn [€468mn/£432mn/$576mn]). Following an open tender by the Greenland government, China Communications Construction Co (CCCC) prequalified to appear on a list of 11 companies or consortia (from Denmark, Iceland, Canada, the Netherlands and the Faroe Islands), that had expressed an interest in the projects (Matzen and Daly, 2018).

Although bilateral relations between Greenland and China had begun in 2012, Chinese interest in Greenland was expressed particularly after Beijing's setting out, in its January 2018 Arctic white paper, ambitions to establish a 'Polar Silk Road' by developing shipping lanes opened up by climate warming (Section 2.6.2, Fig. 2.1), and Chinese enterprises were encouraged to participate in the construction of Arctic infrastructure (Section 2.6.3).

Yet, according to Têtu (2016), Greenland was then still far from being a priority for Chinese investment. Although Greenland's perceived (geo)political stability was an asset, notably because it was attached to Denmark, which was viewed as exhibiting a stable political system, Greenland's attempt to recruit economic development to gain independence from Denmark had the potential to scare Chinese investors away (Têtu and Lasserre, 2017), and Chinese officials were said to be uncomfortable dealing with a self-governing entity that held a clearly stated desire for independence (Sørensen, 2018b).

Kalaallit Airports expected to shortlist five or six companies in the spring of 2018 to begin construction later in the year. The Greenland government said it could finance DKK2.1bn, but would need external funding for the remaining DKK1.5bn. Subsequently Greenland's premier held meetings with the Export-Import Bank of China (Matzen and Daly, 2018). While Greenland's self-rule authority decides on most domestic matters, foreign and security policy remains the responsibility of Copenhagen, and foreign investment in infrastructure projects is a politically grey area.

'We are deeply concerned. China has no business in Greenland,' a 'high-ranking Danish government official' was reported to have told Reuters. 'Denmark has a big responsibility to live up to with regards to our closest ally, the United States.' Soren Espersen, foreign and security affairs spokesperson for the Danish People's Party, part of the governing coalition in Copenhagen, was quoted as saying 'the United States does not want a superpower like China to have such large-scale access to Greenland ... The government in Copenhagen has to stop the Chinese plans, because if it doesn't, the United States will' (Matzen and Daly, 2018). An interesting reflection on both Greenland's relative autonomy and Danish sovereignty (see Chapter 4).

Trump's presidency in 2016 began by continuing a US shift of interest away from the Arctic. But as US–China relations deteriorated and China widened its sphere of projecting political and economic power beyond Asia and into Europe with its 'Belt and Road Initiative' (BRI) and then into the Arctic with its 'Polar Silk Road', Greenland (once again) came to be regarded as strategically important by international actors (Sørensen, 2018a, 2018b; Grydehøj, 2020a).

The Danish prime minister subsequently arrived in Nuuk offering not only support from the Danish government for one-third of the cost of the airports programme but also help to secure further loans (WCN Editorial Team, 2020). This response was interpreted as a result of Nuuk's refusal to exclude CCCC from the shortlist of airport construction contractors in the face of veiled warnings from Copenhagen and Washington (Klarskov, 2018); it was not a case of Greenland rejecting a Chinese bid in favour of a Danish one (Olsen, 2018; Grydehøj, 2020a: 228–229).

Previously, in 2016, the Danish government, allegedly on direct instructions from Washington, had prevented a Chinese investor from buying a former US marine station at Kangilinnguit/Grønnedal in southern Greenland. This appeared to have been the first instance where Copenhagen had directly intervened to prevent a Chinese acquisition in the country (Breum, 2016) (also Box 4.5).

Following Denmark's intervention in offering financial support for the airports development, the short-listed CCCC withdrew its bid, two weeks before the deadline for final submissions, thereby curtailing China's potential participation in this important infrastructure project (Gronholt-Pedersen, Qiu and Heavens 2019).

Aaja Chemnitz Larsen, a member of the Danish parliament representing Greenland and head of Greenland's foreign affairs committee, pointed out that 'In Greenland we don't suffer from China anxiety, like they obviously do in the government in Copenhagen' (Matzen and Daly, 2018). A clearly ironic US State Department official welcomed 'efforts by all countries, including China, to enhance investment in high-quality infrastructure development', but urged that development financing should not be 'predatory or result in unsustainable debt' (Matzen and Daly, 2018).

There followed a US$12.1mn (€9.8mn/ £9.1mn) US aid package for Greenland, followed by a tripartite agreement with Denmark which included provision for the US Thule air base in northern Greenland to become more accessible for civilian use (Humpert, 2020) (Section 4.7). Greenland's strategic importance for the US military and its ballistic missile early warning system had been reinvigorated because of perceived Russian and Chinese commercial and military build-ups in the Arctic (Gronholt-Pedersen and Pamuk, 2020; Section 2.6). An April 2020 statement from the US embassy in Copenhagen opined:

'The United States and the Trump Administration are setting out to wake the West up from our collective complacency before other less trustworthy governments shape the values of the region ...' (Bykova, 2021).

In November 2020, the US Export-Import Bank said it was prepared to invest more than $200mn in Greenland's mining projects, explicitly on the basis that the investment would benefit US interests. While Greenlandic officials welcomed the planned investments, they were wary of likely attached conditions (Bykova, 2021).

At the time of writing it remains to be seen how the Biden administration will take these developments forward.

A Greenland moving towards independence, seeking to develop its tourism and mineral resource 'self-sustainably', is now caught in superpower rivalry over access to and exploitation of Arctic infrastructure, sub-soil and sub-sea wealth (Bislev, Gad and Zeuthen, 2018). Is Greenland in a position – especially if and when it achieves independence – to exploit such a position? Or will it simply shift from one form of dependency to another?

When the airports package – with Danish government financial support – was approved in Greenland's parliament by a small majority, it provoked the pro-independence *Partii Naleraq* to walk out of the coalition government in protest over what it saw as an act that would sustain post-colonial dependence. By contrast, the majority of Greenlandic political actors see the airports development as a means of securing better global connectivity, supporting economic development and assisting the path to independence (Turnowsky, 2018; Grydehøj, 2020a).

Improvement of the country's infrastructure and the time savings for residents, tourists and business travellers could be substantial once the airports package is completed. But in terms of generating economic development, especially in the tourism sector, questions remain. A major issue is to generate sufficient interest in appropriate airlines to establish direct connections to Ilulissat from Europe and the North American east coast in order to generate the induced and, in the long run, wider economic benefits (although accommodation and other provision would need considerable expansion). To this end, it may be necessary for Kalaallit Airports and/or the public authorities of Greenland to be actively involved in route development. Model incentive schemes to stimulate direct international air services include the Northern Norway Charter Fund, which has supported operators establishing air services to airports in northern Norway (Halpern and Graham, 2015).

For its 2019–2023 marketing plan, Visit Greenland (2020e) viewed expansion of Greenland's tourism as contingent upon the airport redevelopments being ready by 2023. The Covid-19 pandemic interceded both in terms of the progress of airport development and in curtailing in-bound tourism.

8.4 Conclusion

Greater engagement between Greenland and international actors is inevitable, as more lustful eyes turn towards the Arctic in general and to Greenland in particular. In terms of tourism, alongside climate change, this process will see a transformation in the territory's relative peripheries and in the country's accessibility. While, in the short term, Danish investment in the Greenland airports upgrade programme appeared intent on excluding Chinese participation, in the longer term Greenland's economic development resources, as yet largely untapped, will raise continuing issues of accessibility, intertwining the explicitly spatial with the crudely geopolitical.

8.5 In the next chapter

In the next chapter, both the symbolism and reality of climate change, and its specific impacts on Greenland's physical environment, are foregrounded. Within this framework tourism in Greenland is highlighted as pursuing two complementary roles: as climate education and as interacting with climate through adventure tourism.

Notes

[1] Even at a scale of 1:3,000,000, a good map of Greenland, such as Joyce and Burns (2013), indicates 12 abandoned settlements, including Ivittuut (Section 5.4).

[2] European cruise tourists' visual experience of Greenland's southernmost tip around Cape Farewell (Kap Farvel/Nunap Isua [Umanarssuaq]) or passing through Prins Christian Sund (Ikerasagssuaq), before undertaking any shore landings, are the exception.

[3] Begun from plantings in 1976, this 200ha arboretum contains around 110 different introduced species.

9

Fulcrum of Climate Change?

'Since the end of the first decade of the twenty-first century, Greenland has been one of the hot spots for climate change science and tourism, which has produced photographic coffee-table books and television documentaries offering snapshots of the change, the end, the last hunters, the outermost vanishing world. This exposure comes at a cost, however, and many people ... express apprehension at being put on display, being portrayed out of context, and, worse still, realizing that neither we nor they have any control over how their identities circulate in forums'
(Flora, 2019: x)

'... the cool nature vision of Greenland is an abstraction, an illusion, an impossibility.'
(Ren, 2012: 154)

9.1 Introduction: remote and frozen?

Greenland's ice and natural environments are in the forefront of global climate change. This chapter examines the physical consequences of this for the territory and considers how, moving towards independence, Greenland's tourism is responding to the challenges raised. It illustrates this by showing how, through the attraction of Greenland's ice, climate change is being mediated for tourist consumption through processes of education and interactive experience.

For some travellers, visiting Greenland may be an expression of individuality, while relatively high costs and difficult access may add to a sense of an exclusive experience, perhaps a sense of initiation. The marketing application of such clichés as 'a once in a lifetime experience' is perhaps intended to replicate or at least allude to the kind of rite of passage associated with images of hardship endured by early (and exclusively male) explorers. An earlier tourism marketing strapline suggested Greenland offered 'experiences from a completely different world' (Kaae, 2006): Greenland as a global traveller's fashion accessory. How far can such fashionability be used to advantage in responding to the challenges of climate change?

In 2012, Ren contended that Greenland is portrayed in popular imagination as a 'colossal, remote and frozen landscape of silence and solitude ... a place of nature ... ice-covered sceneries devoid of people and human structures' (Ren, 2012: 154). Yet, she argued, to obtain access to such environments and the experiences they offer, the traveller to Greenland is required to employ such cultural artefacts as aircraft and cruise ships, travel agents, tickets and documents, local guides and service providers, accommodation, outdoor equipment, maps and positioning technology.

In this way, to facilitate the visions and experiences of a 'cool' and 'natural' Greenland, such manifestations of interposing human impress

 Tourism, Climate Change and the Geopolitics of Arctic Development (D. Hall)
DOI: 10.1079/9781789246728.0009

are necessary: the cumulative effects of which have contributed to climate change. Yet it appears for much of the global tourism and travel sector that climate change remains the proverbial elephant in the room: an elephant that is about to run amok.

Somewhat ironically, Greenland's tourist identity has thrived on the imagery of a frozen empty purified landscape only populated by people entirely conditioned by nature in a closely symbiotic relationship (Ren, 2012, 2016). Yet the imagery of an Indigenous Inuit culture that is closely linked to nature – and thus probably in stark contrast to that of the visitor's own lifestyle – is superficially mediated by Greenland's still familiar European socio-cultural dimensions of (Danish) language and currency (Kaae, 2006). As Smed (2017: 144–145) has also noted, because nature is heavily mediated by culturally constructed images of remote wilderness, peripherality and 'last frontiers', a cultural dimension will be present in any encounter with nature, despite destination marketing that may portray a one-dimensional untouched and 'human-free' landscape.

Yet there is no escaping the reality that this landscape is changing and being changed: while climate change is embracing the planet, nowhere is it being more dynamically expressed than in the Arctic, and especially in Greenland.

The tourism trope of Greenland representing binary opposites to the visitor's quotidian experience is a well-trodden path, with the Arctic climate, ice and snow contrasted to usually temperate home environments; a vast rugged wilderness landscape compared with groomed suburbia; and an unpeopled remoteness in contrast to the high population densities and accessibility of urban society (Kaae, 2006; Wang *et al.*, 2020). The Greenland experience has also been characterised in binary opposition terms to '4S' destination experiences of Europe's south, whereby the imagery of Greenland can be crystallised – if not commodified – as the land of 'many Is': ice, icebergs, icefjords, Indigenous Inuit, inaccessibility, initiation, island isolation and individuality (Kaae, 2006).[1]

In this chapter, the nature and implications of climate and climate change are employed as the context within which to interrogate critical elements of tourism development and concepts of sustainability in Greenland. At least three elements of these interrelationships are highlighted:

- Greenland in general (and Ilulissat Icefjord in particular) as symbols of climate change and exemplars of 'last chance' tourism.
- The paradox that accessible 'ice' representations of climate change such as the Ilulissat Icefjord become tourist destinations by virtue of the co-construction of 'nature' and (global?) culture, yet (local) culture is often marginalised or even excluded from the touristic 'pristine wilderness' narrative.
- While international concern may be focused on Greenland in terms of the nature and effects of climate change, local interest is perhaps more focussed on the opportunities that a changing climate may offer. As a consequence, conceptions of sustainability are seen to be flexible and opportunistic.

Through this framework, interactions between tourism, climate change, and their implications for Greenland's longer term sustainability and understandings of sustainability, are exemplified through the far from mutually exclusive dimensions of:

- tourism as climate education: illustrated by the Ilulissat Icefjord Centre; and
- tourism as climate experience through adventure tourism.

9.2 Climate threat

The world has warmed by about 1°C since pre-industrial times as the concentration of greenhouse gases in the atmosphere has increased, but that warming has not been evenly distributed, with the Arctic warming fastest. Even just in the decade to 2019 Greenland experienced an average 0.75°C temperature increase (Amos, 2019).

Commissioned by the Arctic Council (Chapter 2) almost two decades ago, the Arctic Climate Impact Assessment (ACIA, 2004), conducted by an international team of 300 scientists, highlighted that climate warming effects were particularly intense in the Arctic, where mean temperatures had been rising twice as fast as in the rest of the world, a trend that was likely to accelerate as a result of the accumulation of greenhouse gases in the atmosphere and feedback loops.

Effects identified in Greenland at that time included a 16 per cent reduction in ice sheet cover over the previous 25 years (ACIA, 2004). Now, warm air, retreating sea ice and a darkening of the ice surface caused by algae and other processes (e.g. Tedstone *et al.*, 2017) are combining to heat up Greenland faster than scientific predictions, and the melting of the ice sheet, as the major consequence of this, is accelerating (Graham *et al.*, 2019; Sasgen *et al.*, 2020).

As noted in Chapter 2, the fragile Arctic is sensitive to climate changes, not least the impacts when temperatures rise to above freezing point; and a contraction of Arctic climatic conditions is currently taking place: glacier melt and retreat (Box 9.1), reductions in extent and persistence of sea ice, and of snow and ice cover on land, increasing but wider variations in precipitation, with the generation of shorter and warmer winters, changes in atmospheric and oceanic circulation patterns, and rising sea levels. Along with increasing temperatures, the Arctic Climate Impact Assessment predicted that the Arctic was also likely to experience changes in vegetation, shifts in animal species, coastal impacts, changes in marine transport, a thawing of permafrost, ultraviolet radiation and effects on Indigenous communities (CAFF, 2001) (Box 2.1).

In terms of impacts on tourism, the decline in summer ice is likely to extend navigable waters and lengthen the summer navigation season to result in an expansion particularly of cruise tourism (although this might be countered by increasing numbers and density of icebergs). Rising temperatures will affect the habitats of different species including a decline in trout and other freshwater fish, thereby impacting sport fishing (ACIA, 2004). Such warming-induced potential and current changes which pose both risks and benefits for Indigenous cultures, are not spread evenly. An expanding tourism sector also brings risks and benefits, which may further complicate an already complex situation (Stonehouse and Snyder, 2007).

The 56,000 Greenlanders mostly live in communities perched on the narrow bands of land between sea and ice sheet which, although close, is usually out of sight. The ability of the Arctic's Indigenous peoples to survive has depended on their capacity to adapt to change. Now, they face challenges that may place their cultures and livelihoods at serious risk. At the same time, milder conditions are permitting more farming in the south of the country, stimulating the development of local pastoral economies, including farm-stay tourism (e.g. Steenholdt and Chimirri, 2018).

Box 9.1. Temporary benefits?

Mass loss from the Greenland ice sheet leading to glacier retreat, and an increasing input of glacial melt-water to fjords and coastal waters is currently helping to sustain important fisheries, which generate a significant part of Greenland's national income. With melting of the Greenland ice sheet currently accelerating, this freshwater input is expected to substantially increase in the short- to medium-term. But it remains uncertain how these highly productive high latitude ecosystems will respond to future changes whereby marine-terminating glaciers – which currently play a crucial role in sustaining the high biomass productivity of the fjord ecosystems – retreat to become land-terminating.

This is because marine ecosystem productivity is very differently regulated in fjords influenced by marine-terminating compared to land-terminating glaciers. Rising sub-surface melt-water plumes originating from marine-terminating glaciers force large volumes of ambient deep water to the surface. The resulting upwelling of nutrient-rich deep water sustains high phytoplankton productivity throughout the summer. By contrast, fjords with only land-terminating glaciers lack this upwelling mechanism, and have significantly lower productivity.

Data on commercial halibut landings support the contention that coastal regions influenced by large marine-terminating glaciers have substantially higher marine productivity. These results suggest that a change from marine-terminating to land-terminating glaciers, following ice sheet melt and glacier retreat, could substantially alter the productivity of the coastal zone around Greenland with potentially drastic ecological and socio-economic implications. Such relationships have also been observed elsewhere in the Arctic.

Sources: Fettweis, Franco *et al.*, 2013; Sciascia *et al.*, 2013; Berthelsen, 2014; Bendtsen *et al.*, 2015; Bendtsen, Mortensen and Rysgaard, 2015; ICES, 2015; Meire *et al.*, 2017

If Greenlanders are forced to further abandon some of their communities, fewer opportunities may exist for tourists to experience cultural traditions, and for local tourism stakeholders to prosper. It is therefore in the tourism sector's interests to help address the impacts of climate change on Indigenous communities, and to firmly place socio-cultural sustainability as well as environmental conservation centrally within tourism strategies (Stonehouse and Snyder, 2007).

9.3 The Greenland ice sheet

The Greenland ice sheet is a vast body of ice covering 1.71mn km^2, just over 80 per cent of the island's surface area. As a consequence, it has long been the object of scientific observation and research (e.g. Rignot and Kanagaratnam, 2006; Broeke *et al.*, 2009). It is the second largest ice body in the world, after the Antarctic ice sheet. With a thickness mostly between 2,000 and 3,000 metres (6,600–9,800 feet), the Greenland ice sheet is 2,400km (1,500 miles) long and up to 1,100km (680 miles) wide. Its total volume is approximately 2,900,000cu km, which would raise global sea levels by 7 metres if it melted entirely (Miles, 2019).

Until recently, the ice sheet was generally in a state of balance, with the amount of snow falling in winter being roughly equal to the amount of ice melting in summer. Over the past 30 years this equilibrium has changed, decade by decade, as Greenland has tended to shed more and more ice.

One of the most important recent studies undertaken – the Ice Sheet Mass Balance Inter-comparison Exercise (IMBIE), by a consortium of 89 polar scientists drawn from 50 international organisations – re-analysed data from 11 satellite missions flown between 1992 and 2018. These spacecraft take repeat measurements of the ice sheet's changing thickness, flow and gravity. The IMBIE team combined their observations – 26 separate surveys of Greenland – with the latest weather and climate models (Shepherd *et al.*, 2020). This was the most comprehensive survey yet of the ice sheet over the past few decades, and, crucially, was based on actual observations as well as computer modelling.

In recent decades, the Greenland ice sheet has been a major contributor to global sea-level rise (Shepherd *et al.*, 2012; WCRP, 2018), and is expected to remain so in the future (Pattyn *et al.*, 2018). Although increases in glacier flow (Moon *et al.*, 2012; Enderlin *et al.*, 2014) and surface melting (Trusel *et al.*, 2018) have been driven by oceanic (Straneo and Heimbach, 2013) and atmospheric warming (Bamber *et al.*, 2012; Fettweis, Hanna *et al.*, 2013), the degree and trajectory of today's imbalance remain uncertain (Shepherd *et al.*, 2020).

Greenland is losing ice in two main ways. Reduced surface mass balance has driven 52 per cent of the ice loss owing to increased surface melt-water runoff. The remaining 48 per cent is due to increased glacier discharge – the 'calving' of icebergs (Benn *et al.*, 2017) and then melting where the ice is in contact with the ocean. This has doubled since the 1990s, although between 2013 and 2017, the total rate of ice loss slowed as atmospheric circulation favoured cooler conditions (Bevis *et al.*, 2019), and as ocean temperatures fell at the terminus of Sermeq Kujalleq (at Ilulissat Icefjord/Jakobshavn Isbræ) (Khazendar *et al.*, 2019), the country's most important glacier (Section 9.5). Cumulative ice losses from Greenland as a whole have been close to that predicted by the Intergovernmental Panel on Climate Change (IPCC) high-end climate warming scenario (Shepherd *et al.*, 2020).

The areas that are most sensitive to warming are the peripheral glaciers and edges of the ice sheet, which may lose up to 28 per cent of their mass over the next century. Alarmingly, these appeared to pass through a tipping point in 1997, with a major loss of their surface refreezing capacity at that time (Noël *et al.*, 2017). Further, rainfall events on the Greenland ice sheet are becoming increasingly common, and this liquid water is hastening the melting of the ice (Oltmanns, Straneo and Tedesco, 2019).

Evidence suggests that sea levels rose by 6 metres over 1,000 years during the last inter-glacial period around 100,000 years ago, and that there occurred a near-complete de-glaciation of southern Greenland during the inter-glacial around 400,000 years ago (Fischer *et al.*, 2018). Now, the IPCC (2018) analysis has concluded that irreversible loss of the Greenland ice sheet could be triggered at a temperature increase of 1.5 to 2°C, indicating the urgent global need to reduce emissions (Vincent, 2020: 519).

The 2019 melt season resulted in a net ice loss of about 329bn t from the Greenland ice sheet, and just to emphasise the above, melting was up to seven times faster than in the 1990s (Tedesco and Fettweis, 2020). In 2020 processes were less dramatic, albeit with 70 per cent of the ice sheet experiencing some melting (Scambos, Stroeve and Koenig, 2020).

Understanding the ice changes in Greenland is critical because it is seen as a model for the future behaviour of Antarctica, which contains ten times as much ice but is currently cooler and has not started melting to the same extent (Hook, 2019).

9.4 Global implications

The IMBIE research suggests that mean sea level rises are likely to reach 67cm by 2100, about 7cm more than the IPCC's main prediction. By that time such a rate of rise will put 400mn people at risk of flooding every year, instead of the 360mn predicted by the IPCC (Shepherd *et al.*, 2020). The IPCC (2020) needs to incorporate these updates in its major assessment report (AR6) of Earth's climate in 2022.

While the IPCC's work on climate science is well respected, some voices have been raised that its findings do not take sufficient account of the potential for 'tipping points' – thresholds beyond which climate breakdown accelerates and becomes catastrophic and irreversible – and that new tipping points may be breached sooner than previously anticipated (Harvey, 2019).

Some of the more alarming indicators are the feedback loops that are becoming evident. Firstly, the ice itself is changing. The projected image of Greenland and the Arctic as a pristine white landscape is being compromised by a darkening of some ice surfaces. Large areas of pale grey and smaller patches that are much darker, covered with what appears to be mud or silt, have been appearing and spreading. Initially, it was thought that this 'staining' was caused by a mix of airborne dust and pollution particles transported from distant industrial regions. But it appears that a major cause of the darkening is algae, microscopically small plants that are flourishing in the melting ice (Yallop *et al.*, 2012; Lutz *et al.*, 2014, 2018). By turning the surface grey, or black, from its usual bright white, the algae make it less reflective, thereby absorbing more solar radiation. This accelerates warming and in turn leads to even more melting (Kintisch, 2017; Vincent, 2020: 509–510).

Second, with a diminution of sea ice cover to reflect the sun's rays, the darker ocean is absorbing more heat from the sun, warming the water, and making it harder for sea ice to form. This is one reason why the Arctic is warming so much faster than the rest of the planet. Such 'amplification' represents a vicious circle where warming begets more warming. Oceans have absorbed most of the excess heat – and much of the carbon dioxide – arising from our disruption of the climate to date, but they are reaching the limits of their capacity to do so. Sea level rises are driven not only by melting ice but by the thermal expansion of the seas as they warm.

Other examples of warming feedback loops include the release of CO_2 from exposed near-surface soils (Jex, 2016), changes in cloud cover and the capacity of forests to absorb carbon dioxide.

Greenland is the single biggest contributor of melt-water to the rising ocean and accounts for between one-quarter and 43 per cent of the annual rise (Noël *et al.*, 2017). Although the ice sheet was close to a state of balance in the 1990s, annual losses have risen since then, peaking at around 335bn t per year in 2011. The IMBIE assessment showed Greenland to have lost 3,800bn t of ice between 1992 and 2018, causing the global mean sea level to rise by around 10.6mm. Whereas in the early 1990s, the rate of loss was equivalent to about 1mm per decade, it is now running at roughly 7mm per decade. Greenland's ice contributes directly to sea level rises as it melts because it rests on a large land mass, unlike the floating sea ice that makes up much of the rest of the Arctic ice cap.

Sea level rise adds to the risk of storm surges when fiercer storms, made more likely by global heating, strike coastal regions (Harvey, 2019). Approximately a billion people live less than 10 metres above current high-tide lines, including 250 million below one metre. One estimate suggests that globally, 1 million people will need to be evacuated for every centimetre of sea-level rise (Jonassen, 2020a), posing an existential threat to low-lying islands, river deltas and coastal cities. For many low-lying areas such as Bangladesh, Vietnam, Florida, eastern England and many small island nations, quite starkly, what happens in Greenland could determine their future. One estimate has suggested that flooding resulting from rising sea levels could cost some $14tn (£11.2tn) annually by the end of the century (Hook, 2019). And what of the social and political consequences of mass migrations from perennially or perpetually flooded regions?

The simulation of future Arctic climatic conditions remains complex because of the limited number of direct verifying observations and the general inability of global climate models (GCMs) to describe climatic processes in the Arctic taking place at scales finer than 100–200km (Stendel, Christensen and Petersen, 2008; Schrot, Christensen and Formayer, 2019). The IMBIE exercise therefore underlines the importance of observational satellites, especially those that can monitor events in Greenland higher than 83 degrees north. Only two of the present fleet have been able to do this, and one of those has been operating beyond its design life. Most of the observed changes in Greenland have been in the west, south and east, but impacts are slowly moving north. Discussions between the EU and the European Space Agency have therefore concerned the requirement for the next satellite in the EU's Copernicus programme to fly over higher latitudes (Amos, 2019).

In the following two sections, the dynamic relationships between climate change and tourism development in Greenland, and how they reflect processes of adaption and mitigation in the context of Greenland's ambivalent attitude to sustainability in its path towards independence, are illustrated. The dimensions of tourism as education, and as interactive 'adventure' experiences, are explored. Some of the marketing complications of the latter are taken up in Chapter 10, while a third dimension of the relationship – tourism as offshore gaze – is addressed in Chapter 11.

9.5 Tourism as climate education: the Ilulissat Icefjord Centre (*Ilulissat Kangiata Illorsua/Isfjordcenter*)

Travel and tourism can be a positive activity, it has been argued (Lew, 2018; Galvini, Lew and Sotelo Perez, 2020), if it intentionally and directly enables people to become more aware of, and sensitive to, the planet in an holistic way. While travelling to and within Greenland may contribute to this objective, experiencing the Ilulissat Icefjord (*Ilulissat Kangia*) and being informed by the Icefjord Centre can inculcate an appreciation of the significance of our actions in contributing to the global climate crisis.

While the world's largest national park – in North East Greenland (Fig. 4.1) – is only accessible with special permits, the World Heritage Site (WHS) of Ilulissat Icefjord is probably the most visited tourist attraction in Greenland. Declared a UNESCO WHS in 2004 (UNESCO, 2004/2019)[2] (Box 9.2), the Icefjord is the sea mouth of Sermeq Kujalleq, one of the world's most dynamic glaciers, emptying into a 61km long fjord packed with icebergs slowly making their way to the sea. Rapid melting of the vast glacier is significant for overall ice loss in Greenland and is impacting on global sea levels (Hook, 2019). This spectacular and unique site therefore acts as both an emblem and instructive experience of climate change. The Ilulissat Icefjord offers easy

Box 9.2. Ilulissat Icefjord UNESCO World Heritage inscription summary

Located on the west coast of Greenland (N69 7 60 W49 30 0), 250km north of the Arctic Circle, Ilulissat Icefjord is the sea mouth of Sermeq Kujalleq, one of the few glaciers through which the Greenland ice cap reaches the sea. Sermeq Kujalleq is one of the fastest and most active glaciers in the world. It annually calves over 35cu km of ice,* i.e. 10 per cent of the production of all Greenland calf ice and more than any other glacier outside Antarctica. Studied for over 250 years, it has helped to develop our understanding of climate change and ice cap glaciology. The combination of a huge ice sheet and the dramatic sounds of a fast-moving glacial ice-stream calving into a fjord covered by icebergs makes for a dramatic and awe-inspiring natural phenomenon.

Area: 399,800ha
Buffer zone: 64,890ha

Source: UNESCO, 2004/2019

* Recently it has been producing up to 46 cu km annually.

access for visitors and scientists alike to view the calving glacier front and the ice-filled fjord (Fig. 9.1). As such, just like the Greenland ice sheet itself, the glacier has long been the subject of scientific scrutiny (e.g. Holland *et al.*, 2008). The wild and scenic combination of rock, ice and sea, along with the dramatic sounds produced by the moving sea ice and visiting whales combine to produce a natural spectacle (Kaae, 2006: 104–105).

IMBIE analysis of satellite imagery revealed that the glacier was moving at a rate of 17km a year in 2013, but actually started slowing down around 2016 and began (re-) thickening.

The nearby Icefjord Centre (*Kangiata Illorsua/Isfjordcenter*) opened in 2021 as a source of education and instruction concerning climate change, as a branding exercise and as a narrative of sustainability (Box 9.3). It is architecturally modern but sympathetic (Figs 9.2, 9.4), and provides a complementary attraction to the Icefjord itself, offering both a cultural dimension to the area and culturally mediated interpretations of the natural landscape and the processes moulding it. The adjacent boardwalk to the Icefjord, across otherwise protected terrain (nature and archaeological remains), has clear carrying capacity limits (Figs 9.3, 9.4).

The designation of Ilulissat Icefjord as a WHS predictably accelerated the popularity of the region, assisted by the publication of a well-timed book produced by the Geological Survey of Denmark and Greenland (Bennike *et al.*, 2004). In 2015 the Icefjord received an estimated 37,000 visitors (25,000 foreign, 12,000 local) (BARK Rådgivning, 2016: 4). Ilulissat, with a population of 4,500, is Greenland's third largest town, and, as noted in section 7.5.2, has uniquely experienced accommodation shortages associated with 'overtourism' in the summer peak, reinforced by successive cruise ship arrivals.

The then Greenland Tourism Board's 'absolutely essential requirement that tourism be developed in harmony with, and not at the expense of, the landscape and culture' (Greenland Tourism Board, 2005) would be subsequently severely tested at Ilulissat (Snyder, 2007: 108).

Fig 9.1. A seaward view of part of the Ilulissat Icefjord. Author's own photograph.

Box 9.3. The Ilulissat Icefjord Centre

Design

A simple twisted structure with double-curved surfaces, sitting in the landscape like two conjoined wings. The architects, Dorte Mandrup, who won an international competition for the building's design, refer to it as 'the flight of a snowy owl through the landscape'.

Intended to be a seamless extension of the area's hiking trails, the building is designed such that visitors can move around both inside and outside, including on the roof, where views of the Icefjord and the surrounding landscape can be taken. The exhibits inside the building are laid out like floating islands, and feature holograms and ice cores. The building's open facade 'reflects the desire for a transparent dialogue between the exhibition and its surroundings: between humanity and nature'.

Begun in 2019 and initially planned to have opened in autumn 2020, the visitor centre features an exhibition, café, cinema, activities for children and young people, meeting facilities and a 'laboratory of the future'. It was finally opened in July 2021.

Role

To provide knowledge about the unique nature of the area and the experience it has to offer.

Mission

- To convey knowledge about ice, climate change and local cultural history.
- To provide a meeting place for debaters and researchers from all over the world.
- To create new encounters for tourists and a new gathering point for residents in the area.

Development partnerships

The Danish philanthropic association Realdania had the Icefjord Centre constructed in partnership with the Government of Greenland (Naalakkersuisut) and Avannaata Kommunia local authority.

Funding (in million DKK)

Realdania	108
Central and local government	23
Nordea Foundation	16.4
Augustinus Foundation	3.8
Bloomberg Philanthropies	0.75
OAK Foundation	0.35

Stakeholders' positions

Realdania: active in the built environment aiming to enhance life quality. 'The Icefjord Centre will convey the tangible consequences of climate change to all of us – including tourists from all over the world. That is why we believe that … [it] … will also play a key role in the evolution of Ilulissat as a tourist destination and of Greenlandic tourism in general' (Jasper Nygård, CEO).

Naalakkersuisut: 'The development of tourism is high on Naalakkersuisut's agenda. In this respect, the Icefjord Centre will play a crucial role. The partnership with Realdania will make a hugely important contribution to the development of tourism in Greenland in the coming years' (Jess Svane, Minister of Industry, Energy and Research).

Avannaata Kommunia: The north-western municipality of Greenland, created in 2018 as a result of the division of Qaasuitsup municipality. It comprises four towns and 23 settlements with a total population of 10,600 over an area of 522,700 km^2 (equal to the land mass of Spain). 'The Icefjord will play a major role for the area – also in terms of employment. The project is already creating jobs in the municipality, and we welcome the increased awareness of Ilulissat and Disko Bay, to which it will lead' (Palle Jerimiassen, Mayor).

Source: Ilulissat Icefjord Centre, as viewed 2 August 2019; see also Mairs, 2016; Dorte Mandrup Arkitekter, 2018

Fig. 9.2. The Ilulissat Icefjord Centre as depicted at the construction site (August 2019). Author's own photograph.

Fig. 9.3. Boardwalk through the World Heritage Site (WHS) towards the Ilulissat Icefjord. Author's own photograph.

Fig 9.4. WHS Boardwalk passing the Ilulissat Icefjord Centre, located just outside the WHS boundary, under construction (August 2019). Author's own photograph.

While an underlying premise suggests the conferment of World Heritage status renders sites as successful attractions, according to Olsen (2017), this represents, at best, confusion between cause and effect. While one can point to expressions of great optimism among UNESCO officials regarding tourism potential (e.g. Pedersen, 2002), not all heritages and heritage sites are popular (Robinson and Silverman, 2015: 14), and their economic benefits may be exaggerated (Bowitz and Ibenholt, 2009; Olsen, 2017).

Further, to be nominated for WHS status involves resource-demanding efforts, both before and after sites enter the game of global recognition politics and its shifting alliances within UNESCO (Meskell, 2015: 3): the heavy costs and responsibilities involved in this process are carried by the respective states. Designation can influence changes to how tourists circulate within the area, the time they spend at a location, and, significantly, the nature and degree of visitation at secondary attractions in the locality.

At Norwegian WHSs, Olsen (2017) found large numbers of local people being attracted to visit, the overwhelming majority of whom were young children, who were immersed in the values of the site. This would seem to emphasise the significant educational role to be played by the Ilulissat Icefjord and the Icefjord Centre in heightening awareness of the climate change crisis for both overseas and local visitors, and of its impacts on local culture.

Indeed, the Ilulissat Icefjord Centre's explicit education role and provision both for young people and children and for scientific gatherings (Box 9.2) can play a significant part in climate crisis awareness. However, this may be a higher priority for the Danish philanthropic developers than for local authority and destination marketing organisation (DMO) objectives of generating employment and tourists.

The Icefjord Centre may help to redress the nature–culture balance in the representation of the Ilulissat region's tourism imagery. As Smed (2017) has emphasised, nature and culture are intermingled in the Greenlandic identity. This was represented in Visit Greenland's launch in 2010 of a national brand, 'The Pioneering Nation', created to stress the special relationship between nature and people as the brand identity of Greenland (Visit Greenland, 2016a).

Yet UNESCO (2004/2019) status tends to overshadow cultural elements, both in representations of (the) place and in tourism production. An urge for 'visitability' tends to under-communicate the cultural aspects of the place, through WHS-inspired standardisation and cultural display (Müller and Viken, 2017a: 11). This is because WHS status has a distinct categorisation of heritage (Smith, 2003): sites with architecture, monuments and historical buildings are perceived as 'culture', while wilderness and physical landscapes are 'nature'. Culture is interpreted as material culture, with the heritage list divided into three categories: 'cultural', 'natural' and 'mixed' heritage sites – emphasising difference while acknowledging overlap (www.whc.unesco.org). Clearly some aspects of tourism development are directly influenced by these essentially Western/Southern constructs (Smed, 2017: 142–143).

While the recruitment of nature in Ilulissat for promotional purposes is clear, any dimension relating to cultural aspects is less obvious. In Ilulissat tourism exists as a mostly local business interest, but Indigenous people in Greenland tend to secure a small part of the local tourism environment, which may help to explain the somewhat secondary role of culture in this locality (Smed, 2017).

It has long been recognised (e.g. Pearce and Lee, 2005) that primary motivations within a market segment, such as nature, may dominate secondary motivations, such as culture, to the extent that they may be ignored or effectively marginalised. And where natural heritage takes priority, indigeneity can be a neglected issue in the tourism environment (Smed, 2017).

In visitor surveys, for example, the natural phenomenon of ice has been cited as a key experience, with 'nature and wildlife' also highly favoured experiences. By contrast, Greenlandic culture and interactions with local people were given far lower priority (Visit Greenland, 2013; Smed, 2017).

Researchers found a lack of local involvement in the short season sector in Ilulissat (CRT, 2013), and Smed (2017: 141) queried whether a lack of connection with the Indigenous population could help explain why culture has been secondary in the tourism product, and/or the obverse: because there is a lack of cultural perspective, there is little Indigenous involvement. Challenging the local marketing focus on the Icefjord, Smed has argued that a higher cultural profile could benefit the tourism sector here. Clearly, the Icefjord Centre has an important responsibility and role to play in contributing to this: assisting the integration of the natural and the cultural as part of Greenlandic and overseas visitors' learning processes concerning the local context and global significance of the climate crisis. This will be a challenging yet nonetheless important role.

9.6 Interactive experience: adventure tourism on the ice

Visit Greenland has been looking to establish the country as an adventure tourism destination ahead of expanded international airport capacity anticipated coming into operation in 2023 (Section 8.3).

In its 2016 strategy document, Visit Greenland (2016b: 21) reflexively argued that 'In the eyes of the surrounding world, Greenland is the definition of an adventure destination, and the Greenlandic Pioneering Nation brand is built on values based on the core elements of global adventure tourism ...' In its 2020 marketing plan, Visit Greenland (2020e: 9) clarified its objectives:

> Greenland will be the place for the tourist who wants an authentic and unique experience with varying degrees of physical activity and cultural immersion. This means that we must be skilled at developing our products for this type of tourist and what they demand. At the same time it is necessary that our experiences are developed all year round, so our guests come both in the summer and in the winter, stay longer and spend more money. Not least, it is crucial that our tourists feel that the quality matches the price of the experience. That is why it is important to create a quality scheme to ensure continuous improvement.

The last point acknowledges that while transport costs are likely to be reduced with the new airports' availability, Greenland will still be an expensive destination by global standards.

As to what 'adventure tourism' should mean for a destination, the 2016 strategy document explained that it:

- creates experiences that focus on active participation and on bringing guests close to nature and local culture;

- offers an 'experience universe' with many levels of engagement and activity, from unassisted sightseeing to guided tours and demanding expeditions;
- focuses on a form of tourism that involves immersion at the destination and offers experiences rooted in the landscape and culture of the country; and
- creates products that enable guests to challenge their own limits, be it culturally, intellectually or physically, thus making them feel like a pioneer (Visit Greenland, 2016a, 2016b: 21).

And while the notion of adventure tourism in the Arctic has attracted a level of conceptualisation from a Nordic perspective (Mykletun, 2018), Visit Greenland sees 'adventure' as a mental attitude toward travel experiences, a subjective term relating to how each traveller wishes to shape their experiences. Thus, as a marketing concept underpinning tourism (/tourist self-) imagery, 'adventure' acquires a broad and flexible interpretation, with statements suggesting that simply travelling to and safely arriving in Greenland is in itself (an) adventure.

Importantly, Visit Greenland has recognised a need to strengthen the focus of Greenlandic tourism on responsibility and sustainability, to enable adventure activities, 'that are often sensitive to the effects of nature, environment and culture', to develop without 'straining' the country's resources (Visit Greenland, 2016b: 21). While Visit Greenland's (2016b: 23) understanding of 'responsible tourism' has been spelled out (Section 7.4), 'sustainability' was a little more elusive. And although not explicitly addressed, climate change is the implicit facilitator of expanding adventure opportunities (at least in some circumstances – Boxes 9.4 and 9.5).

In Visit Greenland's (2020e) subsequent five-year marketing plan, released just before the world was made aware of the Covid-19 pandemic, generating increased demand from adventure tourists was one of the country's four 'must wins'. The plan argued that, globally, adventure tourism was the fastest growing form of tourism, and fitted well with Visit Greenland's strategy for 'responsible tourism with a focus on sustainability' (a repeated mantra in danger of losing its intrinsic meaning). The marketing plan contended that from being a niche market only a few years before, adventure tourism was headed towards a market share of 36 per cent of global tourism in 2020 (quoting Technavio.com). Also citing Allied Market Research.com, the plan suggested that global growth in adventure tourism would be between

Box 9.4 Adventure tourism promotions

In seeking to attract adventure tourism and a significant share of the extreme sport tourism market, specific attractions promoted have included:

- Arctic Team Challenge: a five-day adventure race in East Greenland covering over 200km in mountain biking, trekking and canoeing.
- Arctic marathon: taking place in August in the vicinity of Nuuk.
- Greenland Adventure Race: a five-day event in late August over a distance of 200km around Narsaq (South Greenland), involving kayaking, mountain biking and long-distance running.
- Arctic Circle Race: 'world's toughest skiing race' – a three-day event around Sisimiut, on the west coast, with overnight camping. Over 150 cross-country skiers participate.
- World Ice Golf Championships: every April near Uummannaq.

In promotions for post-Covid recovery, a broader brush conception of adventure tourism has embraced cycling, stand-up paddle-boarding, dog sledding, hiking, spotting wildlife 'while experiencing life at the edge of the world', and even staying on a sheep farm. The role of 'ice' would appear to have been downgraded in favour of empty (or at least uncrowded) peripherality, perhaps in response both to the longer-term (physical and psychological) consequences of Covid-19 globally, and to accelerating climate warming and its ambivalent impacts on 'traditional' adventure pursuits (Box 9.5): 'Greenland has never been a destination with mass tourism … [it offers] plenty of unique products … and [has] incredible amounts of space to share' (Por, 2020).

Sources: Burnett, 2001; Snyder, 2007; Visit Greenland, 2016a; Weaver and Lawton, 2017; Por, 2020

Box 9.5. Summary of potential impacts of climate change on adventure tourism in West Greenland (after Schrot, Christensen and Formayer, 2019)

Resort skiing

Resort skiing's vulnerability to climate change depends on temperature, snow cover season length, snow cover thickness and winter precipitation (Scott, McBoyle and Mills, 2003; Scott and McBoyle, 2007). Until 2100 resort skiing in Nuuk (Sisorarfiit resort) and Sisimiut (Solbakken/Tøsebakken) is likely to be within the optimal skiing temperature range of -5.0°C to +5.0°C. Nevertheless, Sisorarfiit resort will be affected by a season shortening of seven weeks, and for Solbakken/Tøsebakken a three-week shortening is predicted by 2081–2100, including a steady decrease in the number of days with more than 20cm snow depth. A predicted 31% increase in winter rainfall by 2081–2100 could compromise ski slope conditions as winter rainfall events (> 5mm) cause a decay of the upper snow cover (Fazzini *et al.*, 2004). Ski slopes in Nuuk's Sisorarfiit resort rely on snow depths of 30 to 150cm because of the complicated topography there. In Nuuk, year-to-year snow scarcity and wind loss have been already disrupting slope management. Recent adaptation actions are labour-intensive and limited: snow guns are used only sporadically, for bottom-layering, while wheel-loaders are used to bring snow from surrounding hillsides.

Ski touring

Ski-touring in Nuuk can be undertaken at Sermitsiaq or Qingaq mountain and Teqqiinngallip glacier near the Kangerlusarsunnuaq ski centre (Visit Greenland, 2017a), but could easily relocate to snow-safe places.

Heli-skiing

As a high profile component of West Greenland's winter tourism (Visit Greenland, 2017b), heli-skiing can be relocated to snow-safe places and altitudes as required. The contribution of such high cost activities to greenhouse gas emissions (Büsser, Stucki and Jungbluth, 2010) should be viewed in light of Greenland's aspiration for sustainable (tourism) development.

Cross-country skiing

Popular and promoted (Visit Greenland, 2017b), this is sensitive to temperatures, precipitation events and snow cover length (Landauer, Pröbstl and Haider, 2012). Track preparation for cross-country skiing requires a minimum snow cover depth ranging between 20 and 30cm, depending on the terrain (Neuvonen *et al.*, 2015). Nuuk, where a minimum snow cover of 50cm is required, is predicted to experience a long-term snowfall decline (27%), a shorter snow cover season length (-13 days with a snow cover deeper than 20cm in 2031–2050 and -79 days in 2081–2100), and increasing winter rainfall (8%). Similar results were found for Ilulissat, where rainfall was projected to increase and winter snowfall decrease in the longer-term. Snow cover was actually projected to increase in Sisimiut (home to eight different cross-country skiing tracks as well as the Arctic Circle Race) (Igloo Mountain ApS, 2017). Only minor snow cover changes were projected for Kangerlussuaq. Cross-country skiing in and around Nuuk, therefore, appeared highly vulnerable to climate change.

Dog sledding

Increased snow melt, more winter rainfall and a shortened snow cover season are contributing to discontinuous snow cover in coastal lowland areas (Callaghan *et al.*, 2011). For Ilulissat, a projected more than four-week reduction of days with snow cover thicker than 20cm and an 8% increase in winter rainfall will have a negative impact on touristic dog-sledding activities. There were reported cancelled or delayed dog-sledding package tours because of 'unusual' spring temperatures in 2019, and dog sledding appears to be a highly vulnerable adventure tourism activity in the future.

Ironically, pre-Covid, tourism providers had been attempting to offset cancelled dog-sledding tours and other cancelled activities with 'climate change' tours, whereby tourists could watch retreating icebergs or, in 'last chance tourism' mode, experience the Greenland ice sheet before it 'disappeared'. Climate warming is rendering winter tourism management more challenging, with activities needing to be interchanged at relatively short notice with assured available alternatives. Increasing tourist demand for dog-sledding tours clearly conflicts with climatic conditions that are shortening the dog-sledding season. As a result, offering dog-sledding tours and needing to feed the dogs all year is becoming increasingly expensive and potentially unviable. Most local sled dog handlers in Ilulissat were abandoning

Continued

Box 9.5. Continued.

multi-day dog-sledding tours during the 2018–2019 winter season and offering single-day tours instead. The reduction in sled dog numbers in recent years is considerable (Section 6.3).

Snowmobiling

Snowmobiles require 15–30cm minimum snow cover (McBoyle, Scott and Jones, 2007). Predictions suggest that warming is resulting in West Greenland's coastal lowlands not reaching these depths in winter, indicating that snowmobiling is highly vulnerable to warming. Using non-snow dependent all-terrain vehicles (ATVs) could serve as a possible adaptation (Scott, Dawson and Jones, 2008). The production of CO_2 emissions again challenges any aspirations to sustainability.

Trophy hunting

Trophy hunting of reindeer or musk oxen around Kangerlussuaq employs dog sleds to stage remote winter landscape experiences (Visit Greenland, 2018). Snow cover decline and sea ice reduction will likely affect access to hunting grounds (AMAP, 2018).

Source: Schrot, Christensen and Formayer, 2019

13 and 21 per cent *towards* 2023, after which it would stabilise at a lower level (Visit Greenland, 2020e: 17; also Lumholt, 2020: 12).

Repeating the Adventure Travel Trade Association definition of adventure tourism as containing the elements of nature, culture and physical activity from very light to physically hard activity, the Visit Greenland marketing plan claimed that adventure tourists 'contribute 65 per cent of their total travel expenses to the local community' (no specific source cited), thereby helping to create growth, jobs and experiences. For these reasons, Visit Greenland (2020e) was looking to attract 15,000 more visitors annually by promoting winter tourism and appealing to those looking to take part in physically challenging activities during their stay that would be 'branded with sustainability'. As noted in Section 7.5.2, Greenland's tourism is characterised by marked seasonality.

When plans for the airport improvements (Section 8.3) were approved, one of the supporting arguments was that they were necessary for the country to expand its tourism sector, as one of three designated key sources of income for the country.

In order to reach a 5 per cent annual growth target for visitor numbers, Visit Greenland was looking to create a 'reason to go' for potential visitors who were not already considering Greenland as a destination. Failure to do so, it was feared, would leave the new airports underutilised (McGwin, 2020b). Thus the need to build Greenland's reputation as a destination for adventure travellers – people whose travel primarily involves outdoor activity, but also elements of culture and contact with local residents –was seen to be a key element in reaching this goal.

Unlike other northern destinations, only 20–30 per cent of visitors travelling to Greenland each year have done so during the winter months. As noted previously, most have arrived in July and August, when numbers have been more than five times those in December (Chapter 7, Fig. 7.1), a pattern reinforced by cruise tourism (Chapter 11). The current air arrival hubs – Kangerlussuaq and Narsarsuaq – are located inland, where weather is more stable, and Visit Greenland is now pursuing a strategy of promoting them as outdoor tourism bases in advance of their future demoted role following completion of new airports at Nuuk and Ilulissat (Chapter 8). Kangerlussuaq was notably marketed as a staycation base during Greenland's Covid-19 lockdown (Chapter 12).

Kulusuk in East Greenland remains the destination for one-day visits from Iceland. East Greenland is also a popular destination for 'advanced' outdoor tourists. 'A few' dare to venture across the ice sheet, usually between Kangerlussuaq and Tasiilaq, on the east coast. This requires a special government permit (Statistics Greenland, 2020: 30). It offers the natural extension to, and is much more challenging than, the hiking route from Sisimiut, on the west coast, to Kangerlussuaq (Dillon, 2019). In combination they form the Arctic Circle Trail.

Greenland often looks to Iceland as a model for development. And following the financial crisis that virtually bankrupted the country, Iceland was seen to bounce back partly because of the successful promotion of adventure tourism (Kirk, 2018), and in generating demand during the winter season (especially from the UK). But Lumholt (2020: 11) points to the fact that unlike Iceland, Greenland has not been able to create low-cost promotions on airline tickets and overnight stays in the traditional off-season. This is largely because, thus far, the country has not been able to attract the operations of low-cost airlines. Lumholt argues that the vigorous marketing of winter experiences is important in generating increased demand outside the summer high season that can at least encourage airlines currently flying into Greenland to create and extend discount promotions.

In the process of promoting and participating in adventure activities, visitors and local tourism entrepreneurs are venturing into areas that were previously perceived to be wild and untouched by humans, thereby rendering such areas increasingly 'cultured' and objects of cultural display; one of the more extreme forms of 'adventure' being heli-skiing or the less adventurous helicopter tour (Smed, 2017: 147). Unfortunately these are hardly emblematic of a (more?) sustainable approach to tourism development.

It is not a little ironic, therefore, that the acceleration of climate change is particularly observable in Greenland's tourism sector. While Demiroglu *et al.* (2018, 2019) have reported on the impacts of climate change on ski tourism and tourists' attitudes in the Nordic Arctic, Schrot, Christensen and Formayer (2019) have examined how (West) Greenlandic winter tourism may be affected by Arctic warming. They modelled temperature and precipitation change over two time 'slices' – 2031–2050 and 2081–2100 – relative to a base period of 1991–2010. They related this to the pursuit of winter adventure tourism activities – resort skiing, ski-touring, heli-skiing, cross-country skiing, dog sledding, snowmobiling and trophy hunting – at recreational locations in or adjacent to Nuuk, Ilulissat, Sisimiut and Kangerlussuaq (Box 9.5). Their model data provided strong evidence for an increasing annual mean temperature of between 1.1 °C and 4.6 °C, and for precipitation increasing by up to 78 per cent in West Greenland by 2100, compared to 1991–2010. Snowfall and snow cover season length would decrease by up to 42 per cent to 2050 and by up to 49 per cent to 2100. Resort skiing and cross-country skiing are affected by snow marginality and wet snow conditions, while ski-touring and heli-skiing demonstrate a higher adaptive capacity whereby they can be relocated to snow-safe conditions. A reduced snow cover thickness particularly threatens dog sledding. In-depth interviews with local stakeholders suggested that they would be challenged in being able to adapt at least some current adventure tourism activities to changing climatic conditions (Box 9.5) (Schrot, Christensen and Formayer, 2019).

9.7 Wildlife tourism

Woodall (2020) for Visit Greenland recognises five types of nature traveller:

- wilderness seeker: e.g. hiking the Arctic Circle Trail;
- extreme adventurer: on a constant quest for remote, challenging landscapes: rock and ice climbing expeditions, skiing or dog sledding across the ice sheet (e.g. Berbeka, 2018);
- nature lover: undertaking day hiking, kayaking, winter ski touring, snowshoeing, snowmobiling;
- special interest adventurer: pursuing angling, photography, bird watching, hunting;
- nature appreciator: enjoying the relaxed simplicity of being near/within beautiful nature.

Although whale watching is one of the Big Arctic Five activities that Visit Greenland (2020d) extols (along with ice and snow, dog sledding, the Northern Lights, and 'pioneering people'), non-consumptive wildlife tourism otherwise appears limited in scope. This partly reflects the longevity of hunting and harvesting traditions.

- Greenlanders have always hunted marine mammals – seals, walruses and whales – for their meat and other purposes. Greenland's 'unofficial national dish' *suaasat*, is a soup of seal or whale meat, onions and potatoes. Another traditional Inuit dish, *mattak*, comprises raw pieces of whale skin with blubber (usually from a narwhal, bowhead or beluga whale);[3]
- while the Greenlandic reindeer is the most commonly found land mammal, particularly on the west coast, it has also traditionally been a main food source, and reindeer hunting remains a popular Greenlandic tradition;

- trophy hunting is a (lucrative) promoted tourism activity, the major targets being musk ox and reindeer (Box 9.5): the activity is claimed (Visit Greenland, 2018) to be 'challenging'. Smaller game such as hare, fox and ptarmigan are also said to be available. Most activity takes place in the Kangerlussuaq area and within the Nuuk fjord system.

Unsurprisingly, therefore, most Greenlandic wild animals have learnt to keep well away from humans, while polar bear and narwhal are mostly to be found in and off the north and north east of the country: the relatively inaccessible and protected North East National Park being the world's largest (Fig. 4.1).

There would therefore appear to be a tension in the notion of Greenlandic wildlife tourism between the 'authenticity' and cultural integrity of retaining Inuit traditions of consumptive wildlife pursuits and the growing global market for (and ethical positions of) non-consumptive wildlife appreciation.

9.8 Conclusion

In responding to Greenland's global climate crisis, the Ilulissat Icefjord Centre represents an important indirect mitigation process of informing and educating foreigners and Greenlanders alike about cause and effect, the latter being illustrated on a huge scale close by. An irony of this process is, of course, that many foreign visitors will have travelled long distances to arrive at this experience, generating substantial amounts of greenhouse gases in the process.

In response to rapidly changing circumstances, Greenland's promotion of 'adventure tourism' would appear to be acknowledging the need for adaptation in casting its net away from a wholly ice/snow focus and in adopting a somewhat liberal interpretation of 'adventure'. As noted elsewhere in the Arctic, such a process might be hastened as increasingly culturally diverse tourism markets seek (ethnocentrically defined) adventure experiences.

9.9 In the next chapter

In the next chapter, within a discussion of tourism imagery, this trend is examined in the context of Asian, and particularly Chinese tourists' perceptions and consumption of 'adventure tourism' in Greenland, understandings which are not necessarily consonant with those of Visit Greenland. Chapter 10 therefore suggests that the DMO's marketing and imagery may require some adaptation.

Notes

[1] With the added word-play that 'Is' is, of course, 'ice' in Kaae's native Danish.

[2] Two subsequent WHS designations in Greenland have been: Kujataa in South Greenland, a large region of fjords and green hills, encompassing five Viking settlements (tenth to fourteenth centuries), declared a WHS in 2017; and the Aasivissuit-Nipisat area, located in the largest ice-free area inside the Arctic Circle, in central West Greenland, where a heritage landscape has 'preserved innovative ancient navigational skills and medicinal use of local flora', and was designated in 2018 (UNESCO, 2018).

[3] Quotas for whaling and fishing exist, and such species as the blue whale are protected. Whale and seal meat can only be consumed locally.

10

Tourism and Imagery: Soft Power, Branding and Cultural Disconnection

> '... places, even the most remote, expand in time and space. They are not stable, but a vibrant field of diverse actors, governance, materialities, interests, memories and mobilities that inflict place conceptions'
>
> (Kramvig, 2019: xix)

10.1 Tourism, soft power and branding

In the previous chapter, while the Ilulissat Icefjord Centre was highlighted as representing an educative role, it retains a (passive) soft power role whereby Greenland is projected as a victim of climate change. The Icefjord Centre also contributes to the branding of Greenland in terms of being in the visually dynamic forefront of climate change. In this way the Centre is scripted to play multiple, overlapping roles representing and projecting Greenland within the retreating ice. This is not value free, and as suggested in the previous chapter, the Danish developers and local stakeholders may not necessarily agree on the priorities the Centre represents. In this way, the intermingling of tourism and geopolitics is expressed: at one level indicating to the world how Greenland sees itself in relation to the climate crisis and being open for business (Covid-19 notwithstanding), and on the other, reflecting different approaches and priorities to the impacts and expressions of that crisis.

Although tourism has been emphasised as being a third economic pillar to support the path towards independence, it also possesses an important soft power role to sustain the journey towards that independence, and beyond. The soft power of victimhood (in relation to the climate crisis) is one dimension. Another is the more explicit attempt at nation and (implicitly) tourism branding, an exercise which has in recent years become more fluid, less fixated on one single ideal, purpose or image. Thus while 'pioneering nation' has been one trope featuring in Greenlandic messaging, extolling tourists also to 'become a pioneer' through pursuing adventure tourism activities, it runs the risk of being variously interpreted.

Tourism, and implicitly (but not wholly)[1] international tourism, can be conceptualised and operationalised as a geopolitical instrument. One geopolitical conception of tourism is through its supposed attribute of 'soft power' (Nye, 1990, 2004; Davis Cross and Melissen, 2013): the ability to achieve goals through attraction rather than coercion (Keohane and Nye, 1998: 98). In the case of tourism, such attraction has several forms. On the one hand it relates to the drawing of visitors to a country or region for them to return home with an empathy with the host country and its objectives, and then to convey that imagery and the narrative reinforcing it to others, in order to generate empathetic associations and perhaps further visits. Soft power can

DOI: 10.1079/9781789246728.0010

also act to attract inward investment for a variety of purposes, for example responding to the imagery of climate crisis victimhood.

Soft power is therefore about the transmission of political and cultural values, beliefs, and agendas: in this case via tourists, tourism marketing and tourism branding, whether related to a destination object, concept or location. Tourism's soft power can support, and be a component of, branding, both sectoral and national.

As a mix of aspirational visions, nation branding can be multifaceted: both outward facing, projecting its imagery in several different directions towards multiple audiences; and reflexively reinforcing, perhaps even reassuring, a domestic audience and its own national self-imagery. As an element of this nation branding process, or set of processes, tourism can become imbued with propaganda value: its symbols and artefacts gather outer and inner meaning and implication, while its stakeholders – tourists, tourism employees, destination residents and host environments – become (perhaps unwitting) actors in the game of image co-construction, projection and reconstruction, processes which have become increasingly played out in virtual space – as 'smart power' (Nye, 2008, 2009): cyber-marketing through destination marketing organisation (DMO), attractions', local authority and travel company websites; quality rating platforms; and, not least, social media posts and blogs.

Globally, a high profile contemporary harnessing of apparent soft power has been the desire to host (and perform well in) tourist-rich 'mega-events' such as athletics, soccer and other large-scale sports tournaments (Grix and Brannagan, 2016). Although Greenland and Greenlanders have not thus far established a significant impression on the global events stage, Nuuk's hosting of the 2016 Arctic Winter Games was significant in attracting favourable global attention, offering opportunities for both (international) prestige and (domestic) reflection and reappraisal (Ren and Thomsen, 2016; Ren and Rasmussen, 2017). Further, the sustaining of some sports events during the worst of the Covid-19 lockdown was seen as an important element in maintaining a sense of Greenlandic identity through domestic recreation and tourism (see Chapter 12).

Generating and exerting soft power has become in itself something of a competition, as reflected in the annual publication of an index ranking of the 'top 30' soft power countries (McClory, 2019). If the goal of soft power is to induce others to tacitly accept, emulate, and aspire to a country's values, beliefs, and agenda, tourism's role, through imagery, propaganda/promotion and direct experience, is to generate empathy and respect for the values projected through the lens of visitors' positive experiences. Yet tourism's soft power role can be something of a blunt instrument, not least because of the many imperfections and inconsistences associated with tourism-related development activity. Like any form of imagery, the positive component of such power may take some considerable time to construct but can be destroyed much more quickly, by, for example, political change, natural disaster, conflict or disease. In the latter case, Greenland's effective isolating of, and itself from, Covid-19, was actually used to advantage by Visit Greenland in its 'Flattening the Curve' imagery and philosophy, both looking forward to post-pandemic possibilities while invoking Greenlandic staycationers as endorsers of those possibilities (Por, 2020).

Tourism's propaganda role can work in two directions. On the one hand, tourism can be employed to present an outwardly positive face to the world of a destination country or region. As already noted, this may be for reasons of generating foreign currency and perhaps to gain benefit from raising the destination's global profile to encourage inward investment and promote an ideology, whether it be pan-Inuit internationalism or 'self-sustainability'.

The destination region or country may seek to generate a positive outward tourism image to counteract or counter-balance imagery which might in other respects be less acceptable, and then can proceed to capitalise on certain aspects of that enhanced, perhaps even falsified, destination image. Or images may be projected by stakeholders such as foreign travel companies that may conflict, or not be consonant with the destination's own (self-) imagery, yet may be relatively successful. For example, while the 'last chance'/'bucket list' imagery and narrative employed by a number of companies operating to Greenland may not be commensurate with Greenland's position in relation to climate change and environmental conservation, it may be effective in attracting those who might otherwise be repelled by 'cold' or 'difficult' experiences. Is this a reason why cruise tourism to Greenland is relatively popular, particularly for travellers from certain cultural backgrounds and demographic characteristics?

Second, tourism's 'reverse' propaganda role can act to diffuse the cultural values of tourist-generating countries within host societies and beyond, perhaps to the further benefit of source countries' political and/or economic interests. For example, a number of commentators in the West might perceive the enormous growth of Chinese outbound tourism as just another dimension of the country's core economic, political and strategic expansionist policies (e.g. Tse, 2013).

In his dictionary summary of tourism and soft power, Ooi (2016) briefly enumerates four 'interrelated approaches' that embrace much of the above. In his critique of Ooi, Hussin (2018) argues that the role of gastronomy has been under-rated and can be an important cultural source of soft power (hence, perhaps, the role of *kaffemik* in Visit Greenland's programme for visitors' encounters with 'pioneering people'). Hussin further notes the omission of international agencies such as UNESCO in their facilitating/gatekeeping role in bringing to the fore for global consumption a country's cultural, socio-political and natural phenomena. Notable in this respect is UNESCO's declaration of World Heritage Sites (WHSs), and those sites' abilities, enhanced through WHS status, to act as image constructors and modifiers. In the case of Greenland's Ilulissat Icefjord, however (Chapter 9), WHS designation has had the power to skew the imagery of a region or country in this case by focusing on ice and 'natural' processes to the virtual elimination of human culture, despite the anthropogenic contribution to climate change processes (Smed, 2017).

However, in a critique of the concept of soft power, Fan (2008) argues that, to some degree similar to tourism, the nature of soft power can be unpredictable, rendering it less efficacious than its proponents might suggest. Indeed, the example of some aspects of tourist behaviour and its negative reflection on their country of origin has been the source of debate both in the academic literature and the popular press. Guo and Zhang (2008), for example, argued that the 'uncivil behaviours' of Chinese outbound tourists had become 'the bottleneck to hamper national soft power upgrade'.

10.2 Projecting Greenland: branding, imagery and marketing reality

Place branding ...

> is a strategic endeavour to make places visible and competitive in the attention economy. Branding creates a visionary platform of becoming; it envisions prosperous viable futures for generations to come ... Place branding is [also] a participatory endeavour, as one activity among many in the remaking of a place through multiple narrative, performative and governance techniques.
>
> (Kramvig, 2019: xix)

Such branding and re-imaging may be especially important in situations where communities may be aspiring to (wider) political and diplomatic recognition at a higher level, such as Kosovo's seeking wider recognition as an independent country (Hall and Brown, 2017), or a former colony or dependent territory seeking to secure political and diplomatic recognition of its independent status, as may be the future case of Greenland.

Nation branding campaigns can therefore reflect processes articulating the meaning of nationhood, and thereby contribute to a (re-)conceptualisation of local and regional geographies (Ren and Gyimóthy, 2013).

Yet the image of a nation is so complex and fluid that it may defy the clarity implied in branding processes (Fan, 2006; Ren and Blichfeldt, 2011), and external images tend to lag behind reality (Fan, 2010). Top-down conceptions themselves may fail to keep pace with the reality of such destination image formation agents as travellers' blogs and social media platforms.

Tourism can hold a significant position in the substantial political, economic and social construction processes required for a society such as Greenland, poised on a trajectory towards independence. Any (re-?) imaging that this requires confronts a number of challenges. First, there is the need to portray an Arctic distinctiveness, or perhaps a distinctive Arctic hybridity integrating nature and culture, and ethnicity and modernity within that culture, inside a frame of sustainability. Second, against the growing yet perhaps not wholly predictable impacts of climate change, a destination is required to project a safe (if at times challenging) and welcoming environment as the context for international tourism promotion. And third, high quality niche tourism products need to be developed and marketed in response to the changing nature of global tourism demand, to reinforce positive national imagery and to help raise per capita tourist receipts (Hall, 1999b: 227–228).

Destination branding's traditional core objective of producing consistent, focused imagery, has usually confronted at least three major challenges:

- sourcing adequate finance to support appropriate marketing and promotion campaigns;
- public or private sector destination marketers being pressured to return short-term results when long-term investment is required to build a consistent brand; and
- perhaps most significantly, the difficulty of developing a coherent brand when a destination is composed of multiple products and environments with diverse stakeholders wishing to promote their own interests and priorities (Hall, 1999b: 230).

Such an approach conceptualises place brands as having the same features as corporate and product brands by acting as catalysts, unifying and bringing coherence to communications, strategies and activities, which are characterised by 'streamlined uniqueness narratives' (Ren and Gyimóthy, 2013: 17) that are implicitly defined geographically. An alternative approach suggests that any differences between place brands and product branding are dynamic and perhaps elusive (Cassinger, Lucarelli and Gyimóthy, 2019: 3). Indeed,

> ... new, emerging approaches to branding, which pursue a more relational and performative (outside-in) strategy ... do not seek to fixate and bombastically communicate 'one clear image', but rather see the construction of place brand stories and images as an ongoing activity, which engage[s] a broad range of stakeholders and is able to take in and address complexity and incoherence (Ren and Blichfeldt, 2011)
>
> (Ren and Gyimóthy, 2013: 19)

The case of Greenland moving towards independence arguably offers exemplification of:

- attempts at national (re-?) branding in response to anticipated new political and economic circumstances;
- the inter-relationships between national identity imagery and tourism branding; and
- the inter-relationships between economic and political re-imaging requirements, especially in relation to foreign investment and de-colonisation (de-coupling from Denmark) processes (after Hall, 2002: 325), and the possibilities of new, potentially significant but geopolitically controversial, tourism markets (Section 10.3).

While earlier chapters have highlighted how Arctic imagery of the natural world can suppress cultural elements of the lived reality, cultural heritage is one, if not *the* key ingredient in Greenlandic-ness and Greenlandic self-image.

But the role of cultural heritage, for example, through statues and commemorations, can act to re-awaken the role of former hegemonic powers and/or highlight contemporary cleavages between cultural groups. Compared to reactions to commemorations of Hans Egede, for example (Box 4.3), does the projection of Norse heritage as tourist attraction in southern Greenland evoke atavistic responses (Chapters 4 and 6)? Can stakeholders ensure that such sites avoid becoming the source of contested and dissonant heritage?

Certainly 'heritage' can occupy an ambivalent role in branding and related processes, drawing on links with the past to justify or reinforce a current trajectory into the future. Any attempt at sustaining an equilibrium between the traditional and the modern in place imagery risks re-inventing and re-interpreting 'heritage'. And 'heritage' is far from being a value-free concept as contemporary economic and political power influence what is preserved/sustained, and how it is (re-)interpreted for the future (Lowenthal, 1997).

In the context of Greenland,

> While the Arctic has for generations been described and represented by people living in the South, the peoples of the Arctic are now to a much larger degree representing themselves, both on the political stage and in the media, art, literature and film
>
> (Thisted, 2015: 23).

Yet, Ren, Gad and Bjørst (2019: 160) suggest that the conditions for, and distribution of agency, are still skewed and complicated, but propel the argument that

> ... the many different institutions undertaking, often independently of each other, the role of cultural or commercial ambassadors for the nation brand, impede the ability to create coherent brand narrative, making national brands far more unmanageable than ordinary brands (Blichfeldt, 2005).

That different Greenlandic brands are enacted into being at state, corporate and market level, is considered to result in a 'polyphonic divergence' (Ren and Blichfeldt, 2011).

In 2010 the national branding campaign of 'Pioneering Nation' was launched, supported by the two narratives drawing on (a) traditional imagery of powerful nature, and (b) the pioneering spirit of the people of Greenland. This is seen as both hybrid and modernistic, linking a close relationship with nature to a modern developing economy, and thus intrinsically part of an internal debate about the nature, direction and end point of Greenland's trajectory. It built on an earlier (tourism) branding exercise that offered promises of the extraordinary – 'Be a Pioneer' – and its precursors 'Once in a Lifetime', 'Experiences from a Completely Different World' and 'Rite of Passage' that conveyed remoteness, exclusivity, individuality and initiation as key allied messages (Kaae, 2006: 59–60).

When the Greenland Tourism and Business Council was split, the 'Pioneering Nation' imagery followed the tourism portfolio and was employed to support the 2016 Arctic Winter Games, an important branding and soft power platform, held in Nuuk (Ren and Thomson, 2016; Ren and Rasmussen, 2017). As part of the media public relations exercise for the Games, in collaboration with Visit Greenland, experiences of nature, local art and Greenlandic gastronomy were offered, encompassing 'hybrid cultural references to Nordic design, Indigenous ethno-symbolism and Greenlandic fashion and produce' (Ren, Gad and Bjørst, 2019: 164) as well as cosmopolitanism, in order to project the ethnic-modern hybridity that Ren and Abilgaard in Chapter 12 consider the fundament of contemporary Greenland logic.

After a new Mineral Resources Act was introduced following the 2009 achievement of Self-Rule, the nation-building and development discourse promoted by Greenlandic politicians witnessed a discursive shift from mining in Greenland towards mining *for* Greenland (Bjørst, 2016). Alongside this re-emphasis, branding initiatives intensified and Greenland's official brand as a *Pioneering Nation* was adopted for the mining sector (Ren, Gad and Bjørst, 2019: 164–165).

This branding application has been regarded as less than appropriate because it summons images of a wild frontier region, positioning Greenland 'as a place on the edge of the world, a place apart and somewhere different' (Nuttall, 2017: 29). For projecting the distinctiveness of a tourism destination, such imagery may possess clear advantages, but pioneering in mining suggests taking risks in an industry that would regard itself as risk averse, and, further, it alludes to the real extra costs involved in undertaking activities on the margins of technology and human abilities.

By contrast, while it remains in the shadow of Denmark, Greenland can draw on Nordic associations of high standards and quality, trustworthiness and stable legal and political systems, a dimension Greenlandic politicians have been able, and willing, to exploit (Ren, Gad and Bjørst, 2019: 167). It became an explicit element of national self-identity and imagery when the Greenlandic minister for independence and foreign affairs, speaking at an Arctic Council conference discussing the UN Sustainable Development Goals, presented three critical narratives (Egede, 2017):

- Greenland was developing a Nordic (explicitly not 'Danish' per se) welfare model that would act as a basis for independence.
- In the face of climate change, Greenlanders, rather than being *pioneers*, had become victims, especially because of the disappearing sea ice constraining traditional subsistence activities (Bjørst, 2012, 2019).
- Somewhat contradicting the above, new climatic and geographical conditions allowed for new investment and business opportunities. This narrative needed to be tempered as it was being presented at a conference discussing Sustainable Development Goals (SDGs).

But as Nuttall (2017) suggests, the ambivalent message was that Greenland was greening, both in terms of climate change transforming Greenland's landscapes from white to green and rendering the territory more accessible, and also, metaphorically, in terms of Greenland's awareness of the need to be environmentally conscious (although also see Chapter 5 regarding Greenland's ambivalent 'sustainable development' stance).

Greenland's 'brand multiplicity' has thus seen different actors draw on 'varying and often diverging discourses, aesthetics and narratives' (Ren, Gad and Bjørst, 2019: 171). Thus place branding in Greenland is undertaken differently in relation to tourism, mining and diplomacy, thereby exemplifying the 'unruliness' of place brands (Ren and Blichfeldt, 2011).

For 'a nation still under construction', the branding of Greenland is viewed as an exercise in identity politics, in which independence from Denmark plays a critical role. While Greenlanders mostly agree that Kalaallit Nunaat should be an independent nation state, the speed of its trajectory is contested, as is the type of nation Greenland should be, and which elements of 'traditional' and modern Greenlandic culture should be granted prominent roles in the process (Gad, 2016) (Chapter 12).

10.3 Imagery and mutual misconceptions: the Chinese outbound market

'Adventure' is clearly an important dimension in the branding of an Arctic country, and, as noted in Chapter 9, Visit Greenland has recently concentrated much of its marketing efforts, promotions, product development and policy support in attracting 'adventure tourists', albeit on the basis that nearly all experiences in Greenland qualify as adventure tourism. Nonetheless, as an active member of the Adventure Travel Trade Association (ATTA), Visit Greenland has adopted the ATTA definition that considers a travel experience to qualify as 'adventure tourism' if it contains two of: immersion in nature, physical risk and cultural exchange (Visit Greenland, 2016a).

At the same time, the global recognition of a burgeoning Chinese outbound market generating high spending and relatively undemanding visitors is a perceived opportunity that Greenland has not been able to ignore, despite limited mutual experience. Given the relatively high costs involved in visiting Greenland, and the skewed nature of the tourism offer available, there has been a clear need for a nuanced approach in marketing Greenland to the Asian market generally and to China in particular. Superficially, aiming to attract Chinese adventure tourists would appear to be an ideal marketing match.

Yet while many observers would suggest that China employs outbound tourism as an exercise of soft power, as noted above, tourist behaviour (from any source society) can be somewhat less than soft power promoters may wish for (e.g. Guo and Zhang, 2008): there may be mutual misconceptions with 'host' societies, and further, as Xu, Wang and Song (2020) suggest, outcomes may depend on global economic power, agenda-setting and qualities located somewhere between hard and soft power, rather than pivoting on the host country's touristic attraction. Tse (2013) recognises that there is ambiguity in China's outbound tourism policy, and that it lies not so much in the consumer area but in politics and the way in which tourism is treated as a form of diplomacy. More explicitly,

> ... outbound tourism development helps to enhance China's voice in the international field of tourism to promote [the] soft power of its culture and values. We should take advantage of the rapid development of the outbound tourism to enhance the Chinese voice in international affairs ...
>
> (Dai *et al.*, 2017: 257)

Xu, Wang and Song (2020) go on to suggest, however, that soft power-building through Chinese outbound tourism may be more effective in Southeast Asia, the Asia-Pacific region, in Africa and South America than in Western countries because of the closer economic relationships, cultural similarities and shared beliefs with the 'East' and the 'South'. This position can only be reinforced with developments such as the establishment of the free-trade Regional Comprehensive Economic Partnership (RCEP) between China and 14 other countries of the Asia-Pacific region that embraces a market of some 2.2bn people (Jolly, 2020), as well as China's 'Belt and Road Initiative' (Sections 2.6.3, 10.4).

Under these circumstances, any evaluation of Chinese tourism to an Arctic society is enlightening, not least in relation to the cultural dissonance that may arise from segmenting such visitors. Surprisingly, the ATTA's (2013) market appraisal report does not include outbound departures from Asia. To fill this gap, Bannikin Travel and Tourism employed the ATTA's methodology to examine the Chinese adventure travel market. They determined this represented 16.3 per cent of Chinese outbound tourists and around US$8.2bn per year. The vast majority of this market, some 88 per cent, are in the 26–39 age group (Bannikin Travel and Tourism, 2019).

The pre-Covid years saw a rapid growth of Asian, mostly Chinese visitors travelling to (European) Arctic destinations (Lee, 2020; Chapter 3). In the last decade, Greenland had seen an increase in Chinese travellers, although somewhat volatile numbers had levelled out to around 1,000 per year (Table 10.1). However, when combining visitors from China with those from Hong Kong (Table 7.4), numbers staying in registered accommodation in Greenland rose from 612 in 2018 to 1,372 in 2019, an increase of 107.8 per cent.

A significant issue with such relatively modest absolute numbers is that just a few large group bookings can make a significant difference, rendering forward planning problematical. Pre-Covid predictions and hopes foresaw a rising number of Chinese visitors. Huang, Tang and Weaver (2017: 89), for example, had suggested that the Arctic was likely to experience a significant increase in Chinese visitors in the following two decades.

In their research on Chinese adventure tourists visiting Greenland, Martin and Ren (2020) argued that focusing on the adventure segment provided an opportunity to offer a more nuanced understanding of 'the Chinese tourist' beyond stereotypical representations, which, like other stereotypes, can result in near meaningless reductionism and abstraction (Ren and Ooi, 2013; Jørgensen and Ren, 2015; Ooi, 2019). There was a need for a more culturally sensitive understanding to better situate and make sense of Asian tourist behaviour, practices and habits; in short, 'the wide variety and range of intertwined social expressions that we have come to understand as culture' (Ooi, 2019: 14). This could be sought through more detailed empirical research.

Table 10.1. Numbers of Chinese visitors to Greenland, 2015–2019. Author's own table derived from Lumholt (for Visit Greenland), 2020: 5 (see also Tables 7.2 and 7.3); Martin and Ren, 2020.

	Numbers of Chinese visitors to Greenland				
Year	2015	2016	2017	2018	2019
Number	1,111	993	1,426	1,015	1,059

Until recently, few people and institutions from China and the Arctic had experienced direct contact with each other, such that pre-conceived Greenlandic images of China and Chinese images of Greenland were likely to be based on (not necessarily accurate) received ideas and information. 'The underlying foundation for studying how Greenland is imagined, represented and constructed as a tourist destination is that places do not just "happen", but are constantly reconfigured through their enrollment [*sic*] into specific discourses, representations and practices of place' (Martin and Ren, 2020: 132). Martin and Ren sought to elicit Chinese travellers' impressions and conceptions of Greenland and their experiences of it through an evaluation of 42 blog posts on the social networking site Mafengwo, employing qualitative and thematic content analysis. They identified a number of recurrent topics and themes, notably: Greenland's destinations status as 'off the beaten track', the difficulty of accessing tourist information, (high) cost, not least in relation to the availability and quality of wi-fi, and the (mis)representation of Greenlandic culture.

Visit Greenland's corporate website, aimed at the travel trade, argues that visitors should not be understood through such traditional categories as age, nationality or employment, but rather through their interests and level of engagement (Visit Greenland, 2016b). To complement this philosophy, the DMO uses a psychographic segmentation model to profile different visitor groups. Eleven psychographic segments range over a level of involvement from 'immersion' and 'over interaction' to 'observation', with a second, focus of interest, axis ranging from 'culture', 'culture and nature', to 'nature' only.

Martin and Ren (2020: 134) view this 'innovative post-demographic' approach as a tool that can reconfigure Chinese adventure tourists in terms of their interests rather than their nationality, 'dissolving the "Chinese" as meaningful category'. In conversations with tourism operators who work with Chinese tourists, however, Martin and Ren were confronted with displays of a 'relapse' back into a nationality-based way of understanding and working with Chinese visitors. Other Visit Greenland activities aimed at helping to understand Chinese travellers also appeared to be based on national culture, thereby undermining the psychographic approach. These included seminars and online material advising tourism and hospitality providers, 'to offer hot water, noodles, slippers and other "comfort" items, which supposedly helps the traveller feel more at home' (Martin and Ren, 2020: 134–135).

Further, although much effort is expended, through business relations and travel road shows, to reach and 'educate' Chinese travel agents, because of a general lack of physical and digital connectivity, they appear to fail to understand 'what Greenland is about'. One apparent consequence of this is the seeming un(der)-preparedness of Chinese visitors to Greenland, as exhibited in the blog posts analysed.

Martin and Ren's research revealed that the main objective of Chinese tourists in winter was to experience the Northern Lights, while in summer calving icebergs appeared to be the main attraction. The clearly expressed interest and main purpose of their visits, as conveyed in the Mafengwo blogs, was predominantly experiencing natural attractions, not culture or physical activities, the two other possible components of 'adventure tourism'.

In those blogs, visitors often appear to have been overwhelmed by the size and force of nature encountered in Greenland, expressing that their experiences were 'pioneering', an emotion consonant with Visit Greenland's brand strapline 'Be a Pioneer'. On the other hand, an expressed lack of access to information in some cases clearly circumscribed conceptions of available activities and experiences, alongside a perceived high cost of wi-fi availability.

Critically, in not experiencing the two other core components of adventure travel as promoted by Visit Greenland – physical activities and interaction with culture – Chinese 'adventure tourism' visitors' experiences diverged from how adventure tourism is typically understood and marketed by Visit Greenland.

Although often ill-equipped for the purpose, Chinese visitors' interactions with some physical risk through, for example, hiking on the ice sheet, were ultimately perceived as rewarding, while accounts reflecting a cultural experience were limited and even culturally insensitive. Thus, for example, some blogs offered advice on where to 'spot' a Greenlandic person: Greenlandic culture being mistakenly understood to be concentrated in specific locations, with Greenlanders 'perceived to be people fitting the stereotypical appearance of an Inuit person and denominated with the derogatory term of Eskimo' (Martin and Ren, 2020: 138). This contrasts starkly with the hybrid concept of culture promoted by Visit Greenland (Ren, Gad and Bjørst, 2019; also Chapter 12).

While personal encounters with the local population might be perceived as awkward or difficult for Chinese because of language barriers, none of the blogs mentioned engaging in such cultural experiences as a Visit Greenland-facilitated *kaffemik* in a local home. Indeed, despite the promotion, in cultural terms, of Greenlandic culinary experiences – locally viewed as representing strong ancestral links (Olsen, 2010) – accounts of such encounters were conspicuously absent from the Chinese blogs, either because they were not considered significant, or, more likely, because they simply did not take place.

While Visit Greenland actively promotes a dynamic view of a Greenlandic culture which is inspired by traditions but does not shy away from progress (Martin and Ren, 2020: 139), this clearly was not conveyed adequately to the Chinese visitors who missed opportunities to learn about Greenlandic culture and engage with it in a meaningful way. The corollary of this was therefore that Greenland failed to be projected as 'more than nature', and for this segment of visitors, at least, the dominant imagery of the Arctic as remote, exotic and empty space was reproduced. Here was clearly a missed opportunity to promote the cultural, gastronomic and artistic products and experiences, that the Greenlandic tourism industry had been working hard to develop and market as part of national imagery.

Martin and Ren (2020: 140) conceptualised the mutual misconceptions characterised by these encounters as being caught between sameness and difference. On the one hand, an understanding of Chinese adventure tourists was caught between sameness in a 'psychographic' sense and difference in a 'national culture' sense, while the Chinese visitors found themselves caught between the positive experiential difference of being 'off the beaten track' and a negative sameness in the experience of comparing prices and commodities – especially wi-fi – with those at home.

Critically, as the commonly used English-language term 'adventure tourism' in Chinese translates as 'nature-based tourism', the multi-layered 'adventure' experiences that Visit Greenland expects and plans for such visitors is not consonant with the largely risk-free nature experience Chinese tourists appear to expect. A clear case of mutual misconception in tourism is thus confronted (Ren, 2010). Interestingly, Rantala *et al.* (2018) observed a 'softening' of adventure tourism and its interpretation in Lapland, which they relate to a 'diversification' of tourist groups.

Defining and performing a destination is a continuous process of evolution and contestation (Edensor, 2001). Greenland, as a destination under construction, can benefit from the way in which Greenlandic and Chinese tourism actors contribute to 'defining and "making sense" of Greenland and of each other in between sameness and otherness' (Martin and Ren, 2020: 140–141). Yet while stereotypical categories of 'the Other' are sustained, opportunities for more meaningful and respectful experiences and encounters may be lost, as will be the opportunities for promoting cultural tourism products and experiences to a potentially growing market, and lost with them, the soft power imagery of a society aspiring to political independence (which may not necessarily appeal to a Chinese audience).

INVITED CONTRIBUTION: 10.4 Understanding Adventure Tourism by Chinese Outbound Tourists

Rong Huang

10.4.1 Adventure tourism by Chinese outbound tourists

The phenomenal growth of Chinese outbound tourism globally, not only in terms of sheer numbers but also the associated level of expenditure ($255bn in 2019; Section 3.7), has attracted the interest of industry practitioners and academics (Huang and Xu, 2018). Indeed, academic attention paid to this phenomenon is reflected in the fact that at least seven papers reviewing research and publication have been published on this subject (Cai, Li and Knutson, 2008; Keating and Kriz, 2008; Keating *et al.*, 2015; Tse, 2015; Jin and Wang, 2016; Law *et al.*, 2016; Jørgensen, Law and King, 2017). However, in reviewing current literature it becomes apparent that little research has addressed adventure tourism undertaken by China outbound tourists (Jin *et al.*, 2019).

In relation to China's stage of economic development, Dai *et al.* (2017) argue that China outbound tourism is still at an early stage. With fast increasing disposable incomes and relaxation of visa restrictions by the Chinese government (Huang *et al.*, 2015; Jørgensen, Law and King, 2017), more Chinese tourists are predicted to be travelling abroad in the future (Huang and Xu, 2018). Chow and Murphy (2008) observed that an increasing number of Chinese tourists are no longer content with sightseeing or shopping but desire some adventurous experience. Furthermore, based on their survey, Jin *et al.* (2019) established that Chinese tourists would prefer overseas locations for these activities because they perceive greater safety and security as well as better service and higher-quality facilities than in China. Hence, Martin and Ren's (2020) research of Chinese outbound tourists in adventure tourism is timely and advances the adventure tourism literature.

However, their observations on China-directed activities by Visit Greenland, and content analysis of travel blogs on the platform Mofengwo, may contribute to a confusing picture of Chinese 'adventure tourists' for two main reasons.

First, the observed activities of Visit Greenland are mostly related to tour operators who organise package holidays for Chinese tourists, while Mofengwo is essentially a travel forum for independent travellers (Mofengwo, 2020). Although there are some similar characteristics between Chinese package tourists and independent travellers (keen to see the outside world and eager to box-tick recommended sites and activities; the importance of word-of-mouth; use of mobile technology: VisitBritain, 2019), more differences can be identified (Chen, Masiero and Hsu, 2019).

Second, as acknowledged by Martin and Ren (2020), there appears to be an insufficient understanding of 'adventure tourists' from China.

- Schneider and Vogt (2012) explain that the term 'adventure tourism' emerged in the literature in the 1990s to describe travel behaviour which engages in activities that have an element of emotional arousal, personal challenge, excitement and risk. Development of the adventure travel sector has historically served visitors from Western cultures (Gardiner and Kwek, 2017); and academic research into adventure tourism has previously concentrated on Western samples (e.g. Gilchrist *et al.*, 1995; Sung 2004). Our current literature lacks an understanding of Chinese adventure tourists. What do Chinese tourists perceive as adventure tourism? Are their views influenced by their national culture? Who influences their decision-making processes for undertaking adventure tourism activities?
- The limited literature on Chinese tourists engaged in adventure activities appears to indicate inexperience and lack of knowledge of Western-style adventure products (Buckley *et al.*, 2014).
- In their research on 'Generation Y' Chinese international students in Australia (or the 'social reform' generation: Egri and Ralston, 2004), Gardiner and Kwek (2017) recognise that a strong sense of familial obligation and distinctive cultural beliefs underpin consumer decision-making.
- Jin and Sparks (2017) argue that perceived operational, financial and physical risks, as well as a lack of perceived importance and value, are major constraints preventing Chinese outbound package tourists from pursuing special interest and adventure tourism activities.

Reflecting the above points, I agree with Martin and Ren (2020) that a more dynamic understanding of Chinese adventure tourism to Greenland is necessary. However, the design and marketing of adventure experiences to Chinese outbound tourists should not only consider their unique cultural understanding of adventure/ risks, but should also take into consideration that this is an early stage in the development of adventure tourism in China and consequently tourists lack familiarity with Western understandings of 'adventure tourism'.

10.4.2 The wider context: Greenland and the Polar Silk Road

As noted in Chapter 2, byproducts of climate warming have included commercial and geostrategic implications for marine transport and sectors such as tourism and mining. Full members of the Arctic Council and observer members (notably China, India, Japan and Italy) have acknowledged the potential opportunities and challenges such climate warming in the Arctic is offering (Assef, 2018). In January 2018, China's State Council Information Office officially released China's first Arctic white paper entitled *China's Arctic Policy* (State Council, 2018) which seeks to justify China's role as an 'important stakeholder' with vested interests in the region (Woon, 2020). To achieve its goals of involvement, China proposed incorporating the Arctic into its 'Belt and Road Initiative' (BRI) and developing a 'Polar Silk Road' (PSR) (State Council, 2018).

Ross (2009) argued that being a strong maritime power is a marker of being a global power. China is developing a comprehensive national ocean strategy to develop and protect its 'blue economy', just as other countries (e.g. USA, UK, Canada, Russia) have done (Gao, Wang and Tang, 2009; Zou, 2012; Cui and Shen, 2019). Although the policy and the Initiative have attracted widespread academic and popular comment (Erokhin, Gao and Zhang, 2018; Lim, 2018; Moscato, 2018; Pelaudeix, 2018; Lino, 2020), one must not forget that China is one of the largest oceanic nations in the world with sea areas under China's national jurisdiction; it has long considered that the twenty-first century is the ocean century, and it is no secret that China desires to become a great sea power (Mallory, 2015). Back in 1996, China drew up its own 'Agenda 21' on the environment and sustainable development, and then a specific 'Ocean Agenda 21' inspired by the 1992 Rio de Janeiro Earth Summit's 'Agenda 21' (SOA, 1996). Hence, extending its BRI to include the Arctic seems to be an obvious move.

Greenland attracts China's various interests, encompassing areas of scientific research, natural resources and related sectors such as tourism and infrastructure, although Denmark and thc USA have their reservations about such developments (Simpson, 2018) (also Section 8.3). For many Chinese people, visiting Greenland is perceived as pioneering activity (Martin and Ren, 2020) because of its physical location and difficulty to reach from China. However, this does not mean that Chinese children have no idea of Greenland. Geography is one of the key subjects for middle school and high school education in China and such education discusses not only the geography of China but also world geography (Guo *et al.*, 2018). Greenland is most likely to be remembered by many Chinese students, if only for examination purposes, as the largest island in the world. Further, Denmark is familiar to Chinese families largely because of its famous writer Hans Christian Andersen (although the Little Mermaid and Legoland are also known).

> Andersen has remained one of the rather few Western writers to maintain and even increase his popularity among Chinese readers to such degree that today, many Chinese may actually have a better knowledge of Andersen than most of his compatriots
>
> (Eric Messerchmidt[2] cited in Li and Huang, 2017).

Greenland's current access difficulties will be ameliorated, and although the China Communications Construction Company (CCCC) eventually withdrew its bid to assist in the expansion of Greenland's airports (Section 8.3), the high degree of Chinese interest certainly stimulated Denmark and the USA into solidifying their support (Gronholt-Pedersen, Qiu and Heavens, 2019; Humpert, 2020), enabling the Nordic Investment Bank (NIB) to provide loans for Kalaallit Airports International (WCN Editorial Team, 2020).

Against the above background, in consideration of the increasing strategic and economic value of Greenland and its natural beauty, China and its consumers are likely to be further engaged in the future development of Greenland. As Blaxekjær, Lanteigne and Shi (2018: 446) argue '... as Greenland continues to debate the possibility of greater autonomy, or even independence from Denmark, China will likely be a major factor in its expanding foreign policy and trade interests'.

10.5 In the next chapter

In Chapter 11, a second major strand of Greenland's tourism development and a third element of tourism's interaction with climate change and 'the ice' is evaluated: cruise tourism. While cruise tourism delivers a significant proportion of all foreign tourists arriving in Greenland, its impacts and dependency dimensions set it apart as a focus of analysis. At the same time, cruise tourism's interaction with elements of climate change – offshore views of glaciers, the blocking of passage by calving icebergs, or the onshore delivery of significant numbers of 'gazers' to Ilulissat Icefjord – provides a range of experiences and opportunities that tourism and environmental planners will need to examine carefully along Greenland's path towards independence and greater accessibility with completion of the airport upgrading programme.

Notes

1 The onset of Covid-19 travel restrictions generated an important inward focus on the (potential) geopolitical role of domestic tourism in reinforcing and enhancing a familiarity and pride in one's own country: a particularly significant role in such a spatially disparate territory as Greenland (Chapters 8 and 12).

2 Director of the Danish Cultural Centre in China.

11

Paradoxes of Cruise Tourism to Greenland

'The only remaining role of real places is fulfilling the few tourist wishes unattainable on the boat' (Supasudaca *et al.*, 2020: 374)

11.1 Introduction

This chapter attempts to evaluate the nature and role of cruise tourism in contributing to Greenland's development and imagery, including the irony of its ships facilitating mass visitor gaze at some of the consequences of Arctic warming to which those very cruise ships may be contributing.

Superficially, Greenland would appear to lend itself ideally to cruise tourism: all settlements and many attractions are adjacent to water, either on the coast or on the shores of fjords, and there is no overland transport. Prime examples of Greenland's major attraction – the ice – can be viewed comfortably as cruise vessels pass the tongues of retreating glaciers, manoeuvre their paths between calving icebergs and linger off unpredictable icefjords, generating multiple images of 'natural' Greenland.

On the other hand, there are few places where large cruise ships can tie up to dry land – Nuuk and Sisimiut being exceptions with adequate berthing space and depth – such that much of the movement of cruise tourists onto shore is accomplished by tender or zodiac (a small inflatable boat), a potentially slow and uncomfortable process. Indeed, deliberate use of the word 'comfort' highlights the paradox of significant proportions of (Western) cruise travellers being elderly and less mobile visiting a destination marketing itself in terms of 'adventure' and often offering challenging environmental circumstances. On the other hand, at least one interpretation of 'last chance' tourism (Section 3.2) possesses a certain resonance for such a market.

In terms of crude numbers, cruise tourism may appear important to Greenland, representing perhaps a third of all foreign visitors (Chapter 7). And yet, while facilitating 'comfortable' viewing and digital capturing of 'the ice' from a safe distance offshore (Chapter 9), it may exert considerable (if individually short-lived) seasonal onshore crowding impacts, while generating often little positive economic benefit (berthing fees aside), and at the same time contributing to both marine and atmospheric pollution through ships' emissions (Section 11.6).

11.2 The global phenomenon

Cruise ship tourism can be crudely divided into niche and mass forms, both of which are experienced in and around Greenland. The difference between these two cruise cultures, notably in tourist expectations and behaviour, and in their inward and outward environments, has long been recognised (e.g. Foster, 1998) alongside the obvious price differential involved (Petrick, 2005).

DOI: 10.1079/9781789246728.0011

Niche cruise tourism, often marketed as an 'expedition' and employing the rhetoric of 'eco-tourism', employs smaller vessels at higher cost, carrying a large crew and staff relative to passenger numbers. Much of the passenger experience hinges on the opportunity to quickly and efficiently land in limited numbers by zodiacs in remote locations and small harbours. A key characteristic of the small cruise ship is that while calling at remoter communities and natural sites that other tourists may not reach, the relatively small numbers of disembarked visitors and the usually relatively short visiting time involved may act to limit both positive and negative local impacts (Thomson and Thomson, 2006).

'Expedition' cruising developed from the 1960s, and was initially characterised by Antarctic cruises combining brief shore visits with environmental and historical instruction (Mason and Legg, 1999; Splettstoesser, 2000; Stewart, Draper and Dawson, 2010). To some extent this category has merged with 'adventure' cruises where, in some cases, participants may camp overnight onshore and pursue such activities as sea kayaking and mountaineering (Berger, 2010), and represent a younger, or at least fitter, clientele.

By contrast, the mass market nature of larger ships, longer disembarking and embarking times and the physical limits to where they can dock, generates a level of self-sufficiency and individual inertia, such that for many travellers the major element of the cruise experience is provided within the ship itself, with destinations often only a secondary consideration (Wood, 2000; Weaver, 2004, 2006; Yarnal and Kerstetter, 2005). Because of the relatively high level of attention and care on board, increasing numbers of elderly people had been opting to travel virtually full-time on such cruises, pre-Covid. For some, serial cruising appears to have offered an attractive alternative to nursing or care homes (Lindquist and Golub, 2004). The onboard presence of this phenomenon also lowers the potential proportion of passengers wishing to disembark at each port of call.

Northern itineraries have been becoming more attractive as cruise operators have faced restrictions at congested southern honeypots such as Venice and Dubrovnik (Corcoran, 2017; Coldwell, 2017). Global growth in the cruise market has placed pressure on operators to seek out newer, exciting destinations, such as Greenland.

Annual cruise ship passenger numbers worldwide rose from 11mn a year in 2009 to 29.7mn in 2019 (CLIA, 2020). In pre-Covid 2019, an estimated 124 new cruise ships – with a capacity of 5,000 passengers or more each – were reported to be under construction or due to be launched in the following few years (Connolly, 2019). In 2020, however, global impacts of the pandemic forced a number of cruise companies out of business (e.g. Cruise Adviser, 2020).

In its strategy towards cruise tourism, Visit Greenland (2016b: 28) recognised the binary nature of the scale and visitor expectations of the cruise sector. Marketing efforts directed at the larger cruise lines have been primarily focused on maintaining and creating contacts, supporting route planning, providing information on navigation rules and other framework conditions, as well as participating in cruise travel trade markets and exhibitions in the USA and Europe. For smaller expedition cruise lines, Visit Greenland's collaboration has encompassed content development, marketing, product development and training, including visits and workshop participation.

As warming proceeds at twice the global average speed to threaten Greenland's unique character, raising the island's high priority 'bucket list' destination status, the already rapidly growing global cruise tourism phenomenon in polar environments was intensifying before the pandemic.

11.3 Cruising, climate change and Greenland

Greenland's cruise and marine tourism both mediates between visitors and the ice, and through pre-cruise literature and onboard information may inform and thereby mediate in the cultural construction of the country's climate change imagery. This can be reinforced by:

- delivering visitors to and around the Ilulissat Icefjord with a local guide;
- generating small-boat excursions whereby visitors gaze at the Icefjord and icebergs (as well as Arctic maritime wildlife), while perhaps engaging with both local boatmen/women and the surrounding marine fauna;

- viewing melting tongues of retreating glaciers when passing through Prins Christian Sund, along coasts and fjords;
- being blocked from entering fjords and harbours by groups of giant icebergs.

Thus, the consequences of climate change can act as both a driving force in the growth of cruise tourism to and around Greenland, and as a constraining factor in the production of Arctic maritime obstacles that can intercede to deny visitor experiences.

In 2016 the Greenland government launched its 'What it Takes' strategy, focusing on creating more favourable conditions for tourism development. This included a change in the structure of cruise ship fees, improving landing infrastructure for cruise ships, promoting private investments and supporting the start-up of visitor centres (Visit Greenland, 2020e: 8). Also in 2016, Visit Greenland (2016b: 28) set a growth target for cruise passenger numbers arriving in Greenland at 3.5 per cent per annum. Having levelled out over the previous few years, following successive years of decline in the wake of the global economic depression, such numbers did in fact almost double between 2016 and 2019, although the growth between 2018 and 2019 was a modest 1.9 per cent (Fig. 11.1, Table 11.1). However, when the numbers of passengers on each port call are totalled, the 2019 figures actually represent an increase of 10.6 per cent (Table 11.3) (in ports where calls are registered by Statistics Greenland).

'Cruise passenger numbers' can therefore represent a crude and blunt instrument for assessing the actual numbers of cruise tourists disembarking at any given port of call and the impacts – both positive and negative – they may exert. As noted above, there may be categories of 'mass' cruise passengers who may disembark only rarely, perhaps when the ship is actually able to dock at a destination, or may not leave the ship at all.

While the impact of cruise tourism around the world can be partly mitigated by the diversified economies of large port cities and the multiple attractions of their hinterlands, given the small scale of settlements and physically restricted areas of movement in Greenland, the island does not easily fit preconceived models relating to the logistics and onshore impacts of cruise visitation. Yet those impacts, assuming they continue on an upward post-Covid trajectory, are likely to contribute a significant element to Greenland's social, political, economic and environmental transformations: transformations that are inextricably intertwined with climate change.

As indicated in Table 11.1, for the smallest ('expeditionary') class of vessel (up to 250 passengers), while ship numbers dipped and then

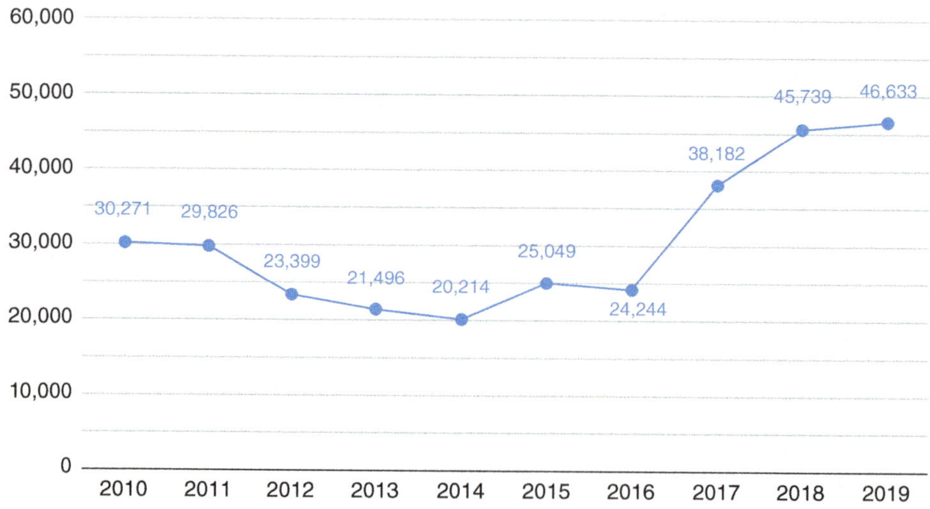

Fig. 11.1. Cruise passengers to Greenland, 2010–2019. Sources: Lumholt (for Visit Greenland), 2020: 12, after https://bank.stat.gl/tuekrp. Figure courtesy of Statistics Greenland and Mads Lumholt.

Table 11.1. Greenland: numbers of cruise passengers and ships by size of ship, 2016–2019. Author's own table derived from an amalgamation of Naatsorsueqqissaartarfik, 2018: Table 2; 2019: Table 2; 2020: Table 2; and the author's additional calculations.

Size of ship (passengers)	2016		2017		2018		2019	
	Passengers	Ships	Passengers	Ships	Passengers	Ships	Passengers	Ships
1–250	5,302	69	6,542	54	6,395	62	7,462	69
251–500	2,699	14	4,715	17	7,111	27	6,239	24
501–1200	10,938	17	11,962	20	9,567	16	10,649	19
1200+	6,466	4	14,963	10	22,666	14	22,283	13
Totals	25,405*	104	38,182	101	45,739	119	46,633	125
Annual % increase	11.3	13.5	33.5	–3.0	16.5	15.1	1.9	4.8

*The official figure for 2016 – when registration procedures changed – is higher than that in Fig. 11.1 and in Table 11.5, also derived from official sources.

recovered over the 2016–2019 period, passenger numbers from 2016 to 2019 increased by 40 per cent within the same number of ships, suggesting an overall increase in size of this type of vessel visiting Greenland. On the other hand, the overall proportion of cruise passengers travelling in this category, who are likely to be more environmentally conscious, actually declined from 20.1 per cent in 2016 to 16.0 per cent in 2019. This reflected the fact that for the three largest vessel classes there was a growth both in ship and passenger numbers, with the largest vessels providing the greatest increase in numbers of visitors: those with more than 1,200 passengers raising their proportion of total passengers from 25.5 per cent in 2016 to 47.8 per cent in 2019, and as a proportion of total vessels from 3.8 per cent to 10.4 per cent. As can be seen in Table 11.1, this meant that in 2019 the thirteen ships in the largest size group carried almost three times the total number of passengers in the 69 vessels of the smallest size group, and that the 32 ships with a capacity of more than 500 passengers carried 70.1 per cent of all cruise passengers to Greenland: clearly a predominantly 'mass' experience.

11.4 Greenland port calls

The number of port calls for the three largest vessel groups rose from 136 in 2015 to 230 in 2019, representing a growth of 69 per cent. Table 11.2 indicates the total number of port calls for the period 2015–2019. The striking trend that emerges from comparing Table 11.2 with Table 11.1 is not only the growing proportion of 'mass' cruise tourists, but the concentration of them in a small number of ports. For 2016 four ships in the largest category called at a total of just seven ports; in 2018, fourteen ships made just 16 port calls; and in 2019 13 vessels made a total of just 20 port calls.[1] Table 11.3 expresses the result of this logistical/temporal concentration in spatial terms: in 2019 just 5 ports out of the 19 registered – Qaqortoq, Nuuk, Ilulissat, Nanortalik and Sisimiut – were host to vessels with 69.0 per cent of all the passengers in cruise ships visiting Greenland; and the most favoured port Qaqortoq, 'capital of the South', alone hosted vessels with 23.1 per cent of total cruise passengers (Fig. 11.2).

Table 11.2. Greenland: numbers of cruise ship port calls, 2015–2019. Author's own table derived from Lumholt (for Visit Greenland), 2020: 13.

Year Size of ship (passengers)	2015	2016	2017	2018	2019
1–250	191	169	151	186	167
251–500	81	78	84	133	117
501–1200	48	67	61	74	93
1200+	7	7	8	16	20
Total number of port calls	327	321	304	409	397
Of which, calls at uninhabited locations	7	9	15	19	1

Further, the 2019 figures for cruise ship passenger arrivals by month (Table 11.4) show a

Fig. 11.2a, 11.2b. Interior (a) and exterior (b) views of Qaqortoq tourist information centre, one of the largest in Greenland. Author's own photographs.

clear shift towards a seasonal concentration, particularly of passenger numbers, with larger vessels dominating arrivals in August – the month with the lowest risk of disruption from weather and water conditions – imposing more obvious mass impacts on port destinations. By contrast, only smaller vessels, with a considerably lower passenger-to-ship ratio, visited in the 'shoulder' months of May, June and October, thereby helping to diffuse impacts temporally as well as spatially, to less visited locations, and with smaller numbers of cruise visitors.

The theme to therefore emerge from interrogating these official figures would seem to be that pre-Covid cruise tourism to Greenland was adopting growing mass tourism characteristics in terms of (i) larger ships, (ii) a greater proportion of passengers arriving in the largest ships, (iii) a spatial concentration whereby just five ports hosted ships with more than two-thirds of all cruise passengers and (iv) a temporal concentration whereby in 2019, 97 of 125 vessels (including 11 of the largest 13 ships), or 78 per cent of the total, visited within the two months of August and September.

Destination ports and governments may often be pressured by cruise companies to keep their fees low (Klein, 2010). On 1 January 2016, as noted earlier (Section 11.3), a new cruise port charge structure was introduced in Greenland which meant that rather than cruise companies paying DKK525 (€68.25/£63/$84) per passenger once, each port call would now entail a port charge of DKK1.1 (€0.14/£0.13/$0.18) per gross tonnage. This made it significantly cheaper for especially large ships to call in Greenland, and has in some cases provided savings of up to 93 per cent in port charges (Lumholt, 2020: 12). This appears to have been a significant contributory factor in the trend towards 'mass' cruise tourism noted above. But was it intentional?

The calving of icebergs from melting ice sheets can provide a spectacular experience, cruise ships often being able to provide their travellers with grandstand views of such processes. But those same icebergs can also act as a physical barrier to accessing other attractions and destinations within the country. For example, for the first half of the summer season in 2019, passage through the impressive Prins Christian Sund (Ikerasagssuaq) in the extreme south of the country – which affords magnificent views of glacier tongues both marine-terminating and retreating – was blocked for large vessels, while much further north, the town of Uummannaq was rendered inaccessible as large icebergs prevented access into its harbour from 'Umanak (*sic*) Fjord' (Fig. 11.3; Table 11.3).

While for cruise passengers, confrontation with such phenomena 'plays into the narrative of the Arctic as a remote and challenging place' (Bislev and Smed, 2018: 36), onshore disruption caused by the late cancellation of ship and passenger visits in such circumstances (icebergs/weather or other unanticipated scheduling problems) may result in local operators and guides being left with unrecoverable expenses, and local residents and traders facing disappointment and frustration. With the likelihood of intensified iceberg calving from glacier fronts in coming years, and an increase in difficult sea conditions with more storm surges following the diminution of sea ice (Lamers and Amelung, 2010), such

Table 11.3. Greenland: numbers of cruise passengers by port, 2016–2019. Author's own table derived from an amalgamation of Naatsorsueqqissaartarfik, 2018: Table 3; 2019: Table 1; 2020: Table 3; and the author's additional calculations.

Year		2016	2017	Change 2016 to 2017		2018	Change 2017 to 2018		2019	Change 2018 to 2019	
Region/port	Approx. population size			Nos.	%		Nos.	%		Nos.	%
North											
Aasiaat	3,100	963	1,072	109	11.3	1,133	61	+5.7	536	–597	–52.7
Ilulissat	4,700	12,577	15,560	283	2.3	12,914	–2,646	–17.0	14,762	1,848	+14.3
Qasigiannguit	1,150	449	739	290	64.6	693	–46	–6.2	529	–164	–23,7
Qeqertarsuaq	850	3,117	6,154	3,037	97.4	4,336	–1,818	–29.5	4,895	159	+3.7
Qaanaaq	650	396	507	111	28.0	713	206	+40.6	795	82	+11.5
Upernavik	1,150	2,000	903	–1,097	–54.9	812	–91	–10.1	595	–217	–26.7
Uummannaq	1,300	3,579	2,618	–961	–26.9	4,747	2,129	+81.3	2,717	–2,030	–42.8
Arctic Circle											
Kangerlussuaq	500	5,939	9,384	3,445	58.0	8,904	–480	–5.1	7,953	–951	–10.7
Maniitsoq	2,650	1,460	80	–1,380	–94.5	717	637	+79.6	1,736	1,019	+142.1
Sisimiut	5,600	10,657	11,686	1,029	9.7	13,660	1,974	+16.9	11,629	–2,031	–14.9
Capital region											
Nuuk	18,500	10,849	14,975	4,126	38.0	14,683	–292	–2.0	18,172	3,489	+23.8
Paamiut	1,500	2,961	3,327	366	12.4	2,255	–1,072	–32.2	5,853	3,598	+160.0
South											
Nanortalik	1,300	5,879	5,963	84	14.3	7,719	1,756	+29.4	11,666	3,947	+51.1
Narsaq	1,500	2,601	2,772	171	6.6	550	–2,222	–80.2	2,705	2,155	+391.8
Narsarsuaq	130	2,205	2,882	677	30.7	2,281	-602	–20.9	2,968	687	+30.1
Qaqortoq	3,200	12,086	23,837	11,751	897.2	28,025	4,188	+17.6	28,214	189	+0.67
Qassiarsuk	40	–	586	–	–	597	11	+1.9	630	33	+5.5
East											
Ittoqqortoormiit	450	1,916	1,649	–267	–13.9	1,754	105	+6.4	2,860	1,106	+63.1
Tasiilaq	2,000	3,971	5,302	1,331	33.5	4,074	–1,228	–23.2	3,099	–975	–23.9
Totals		**83,605**	**109,996**	**26,391**	**+31.6**	**110,567**	**571**	**+0.5**	**122,314**	**11,747**	**+10.6**

Table 11.4. Greenland: seasonal distribution of cruise ships by passenger capacity, 2017–2019. Author's own table derived from an amalgamation of Naatsorsueqqissaartarfik, 2019: Table 1; 2020: Table 1; http://bank.stat.gl/TUEKRK.

Ship Capacity	Total number of ships			May			June			July			August			September			October		
Year: 20	17	18	19	17	18	19	17	18	19	17	18	19	17	18	19	17	18	19	17	18	19
1–250	54	62	66	2	2	5	2	4	4	7	8	5	21	23	26	20	23	27	2	2	2
251–500	17	27	23	–	–	–	2	3	–	–	5	6	9	14	13	6	5	5	–	–	–
501–1200	20	16	18	–	–	–	–	1	–	7	4	4	9	8	11	4	2	4	–	1	–
1200+	14	14	13	–	–	–	–	–	–	2	4	2	3	4	7	5	6	4	–	–	–
Totals	**101**	**119**	**125**	**2**	**2**	**5**	**4**	**8**	**4**	**16**	**21**	**17**	**42**	**49**	**57**	**35**	**36**	**40**	**2**	**3**	**2**

occurrences, and the disruption and uncertainty they can cause for local communities, are likely to increase as a result of climate change.

Thus statistics noting annual change in the arrivals of cruise passengers at individual ports may reflect a number of environmental, economic and structural factors. While the significant increase in numbers at Uummannaq for 2018 and the equally significant reduction for 2019 may reflect variations in accessibility to the port because of its approaches' susceptibility to large icebergs (Fig. 11.3), other changes, such as the reduction in Sisimiut's arrival numbers, may represent apparent arbitrary route decisions made by cruise company schedulers.

Longer-term deliberate decisions over whether to include a port in a ship's itinerary can reflect a range of economic, environmental and altruistic factors varying from the costs of port facilities and provisioning, the level of profit derived from selling onshore excursions, to passenger feedback from previous visits. They highlight elements of the asymmetrical nature of power relations between cruise companies and land-based communities. In particular, the instability of such figures as exhibited in Table 11.3, and the uncertainty they engender, make it difficult for local tourism planners, entrepreneurs and communities in a destination such as Narsaq – where passenger arrivals declined by 80 per cent in 2018 then increased by 390 per cent in the following year – to plan ahead with any confidence.

In terms of regional change, and only partly confirming the trend of spatial concentration noted above, 2019 saw a significant growth in the number of passengers calling in the ports of the Capital Region (41.8 per cent) and in South Greenland (17.9 per cent). By contrast, a decline in numbers was experienced in North Greenland (–2.1 per cent) and in the Arctic Circle Region (–8.4 per cent). East Greenland showed virtually no change (Lumholt, 2020: 14).

What is instructive from Table 11.3 is the comparison of port towns' populations with the *potential* cruise passenger numbers coming ashore there. By global standards, all Greenland's port destinations are small, with the capital Nuuk, by far the country's largest settlement, having a population of just over 18,500. And while the country's main tourist centre, Ilulissat, with a population of around 4,500, received cruise passenger numbers in 2019 more than three times its resident population size, Kangerlussuaq (population less than 500), currently Greenland's main air transit point (see Chapter 8), hosted cruise visitor numbers some 16 times the magnitude of its resident population, while roughly the same disproportion had the opportunity to descend on Narsarsuaq (population less than 200). Kangerlussuaq's numbers are inflated by its role as a start and end point for some cruises, with passengers flying in or out as appropriate. Passenger numbers passing through Narsarsuaq's airport will also have influenced the figures there.

As port charges were changed, so also, in January 2016, the registration of cruise guests by nationality began, although the process was initially far from perfect, as the large numbers included in the unregistered category in Table 11.5 indicate.

While cruise passenger numbers from Germany have dominated the figures for some time, in 2019, a 21 per cent increase in US citizens and a 4.6 per cent decline in German travellers narrowed the gap between the two groups to just 3.2 per cent. At the same time 2019 saw the greatest decline in numbers from the UK (–22.9 per cent) and

Fig. 11.3. Large icebergs obstructing the seaward entry to Uummannaq (August, 2019). Author's own photograph

Australia (–51.1 per cent). The number of passengers from Canada, by contrast, represented a growth of 35.3 per cent, and from Switzerland 11.5 per cent (Lumholt, 2020: 15). The major, fairly obvious, conclusion to be drawn from these figures is that the pre-Covid North American market was gaining in strength, in 2019 representing just over a third of the total registered nationalities. Numbers of cruise tourists from most registered European countries – with the notable exception of the UK – appeared to be showing steady, if largely unspectacular growth, while, for the 2016–2019 period at least, the Chinese market looked, paradoxically, reasonably static if a little erratic, this contradiction presumably reflecting the relatively small numbers involved.

But, as with so many Greenland tourism statistics, it bears repeating that because overall numbers are relatively small, small changes can appear to produce a disproportionate effect.

11.5 Dependenc(i)es

Twenty years ago, Johnson (2002) was pointing out the disparities in cruise tourism destinations' control of, and interface with, cruise companies, suggesting that the externality costs of this sector fell unequally on smaller, less developed destinations.

Supasudaca *et al.* (2020: 363–364) warned that while cruise tourism can act to add economic value to local cultural heritage, there is always the danger that because of its monopolistic position and arbitrary exogenous decision-making, the cruise industry could fake, side-step or totally abandon local cultures and economies. The pressures of cruise lines' requirements can thus distort local cultural representation through commodification and re-interpretation, in a manner that may resemble the practices of colonial times, a theme all-too sensitive in the Greenlandic context (Section 4.3). Supasudaca *et al.* therefore postulated that the current model of cruise tourism can contain a pattern of development with trajectories that are inherently difficult to reverse, contributing to path dependencies (e.g. Sorensen, 2015). These dependent paths may have been established in colonial times and become entrenched in subsequent patterns of development, whereby cruise tourism destinations may be perpetuating and reinforcing spatial, economic and even political imbalances deriving from colonial experiences.

Table 11.5. Greenland: numbers of cruise passengers by nationality, 2016–2019. Author's own table derived from an amalgamation of Naatsorsueqqissaartarfik, 2018 Figure 2; 2019: Figure 2; 2020: Figure 2; Lumholt (for Visit Greenland), 2020: 15; http://bank.stat.gl/TUEKRL; and the author's additional calculations.

Year	2016 (official figures subject to revision)		2017		2018		2019	
Nationality	No.	% of total	No.	% of total	No.	% of total	No.	% of total
Germany	7,084	29.2	13,418	35.1	14,189	31.0	13,658	29.3
USA	1,309	5.4	9,543	25.0	11,074	24.2	13,235	28.4
UK	2,051	8.5	3,272	8.6	5,096	11.1	3,925	8.4
Canada	1,043	4.3	2,153	5.6	2,211	4.8	2,989	6.4
France	574	2.4	1,504	3.9	1,907	4.2	2,011	4.3
Australia	485	2.0	1,230	3.2	3,591	7.9	1,756	3.8
Switzerland	533	2.2	1,080	2.8	1,093	2.4	1,241	2.7
Denmark	357	1.5	540	1.4	505	1.1	738	1.6
Netherlands	122	0.5	422	1.1	660	1.4	707	1.5
Austria	256	1.1	586	1.5	467	1.0	671	1.4
China	638	2.6	690	1.8	411	0.9	651	1.4
Belgium	108	0.4	681	1.8	539	1.2	445	1.0
Italy	86	0.4	82	0.2	119	0.3	366	0.8
Norway	121	0.5	194	0.5	230	0.5	234	0.5
Spain	40	0.2	68	0.2	101	0.2	165	0.4
Russia	157	0.6	46	0.1	71	0.2	145	0.3
Japan	97	0.4	58	0.2	23	0.1	81	0.2
Other countries	748	3.1	1,078	2.8	1,930	4.2	1,981	4.2
Unregistered	8,435	34.8	1,537	4.0	1,522	3.3	1,715	3.7
Totals	**24,244**	**100***	**38,182**	**100***	**45,739**	**100***	**46,633**	**100***

*The sum of percentages does not equal exactly 100 due to rounding.

How applicable might path dependency conceptualisation (and/or its variations such as path plasticity) be to contemporary, and indeed future, Greenlandic experience?

Although relatively little work appears to have been undertaken in this area, James, Olsen and Karlsdóttir (2020) examined stakeholders' perceptions of the sustainability of cruise tourism in the southern Greenland cruise port town of Qaqortoq[2] (Table 11.3). As noted in Section 11.4, in 2019 Qaqortoq attracted almost a quarter of all cruise ships visiting Greenland.

In investigating how stakeholders are managing the development of the sector in the context of imbalances in power between their place-bound local role and global cruise lines, the researchers found that such stakeholders conceptualised relations in terms of trade-offs and competition, but lacked an holistic perspective and concern about economic dependence on the cruise companies. The researchers concluded that intergovernmental agreements are needed to address regulatory issues and that national co-ordination could help to improve collaboration between destinations (and, implicitly, strengthen bargaining power) (James, Olsen and Karlsdóttir, 2020).

As a vertical 'industry', tourism differs from fishing or mining, with many more business, economic and cultural sectors affected. This can result in not simply monopolistic control of one product but of a concentration and deformation of the market – both sectorally and geographically defined – by a small number of global companies. In this way historical patterns of dependency can return in new forms while unequal power relations are sustained, with major economic decisions remaining in the hands of exogenous actors, whose prime interests may not correspond to local agendas (Supasudaca *et al.*, 2020: 368–369).

In sustaining path dependencies, what may appear as dynamic processes could actually consolidate into a static relationship, rendering change increasingly difficult (Supasudaca *et al.*, 2020: 377). However, long periods of path dependent institutional stability and reproduction can be punctuated by brief phases of institutional flux – referred to as critical junctures – during which more dramatic change is possible (Capoccia and Keleman, 2007: 341). Thus path dependency can have both enabling and constraining effects (Brouder, 2014a, 2014b; Brouder *et al.*, 2016).

Changes in orientation creating new paths might be stimulated through formal or informal policy measures, grass-roots collaboration and innovation-led mechanisms (e.g. Chimirri, 2020), that could, however, lock the destination into a new dependency (Sanz-Ibáñez, Wilson and Clavé, 2016).

Path plasticity (Halkier and Therkelsen, 2013; Strambach and Halkier, 2013) recognises the importance of (implicitly endogenous?) forms of knowledge not seen as central to the emerging knowledge economy. It provides opportunities for creativity and policy entrepreneurship, highlighting that some options are more likely to succeed than others. It challenges policy-making by combining different learning processes, resources from different sectors, and reaches outside the geographical area. Thus path plasticity is a more fluid, flexible notion which might imply an increased level of resilience to path dependency and future negative path shaping effects. But its influences may be smaller and slow.

So, in Greenland (Chimirri, 2020) it is important to identify the brokers in knowledge networks who may contribute to a (cruise tourism) destination's path plasticity, by, for example, offering the opportunity for new activities, excursions and co-operative relationships with other destinations. The latter course would, however, appear to be limited for some time to come, given Greenland's often large distances between port destinations and their lack of landward connections. At the present time, such co-ordination would necessitate building on relatively slow water connections (or more expensive air links), between, for example, Qaqortoq, Narsaq, Narsarsuaq and Qassiarsuk, which are connected by a summer coastal ferry service.

11.6 Cruise tourism impacts

11.6.1 Onshore impacts

As one of the fastest growing global economic sub-sectors, cruise ship tourism and its impacts have resulted in cultural and spatial transformations globally (Hein, 2016; Supasudaca *et al.*, 2020). Despite the voluminous academic literature on such impacts and cruise tourists' onshore perceptions and behaviour, the appropriateness of such literature to the Greenland situation is severely limited. Debates on cultural impact and heritage consequences, for example, are meagre (Hein, 2016; Epler Wood, 2017). And in a post-Covid-19 world, how many of our concepts and practices will be relevant? How far can many of the received wisdoms noted below be sustained where social distancing, facial masking and other personal and collective risk avoidance constraints may need to persist to varying extents? Not least is the consideration that 'Cruise ships are often settings for outbreaks of infectious diseases because of their closed environment, contact between travelers from many countries, and crew transfers between ships' (Moriarty *et al.*, 2020: 347). The literature on tourist crowding for example, as applied to cruise visitors (e.g. Sanz-Blas, Buzova and Schlesinger, 2019) draws on earlier concepts (Machleit, Eroglu, and Mantel, 2000; Kim, Lee and Sirgy, 2016) that distinguish between human crowding ('full of people') and spatial crowdedness ('restricted movement'). Will such hitherto somewhat blurry distinctions become sharply clarified for tourism planning and management in the post-Covid traveller's world in which there is likely to be an increased awareness of 'space' (Galvani, Lew and Sotelo Perez, 2020)? Will 'adventurous' tourists become more risk averse?

The Association of Arctic Expedition Cruise Operators (AECO) has been working to develop site-specific guidelines/codes of conduct for cruise visitors, following the production of community guidelines in Svalbard, where AECO is based. A local guide/code has been developed for Sisimiut in collaboration with Arctic Circle Business and the local museum, and this was to be followed by a similar guide for Nuuk (Ren, Chimirri and Abildgaard, 2020).

In looking at the dispersal of onshore cruise tourists away from congested honeypots – 'in the case of cruise tourism, some sustainable concepts will be less exact and create trade-offs' (UNWTO, 2016: 59) – Mangano and Ugolini (2020) argue that if the cruise industry's aspirations for responsible tourism (CLIA, 2020) are to be realised, the interaction between cruise companies and local stakeholders is essential to secure a diversification of onshore excursions and a programme of staggered arrivals and departures for the cruise ships themselves (Fig. 11.4). Short excursions and/or limited free time ashore constrain the development of any deep and intimate connection between visitors and hosts that could otherwise generate a unique experience.

For cruise visitors to better engage both economically and socially with host societies, the need for sufficient length of onshore stays would appear necessary. The inclusion of land-based overnight visits would substantially increase their value for local economies and provide greater opportunity to interact (Mangano and Ugolini, 2020). It would also raise costs for visitors and complicate cruise lines' logistics, not least in extending potential economic leakages from the almost ring-fenced nature of discretionary cruise tourist spending. Taking meals onshore is perhaps the next best option both in terms of generating local income and, literally, tasting local (food) culture (Domenèch and Gutiérrez, 2019).

Cruise tourism impacts depend upon how such visits are prepared, carried out and how well informed and aware visitors are of the ecological and socio-cultural vulnerability of the areas visited (e.g. Walker and Moscardo, 2002). In Greenland, while the benefits of cruise tourism visits to local communities appear to have been limited and not well documented, the timings of cruise passenger arrivals may not always be well communicated for onshore advantage, a characteristic reported from other Arctic destination regions (Kaae, 2002, 2006; Thomson and Thomson, 2006) (Figs. 11.5 and 11.6). In a somewhat rudimentary form, the information board for local people illustrated in Fig. 11.7, located in the central area of Nanotalik town, is perhaps representative. It indicates that over a period of three days in August 2019, four cruise ships were arriving with a total number of passengers of just over 40,000, who, on average, could spend a maximum potential time onshore of just under seven hours, albeit reduced with the bottleneck process of tendering to and from each ship.

Apart from Jaakson's (2004) conceptual recognition of the cruise tourist 'bubble' in relation to landside activities – not to mention Curson's (2009) pejorative description of 'pack behaviour ... producing more than a ripple of unease to run through the local population' (cited in Klein, 2010: 66) – relatively little conceptually framed detailed empirical research on cruise tourists' impacts on, and interactions

Fig. 11.4. Cruise liners *Pacific Princess* (680 passengers) right, and *Marco Polo* (848 passengers) left, moored off Nanortalik harbour. Author's own photograph.

Fig. 11.5. One of only two stallholders in Qeqertarsuaq (Disko Island) selling local items during an early Sunday morning visit of cruise line passengers. Author's own photograph.

Fig. 11.6. Cruise line passengers queuing in Qeqertarsuaq (Disko Island) for tenders to return them to their ship (MS *Marco Polo*) on a Sunday morning just as the town's museum, left, is opening for the day. Author's own photograph.

Fig. 11.7. Information board indicating the timings and potential numbers of cruise ship passenger arrivals, Nanortalik. Author's own photograph.

with, 'host' populations appears to have been published (Supasudaca *et al.*, 2020); and certainly little that might appear directly relevant to the Greenland context.

Equally, tourism policy-makers and planners have tended to overlook the culture and heritage values of the local population in such interactions, the dominant discourse often adopting a short-term perspective focused on tourist arrivals' numbers and expenditure levels, 'leaving aside local actors, their agendas ... and their specific identity concerns' (Supasudaca *et al.*, 2020: 365).

Because cruise tourism has a different pattern to land-based tourism, with large numbers of visitors making short-term visits, it can exert a disproportionate impact on more remote coastal communities and archaeological sites. While some Arctic communities, such as Kodiak and Tenakee Springs in Alaska, and at heritage sites in Canada, have expressed opposition to cruise arrivals (Klein, 2010), little such opposition has been reported from destinations in Greenland.

The wider constraints on visiting Greenland previously discussed (Chapter 7), such as cost and accessibility, have prevented a descent into conventional mass tourism, a circumstance which, in its post-Covid marketing, Visit Greenland now emphasises as a positive attraction (e.g. Por, 2020). This seriously overlooks, however, the growing neo-mass character of much of the country's cruise ship tourism.

The relative compactness of, and lack of road connections between, Greenland's towns and villages, provides the ideal context for extreme examples of the cruise-line tourist 'bubble' (Jaakson, 2004) phenomenon and its impacts, whereby visitors' onshore activities are spatially concentrated within a relatively small area.

Greenland often looks to Iceland as a model for tourism development, and that country's 'Golden Circle' day tour from Reykjavik, taking in the impressive Gullfoss waterfalls, the excitement of spouting Geysir and the geologically and historically important site of Þingvellir, has no obvious equivalent in Greenland, if only because no major inland circular routes are yet possible.

Apart from walking routes, to, for example, the Ilulissat Icefjord from Ilulissat town (Section 9.5) or to the 'Valley of the Winds' (Blæsedalen) from Qeqertarsuaq – both essentially linear there-and-back treks[3] – or just exploring urban Nuuk, Greenland 'onshore' excursions tend to be expensive because they require small boats or flights (e.g., to and/or over, the ice sheet), and the cruise line will be extracting its 'facilitating' fee.

The World Travel and Tourism Council (WTTC, 2004) has argued that destinations have the opportunity to 'convert' cruise visitors into returning longer-stay tourists. But the theory of conversion seems to contradict the pre-Covid trend of tourists actually spending less time onshore, thereby diminishing the chances of the destination promoting itself. Further, this limited window of opportunity has been exacerbated by cruise companies exerting increasing control over onshore excursions (Supasudaca *et al.*, 2020: 370) in their asymmetrical power relationships with destinations.

As some cruise ships have become larger and more self-sufficient, it has been suggested that they even call into question the relevance of destinations, whereby the ship itself becomes the destination (Wood, 2000; also Rodrigue and Notteboom, 2013). This would seem to apply no less when tourists can view icebergs, witness multiple examples of glacier tongues and obtain good views of ice and rock landscapes from the comfort of a lounge or sun-deck on the ship: thereby further marginalising/eliminating the human presence in the object of the gaze.

11.6.2 Offshore considerations

As well as their landward impacts, cruise ships also represent a source of marine and atmospheric pollution. On average a cruise ship passenger generates about eight gallons of sewage per day, and a cruise ship produces more than 95,000 litres of oily bilge from engines and machinery (Connolly, 2019). In the absence of land-based energy supplies, cruise ships' engines need to keep running in order to supply the ship with energy, in particular electricity. Dockside electricity supplies are expensive and rare, and because many of today's ships still rely on heavy fuels, ship engines are a significant source of air pollution. While this problem is not restricted to cruise vessels, because of their high energy needs and requirement to stay close to the shore, cruise tourism ships are a particularly significant source of air pollution for coastal communities. Among the most problematic pollutants emitted by ships are sulphur oxides (SO_x) and carbon dioxide (CO_2) (Kirchner, 2018: 73). Ships often need a sudden surge of energy while manoeuvring close to shore or when energy-intensive systems are switched on. Conventional ships need to start an additional motor or generator, only to turn it off again a short time later. This process drives up fuel consumption and pollution, often adjacent to shore communities.

Pollution impacts on marine and coastal environments have generated opposition to Arctic cruise tourism, notably from European conservationists (e.g. *Neue Osnabrücker Zeitung*, 2019; *HLN*, 2019). A number of cruise companies visiting Greenland and the Arctic, however, have been moving towards more sustainable hybrid forms of propulsion and instilling greater environmental awareness among their passengers and crew while on board as well as on land (Wüst, 2019). Nonetheless,

> Even if the cruise industry implements more sustainable practices (e.g. advances wastewater treatment systems, a total ban on the discharge of liquid and solid wastes), the risk of vessels colliding with icebergs or running aground remains ... The potential spill of large quantities of oil, fuel and other hazardous substances from a cruise vessel could potentially have devastating consequences for both marine and coastal environments
>
> (Lück, 2010: 128)

This, of course, applies not just to Arctic cruise ships.

The international shipping industry did not play a role in the 2015 Paris Climate Accords, although concern over emissions led to the International Maritime Organisation (IMO) establishing specific Sulphur Emissions Control Areas (SECAs) in the Baltic and North Seas and along much of the US and Canadian coasts. None of these include areas north of the Arctic Circle, and while January 2020 saw global limits of 0.5 per cent sulphur m/m in ships' fuels established, this was still a level five times higher than limits imposed in the Baltic (Kirchner, 2018).

As noted in various sections of this volume (2.6, 3.7, 5.4, 8.3, 10.3–10.4), the multiple roles of China and Chinese actors in the Arctic, and the perception of those roles from within the Arctic and from the sources of Arctic governance outside, have generated debates and narratives along a number of often intertwining dimensions and at different scales. The role of China and Chinese citizens in Arctic tourism development is the source of several of those intertwinings, including the relationships between Arctic shipping, cruise tourism and pollution control.

While China's direct role in Arctic cruise tourism remains limited, the significant numbers of Chinese visitors (Chapters 3, 7 and 10) provide China with an interest in the region and (potentially) the well-being of its citizens. In comparing the narratives of Danish and Chinese passengers on a cruise to West Greenland, Bislev and Smed (2018: 39–40) observed that

> Understanding these different narratives and their sources are extremely important in discussing Chinese tourism as a resource to Greenland and in understanding why Arctic nature is perceived as a resource to the busy Chinese tourists escaping from polluted and crowded mega-cities in China ... the wider consequences of these impressions may reach beyond the scope of tourism [and] can become quite significant in a wider context of global understanding of Arctic places and spaces.

Within such understandings, Kirchner (2018: 72) suggests, ideal(istica)ly, that states, like China, should not overlook the impact that overseas tourism and tourists has on local communities, and

> While the sovereignty of the receiving states prevents tourists' home states from taking direct action, a cooperative effort aimed at safeguarding the rights and interests of local communities in tourism regions is in the interest of all sides. The sustainability of Arctic tourism can benefit from the involvement of the home countries to tourists who visit the Arctic.

While, of course, this objective can be sought through raising tourists' awareness of local environmental and cultural conditions and the needs and sensibilities of local communities, prior to departure, it could also be sought in the diplomatic arena by influencing international legal frameworks relevant for tourism activities. Kirchner (2018) therefore postulates that China has the potential to play several roles in contributing to the protection of Greenland's coastal communities' right to live in a healthy environment. He argues that China, as a major maritime nation could, for example, encourage international co-operation in relation to the IMO's work to support the establishment of a SECA for the Arctic region, while also developing the technological capacity in China's shipbuilding industry[4] to meet the needs of more environmentally conscious cruise ship companies. 'While such action might not provide immediate benefits for China, it could increase acceptance of Chinese tourism-related activities by local communities in the Arctic' (Kirchner, 2018: 72).

A potentially interesting deployment of soft power, perhaps, but China's response to the 2015 Paris Agreements has been hardly reassuring (Harris, 2017). And certainly, Kopra and Nojonen (2020: 18) argue that China needs both to critically assess its environmental footprint in the Arctic and to raise the level of ambition of its climate mitigation approach in order to fulfil its polar climate responsibilities.

11.6.3 Further considerations

An holistic approach to cruise tourism impacts should consider not only local destination impacts and en route maritime effects, but also the contributions of transport used by passengers travelling to and from their cruises' starting and finishing ports (Amelung and Lamers, 2007; Lück, 2010).

Finally, but not least, the pay and working conditions of crews employed in cruise ships – international minimum standards often being circumvented by cruise companies' use of flags of convenience – continues to leave much to be desired. This was further highlighted in April 2020 when the crews – numbering over 100,000 – of at least 50 cruise vessels were virtually abandoned in empty, potentially infected ships, after the Covid-19 pandemic took hold (McCormick and Greenfield, 2020; US Coast Guard, 2020; also Higgins-Desbiolles, 2020).

11.7 Conclusion

Pre-Covid, cruise tourism appeared to have become a significant factor in Greenland's tourism development path, albeit increasingly spatially and temporally concentrated and increasingly developing mass characteristics: all elements contrary to Visit Greenland's development objectives (2020e; Por 2020).

How relevant are cruise tourism research literature observations to the Greenland post-Covid context? Even before the pandemic, Chiappa, Lorenzo-Romero and Gallarza (2018) could argue that studies applying community-based tourism approaches to cruise tourism activity were highly site-specific and hardly generalisable. Apart from James, Olsen and Karlsdóttir (2020), noted earlier, none appear to have drawn on Greenlandic experience.

An emphasis on viewing the ice both off- and onshore reinforces the false imagery of a relatively un-peopled environment, lacking any cultural dimension. This may be further reinforced with the increasing size of cruise ships and their pursuit of self-sufficiency at the expense of deeper engagement with local people and cultures on shore.

As highlighted in Chapter 1, imagery of the Arctic is co-constructed, derived from experience, both first- and second-hand: from friends and relatives, (social) media images and narratives, tourism marketing and less explicit means of employing tourism in a soft power role (Chapter 10). In relation to imagery underscoring and derived from cruise tourism to Greenland, Bislev and Smed (2018: 34) allude to this soft power dimension:

> ... different perceptions of the Arctic are coming into play, both in terms of the perceptions being brought into the Arctic, e.g. when tourists are visiting, but also when bringing perceptions 'out' of the Arctic to live and breathe in the rest of the world.

At the time of writing, the post-Covid recovery role of cruise tourism in Greenland's development trajectory towards independence was difficult to predict. Undoubtedly it will remain an important element in the tourism mix, and may be further enhanced through the generation of fly-cruise packages when greater accessibility to Greenland's tourism centres is achieved through the airports upgrading programme. Two dimensions appear critical:

- The need – at some cost – to upgrade port infrastructures to permit ships to dock in more locations rather than their current requirement to tender passengers ashore, which can act as a significant constraint to the disembarking of less mobile but potentially high spending passengers.
- Greenlandic authorities, either individually or collectively, need to find ways of strengthening their bargaining positions in relation to decisions taken over the cruise companies' logistics and onshore programmes. In the early stages of the pandemic, Renaud (2020) could argue that as cruise lines had been weakened by the curtailment of activities, destinations or groups of destinations would have more bargaining power post-Covid. Would this 'window of opportunity' be sustained?

11.8 In the next chapter

In the next chapter, acting as both the culmination and conclusion to Part 2 of this volume, Carina Ren and Mette Simonsen Abildgaard explore Greenlandic futures at the intersection of tourism development and possible independence.

Notes

[1] Superficially there would appear to be a discrepancy in the two official sets of data for 2017, with ten ships making a total of only eight port calls. But it could be the case that some were unable to enter port because of icebergs, bad weather or other constraints.

[2] As part of the Arctic Cooperation Programme project 'Sustainable business and demography: exploring critical links between gender, youth and small-scale business development in fisheries and tourism in South Greenland'.

[3] Notable circular walking routes include the two- to three-hour trek around Lake Tasersuaq, just outside Qaqortoq, where loons (great northern divers) may be heard and possibly seen; and various extended routes taking in the Icefjord and Semermiut pre-historic village from Ilulissat town.

[4] As noted in Section 2.6.3 Chinese shipyards launched their first polar ice-breaker in 2018, designed in collaboration with the Finnish company Aker Arctic (Eiterjord, 2018).

12

Greenlandic Independence and Tourism Futures – Exploring *Modern* and *Ethnic* Logics

Invited contribution

Carina Ren and Mette Simonsen Abildgaard

12.1 Introduction

In this invited contribution, we explore Greenlandic futures at the intersection of possible independence and tourism development.

As noted earlier, alongside fishing and mining, tourism is one of three 'pillars' considered fundamental to sustaining a future self-sufficient Greenlandic economy. With Greenland slowly moving towards independence from Denmark, the future trajectories of tourism and statehood in Greenland are thus inextricably intertwined – politically as well as economically. To further investigate tourism as an arena for discussing independence, we begin by identifying and describing the 'identity logics' underpinning political debate around independence in Greenland and trace these in our analysis into current tourism activities in Greenland. These logics are analytical constructs that we do not seek to find empirically in their pure form, but rather use as tools to understand statements and events in the Greenlandic independence process. The analysis is based on a review of current and former Greenlandic tourism strategies and campaigns together with a series of qualitative interviews and a workshop with Greenlandic tourism actors.

In the first logic – *modernity* – Greenland is following a path to statehood, where tourism is inscribed in the building of a strong national identity, enabled through infrastructure development on and mobility within Greenlandic territory. In the second, *ethnic* logic, a path that leads to strengthened Inuit identity, tourism is about immersion in Indigenous tradition and ethnic expressions. In the analysis, we trace the modernity and ethnic logics through two tourism trajectories. We emphasise that the logics are abstractions that do not exist empirically in their pure form, as they intersect and entangle in practice. But they are instructive in being distilled as each holds distinct answers to how tourism in Greenland can and should play out in the future.

While the two dominant logics underpinning independence in Greenland are identifiable within both political and tourism arenas, we argue that emerging issues and concerns in tourism, such as tourism as a colonial undertaking, can pave the way towards the crafting of new identity positions.

12.2 Modernity and indigeneity: a view from the arena of politics

Since World War II, Greenland has been on a path of increased independence from Denmark (Chapter 4). The process has not been straightforward, and Greenland has so far enacted a policy of 'almost sovereign, but not quite' (Adler-Nissen

DOI: 10.1079/9781789246728.0012

and Gad, 2017: 188). In this political arena, both the logic of ethnicity and the logic of modernity have underpinned central priorities and decision-making in Greenland's independence process.

In the context of politics, the modernity logic entails working towards formally becoming a state in the traditional sense through full Westphalian sovereignty, i.e. exclusive territorial sovereignty. In foreign policy, this means securing authority and 'a seat at the table', both figuratively and literally, in forums such as the Arctic Council (Chapter 2), where Greenland as part of the Kingdom of Denmark shares one seat with Denmark and the Faroe Islands. These efforts were highlighted in 2011, when the new Swedish Arctic Council leadership replaced the three chairs habitually placed at the table for Denmark/Greenland/Faroe Islands with a single chair – for the Danish representative. The representatives from Greenland and the Faroe Islands were placed away from the table, behind the Danish chair, in this way signalling that sovereign states hold a primary position in the council (Jacobsen, 2019). The response from Greenland's premier, Aleqa Hammond, was to challenge this division between council members, and Greenland's participation in council activities was suspended until a solution had been found in which Greenland regained full rights to participate in council meetings (Jacobsen, 2019). Today, Greenland, Denmark and the Faroe Islands internally decide who has a seat at the table prior to meetings where there are insufficient available seats. However, in the same process, Greenland (and the Faroe Islands) lost the right to display their flag at council meetings. As remarked by Gerhardt (2018: 120), 'if the paradiplomatic protest was about asserting ones [sic] nationalistic right to be at the table, based on a recognition of quasi-statehood, then the loss of the flag, the ultimate symbol of that state, appears to be a loss'.

Continuing these considerations of the role of the flag as a symbol of statehood, other examples of the modernity logic at play within the political arena are the ongoing efforts in Greenland to assemble what Löfgren (1991: 104) has called a 'national inventory', that is, elements that are widely considered to make up 'a proper nation'. We could see the Greenlandic decision on the design of *Aappalaartoq*, the Greenlandic flag adopted in 1985, as an example of this assembly. Also, Greenland has had, since 1979, two national anthems, not to mention the world's largest national park. The examples of the Arctic Council and the national inventory showcase how modernity logics in Greenland's politics are underpinned by traditional Western concepts, in the case of Westphalian concepts dating back to the seventeenth century, while national inventories were established by Western nations in the nineteenth and early twentieth centuries (Löfgren, 1991). While such an effort to 'piece together' nationality may be downplayed as artificial, it is worth remembering that Western nations have had a long time to establish themselves as natural fact, thereby rendering invisible the comprehensive work entailed in proclaiming their identity as sovereign nations.

Alongside state-oriented decision-making forums such as the Arctic Council, assemblies such as the Arctic Peoples' Conference in 1973, the First World Council of Indigenous Peoples in 1975 and the first Inuit Circumpolar Conference (now Council) (ICC), in 1977 have increasingly enabled communities other than sovereign states to have a voice in international politics (Chapter 2). As part of this development, a political logic of ethnicity has come to supplement that of modernity, focusing on the Greenlandic population as Inuit and Indigenous, on their connectedness to the Arctic region, its peoples and traditions. The logic of ethnicity also encompasses a goal of achieving increased independence and representation in matters relating to Greenland's interests, not necessarily with an emphasis on formal state formation, but rather a process of decolonisation.

A balance between the two logics in the arena of politics has not always been easy to obtain, and in some instances modernity and ethnicity have been positioned as opposites. According to Kleivan (1992), Greenland's initial response to the potential collaboration and engagement in forums such as the World Council of Indigenous Peoples was hesitant because it was perceived as a goal that at the time competed with preparing for and achieving Home Rule (as a greater degree of independence from Denmark).

To understand this hesitation, it is worth looking deeper at the concept of indigeneity

(Chapters 1 and 6) and its paradoxical relationship with independence. The United Nations (2007) declaration regarding the rights of Indigenous peoples includes no definition of the term, but implies an understanding of a people governed by a quasi-colonial power by insisting on the right to self-government, and by recognising a distinction between, among others, the politics and culture of Indigenous peoples and that 'of the State':

> Indigenous peoples have the right to maintain and strengthen their distinct political, legal, economic, social and cultural institutions, while retaining their right to participate fully, if they so choose, in the political, economic, social and cultural life of the State.
>
> (United Nations, 2007: Article 5)

Indigeneity is thus posited to exclude majority and post-colonial status. With no majority Indigenous nation state currently in existence, there is no pre-existing model to draw on in international politics. In fact, the principles of statehood and Indigenous rights are often perceived as mutually exclusive; an opposition between capitalist economic practices and Westphalian logics of territorial sovereignty versus Indigenous approaches to territory and ownership (Strandsbjerg, 2014). For Greenland in particular, an issue that well exemplifies this rift is that of 'private property' as it pertains to land, as Greenland has a long tradition of collective land ownership. Before the colonial period, the determination of territorial rights was based on criteria such as regular use and the investment of work or materials towards more permanent establishments (Brøsted, 1986), and these principles are to some degree carried over in contemporary Greenland, where there is no private ownership of land, only of the structures on it.

As the Greenlandic politician and former prime minister, Kuupik Kleist, explained, such collective land ownership practices do not translate well into politics at an international level:

> It was extremely difficult to be taken seriously in international political fora if he [Kuupik Kleist] were to make the case for alternative arrangements of sovereignty and the relationship to geographical space. As such, he suggested a structural necessity to adapt to established conventions in international politics when making the case for Greenland's interests.
>
> (Strandsbjerg, 2014, 271)

As the example of land ownership as well as the negotiations about representation in the Arctic Council above show, there is no shortage of instances in foreign policy where being cast in the role of Indigenous rather than a sovereign nation has diminished Greenland's influence and opportunities for representation.

Because of such experiences, the logic of ethnicity is often downplayed in Greenland's current politics as they relate to matters of independence. Still, there are arenas where ethnicity is playing an increasingly central role. One instance is the politics of sustainability, where Greenland has followed a 'double climate strategy': on the one hand acting as a nation that supports CO_2 reduction on a global scale, while on the other stepping out of, and finding exemption from, climate deals, as Greenlandic politicians draw on their Indigenous status to take a position as victims, rather than producers of CO_2 emissions (Bjørst, 2011, 2019) (Chapter 5).

Another example is international trade, where seal products have been banned from import and commercialisation in the EU since 2009. However, as an Indigenous Inuit culture whose hunting traditionally includes seals, Greenland was able to obtain an exemption from the ban in 2015 (https://ec.europa.eu/environment/biodiversity/animal_welfare/seals/seal_hunting.htm).

In the next part of this contribution, we attempt to follow these well-researched identity positions and their connection to nation building into the not so well-travelled landscape of tourism. After a short reprise of tourism development in Greenland, we zoom in on the logics of territory and ethnicity in the development of Greenland as a tourism destination.

12.3 Tourism in Greenland: towards the *modern* or the *ethnic*?

Until 1953, when the status of Greenland changed from that of a colony to a county within the Danish Kingdom, people wishing to visit Greenland had to seek permission to enter the country from the Danish Ministry of Greenland.

Travellers to Greenland would typically venture there as part of the colonial government (trade and commerce, teaching) or through missions or expeditions. The basis for tourism in a more traditional sense was only laid shortly after 1953 as charter flights were initiated to Kulusuk (East Greenland) and Narsarsuaq (South Greenland) on airstrips laid by the US army during World War II (Hegelund, 2009) (Sections 4.4 and 8.2).

Since then, tourism numbers have slowly increased, although not at the rates projected in reports published on tourism development (Ministeriet for Grønland, 1973; Grønlands Landsstyre, 1998; Udenrigsministeriet/Udenrigsdirektoratet, 2011; Instituttet for Fremtidsforskning, 2013; Rambøll Group, 2014). Nor has tourism developed in such drastic ways as seen elsewhere in the Nordic Arctic regions or Alaska (Chapter 3). This is partly explained by the short tourism season, a lack of infrastructure, the current limited capacity, the low standards of customer service, low growth rate, a lack of package tours and the low average of overnight stays (only an average of four) and spending per tourist (DKK1,100/€143/US$160/£132 per day) (Rambøll Group, 2014) (Chapter 7). While this has arguably improved during recent years, the cited report also mentions low priority and lack of concrete initiatives by the Government of Greenland, the lack of online destination information, and the difficulties with internal and external co-ordination in the Greenlandic tourism sector (Rambøll Group, 2014: 40–52; Bjørst and Ren, 2015: 97). However, with infrastructure development and the current construction of two international airports in Ilulissat and Nuuk, many believe they are now looking at a 'game changer' in Greenlandic tourism (informal conversations and formal presentations on infrastructure projects).[1]

The tourism offer in Greenland currently centres around adventure tourism and the 'Big Arctic Five' experiences marketed by Visit Greenland, which encompass dog sledding, whale watching, the northern lights, ice and snow, and the 'pioneering people' of Greenland (Visit Greenland, 2020d) (Chapter 9). In combination with sailing, hiking, fishing and hunting, as well as cultural offers such as *kaffemik* home visits and visits to handicraft workshops, this makes up the product offered to the near 100,000 people visiting Greenland annually (Ren, Chimirri and Abildgaard, 2020).

Nowadays, Greenland increasingly positions itself within a global adventure market. A noteworthy major difference compared to other Arctic tourist destinations is how Indigenous culture plays a much less prominent role in promoting, developing and articulating tourism. This contrasts with destinations in Arctic Sweden, Norway and Finland, where Sámí tourism has been extensively developed and marketed (Ren *et al.*, 2021). This could be explained by a low level of tourism, which has not (yet?) resulted in an extensive commodification of cultural products (Chapter 6), and also by the situated nature of the Indigenous and local culture in which cultural artefacts and performances hold different weight, value and presence in the everyday.

12.3.1 #staycation – Greenland as desti-nation

As the Covid-19 virus brought tourism mobility to an abrupt halt in the spring of 2020, concerns over how to reduce its massive impacts on tourism-dependent business and communities were voiced on a global scale. On both national and local levels, destination management organisations (DMOs) and tourism authorities initiated activities to reposition tourism to cater for the only available market at the time: the domestic.

Greenland was no exception to this movement. As the isolated island nation shut down, the national DMO, Visit Greenland (2020a, 2020b), developed the *Nunarput Nuan* ('our Wonderful Greenland') campaign, which encouraged Greenlanders to (re)discover their island nation. On the travel trade website of Visit Greenland, CEO Julia Pars explained

> Greenland is beautiful, safe, large and varied. Many adventures are hidden, which even we, who live here, do not know yet. With Nunarput Nuan, we want to emphasize that you can easily staycation and still have new world-class experiences – also in the company of friends and family that you have not seen for a long time.
> (Visit Greenland, 2020a)

Alongside this national campaign, other local activities were promoted under #staycation. On the website of the large tour operator Guide to Greenland, a number of tours were offered under that heading accompanied by: 'Living in Greenland? Spend your holidays exploring the country with one of the many "Staycation" offers from all over Greenland. With no guests from outside, now is the best time to explore and support local tourism' (guidetogreenland.com, 2020).

Offers included new products such as a weekend 'city trip' by boat from Nuuk to Sisimiut or to Maniitsoq or even an extended weekend in Kangerlussuaq, otherwise primarily known for its status as Greenland's main flight hub, from where domestic and transatlantic travellers are distributed along the west coast. As currently one of only two national airports where international flights can land, it is an obligatory transfer for many visitors (Chapter 8). The town of Kangerlussuaq is a relatively recent settlement built around an airfield established by the US military in 1941 (Section 4.4). As the only inland town in Greenland, Kangerlussuaq has a stable and dry sub-Arctic climate, with favourable conditions for air travel (Ingeman-Nielsen and Foged, 2016).

In a national newspaper campaign, and challenging many domestic travellers' perceptions of Kangerlussuaq as a tedious and generic 'non-place', the settlement was promoted as 'Kangerlussuaq – much more than you think' (Sermitsiaq, 2020). Until recently, few visitors considered Kangerlussuaq as a destination in its own right. As illustrated by Visit Greenland's suggestion that tourists 'take advantage of the wait in Kangerlussuaq' to experience local sights such as the nearby ice sheet, perhaps by air safari (Visit Greenland, 2020c), stays in Kangerlussuaq are often perceived as something to get over with as quickly as possible. In the wake of Covid-19 however, the development of tourism products and experiences in Kangerlussuaq was responding to the frustration of many local tourism businesses to finally have the town recognised as something more than a waiting room. It seemed that Kangerlussuaq was finally receiving attention and credit as a destination in its own right.

As opposed to other staycation destinations under Covid-19, Kangerlussuaq is a genuinely emerging domestic destination. While local tourism businesses are usually fully occupied with foreign tourists as Greenlandic tourists use the hub to access summer destinations in Europe or elsewhere, both have been forced to reassess and embrace each other. Thus, the staycation policy in Greenland has not only resulted in stimulating a higher proportion of locals in already known and recognised destinations, such as Ilulissat and South Greenland, but it has also played a part in crafting new ways of seeing and visiting a place as a tourist destination in its own right rather than just as a transfer point.

Beside the opportunity to travel with and visit friends and relatives, Visit Greenland CEO Julia Pars pointed to other reasons to undertake domestic tourism in the time of lockdown: to establish a deeper connection with the home country, and to help local businesses and workplaces to survive (Visit Greenland, 2020b). Interestingly, this explicitly allocates a role for tourism beyond the hedonist or pleasurable motivation of travellers and aligns with Franklin's exploration of the foundation role of tourism. Franklin (2003: 2) argues that tourism should be perceived as 'a central component of modern social identity formation and engagement, rather than something shallow and insignificant that takes place on the social margin'.

Franklin also points to the relations between tourism and attempts to establish social order in modern societies. Thus, 'although tourism is perceived as a domain of escape and freedom, we have always to understand that this takes place against a background of legislations, controls, subsidies, policies, nationalisms and controls and manipulations of public spectacle, ceremony and building' (Franklin 2003: 16). According to Franklin (p. 22), even for Thomas Cook, one of the founders of mass tourism,

> extending the numbers and sorts of people who had access to travel and a world beyond their home was seen as a positive, democratising project that could produce a more evenly educated civil society with more equal life chances and a society more tolerant of others; a civil society that could more easily take part in national life.

In an extension of this argument, in which tourism is seen as a nation-building activity, Franklin (p.16) establishes a close relationship between tourism and the state:

> The role of the state as an ordering vehicle for modernity and the cultural processes of nation formation are critical to understanding how modern tourism came about. In a variety of ways nation states also provided the conduit for the emergence of forms of governance and ordering strategies through the domains of sport, leisure and tourism

In his examples, Franklin situates the simultaneous emergence of modern tourism and instances of nation-building in the late nineteenth century. For the citizens of the nation state, an important aspect of performing citizenship was to travel in one's own country, paying homage to important national sights, much along Löfgren's (1991) idea of the national inventory. A central requisite of nation-state tourism is thus the identification of national sites that belong on the 'bucket list' of its citizens. Drawing on Franklin's suggestion of tourism as a tool for ordering modernity and the nation, the current Greenlandic 'destinisation', as exemplified with Kangerlussuaq, also invites us to think about the relationship in Greenland between the development of tourism and the forging of a modern national identity in the building of a 'desti-nation'. The development of national sites and their cultural history is a further indication of Greenland slowly becoming a 'desti-nation' in itself.

The modernity logic in Greenlandic tourism is at present imagined and enacted in a post-colonial space where Greenland is, on the one hand, on a path towards independence, competing with other Arctic states for visitors through the making of an Arctic destination open for tourism. In such a space, tourism, on the other hand, functions as a national project, adding to and supporting ongoing efforts to discover and construct the spaces of a future independent Greenland. But where, we ask, looking back at the Indigenous identity position in the political arena, might we locate an ethnic logic in tourism development?

12.3.2 Another trajectory: Greenland as an ethnic destination?

Regardless of whether the tourism offer is directed towards foreign, or increasingly domestic, tourism, the official tourism strategy is primarily concerned with adventure tourism, a strategy similar to other North Atlantic and Arctic destinations. This focus, in combination perhaps with the dominant political perception of tourism as an economic tool to further independence from Denmark (Bjørst and Ren, 2015), has meant that little attention has been given to the role of culture in either marketing or developing tourism (Ren, Chimirri and Abildgaard, 2020).

However, as part of Greenlandic culture, Inuit heritage is receiving growing public interest in related fields such as arts and crafts, music and events. Reflecting the limited interest of tourism providers, almost no academic literature specifically looks at the relationship between Indigenous culture and tourism in Greenland. An early exception to this was the short article 'Tourism to Greenland: renewed ethnicity?' by Valene Smith (1982), an early pioneer of tourism anthropology, in which she discussed the role of ethnicity in Greenlandic tourism. In her view, tourism was directly linked to an ethnic revival in Greenland: 'Unlike elsewhere, Greenlandic tourism is a conscious indigenous strategy and may well be an added stimulus to national ethnicity' (p. 26).

Interestingly, Smith (p. 26) pointed to how 'Growth of the tourist industry may well intermesh with the process of modernization associated with culture change, but here, as in a few other areas, it may also serve to reinforce ethnic identity'. Concerns about the consequences of tourism was reported (although unsourced) by Smith (p. 27), noting how

> ... some Kalatdlits [Greenlanders] are wary that tourist expectations – 'to see Natives' – will make them objects of curiosity. In addition, they fear that the control of tourism will pass to tour operators and non-Greenlandic carriers who will derive the largest economic benefits.

In spite of this, Smith (p. 27) saw tourism as 'a viable new industry with advantages, including a renewed sense of ethnic identity, if controls are instituted to maintain standards of authenticity'. Optimistically, she concluded: 'If this new industry provides a cash income and also generates new pride in their heritage, why NOT tourism?' (p. 27).

While tourism in Greenland failed to grow in the 1980s and 1990s as projected

(Box 7.1, Table 7.1), the current building of two international airports could herald a steep rise in visitor numbers in the years to come (notwithstanding Covid-19). Looking at the Nordic Arctic, where Sámí tourism has developed intensely (Pettersson and Viken, 2007), the question is whether and how an intensification of tourism in Greenland might lead to a rise in 'ethnic' tourism, and which tourism futures this might create. According to Olsen (2016), Indigenous tourism research has identified three main conflict areas: internal conflicts over Indigenous identity, the use of Indigenous culture in destination marketing, and land-use conflicts.

So far in Greenland, the level of such conflicts has remained low, although conversations with tourism stakeholders reflect concerns about tourists 'othering' the Greenlandic population in a search for 'traditional Inuit culture', as detected by Smith in 1982. This was expressed by a regional DMO representative, when discussing how to represent Greenland as an ethnic destination:

> I know there is a demand from the West or people that are coming here to experience the so-called Greenlandic culture. We have various offers here in Nuuk in what you would call traditional Inuit culture. However, this is not something that we encourage or that we especially focus on to preserve or to make sure that it is there. I feel confident enough that the operators are offering it. I do not need to make sure that it is happening.

Another DMO representative explained that part of the apprehension towards products based on traditional Inuit culture stems from an interest to reflect a majority Indigenous population living in modern cities with a culture that is 'very much impacted by the fusions that Greenlandic people have met for centuries'. According to her,

> ... it is not up to us as a DMO to define what Greenlandic culture is. We are never going to do that. It is very much up to the stakeholders, tour operators, people that develop products and so on. ... Many of the visitors coming to Greenland are coming to in a way experience this authentic or some sort of Indigenous life in Greenland. My personal opinion is that it will be hard to find for them. The Indigenous population in Greenland is the majority, it is not the minority as e.g. in Sweden, Norway, Canada or Alaska. We are the majority. So of course it is reflected in all parts of life.

Ongoing work in tourism reflects this will to represent a culturally fused Greenland to tourists. In this way, the logic of ethnicity in Greenlandic tourism actually resembles how tourism actors have framed the logic of modernity, by enacting it in a post-colonial setting. Greenland does not have a minority culture whose traditions are perceived as threatened or as risking exotification as they are consumed by its majority visitors, but rather as a majority population that is interested in inviting in foreign tourists who can experience – and consume – its contemporary culture. As an example of this, the business council of the capital municipality, Sermersooq, highlights the commercialisation of culture and arts as one of its main priorities (Møller-Sørensen, 2018).

While there is still limited interest from the point of view of DMOs in preserving or defining Greenlandic culture through tourism, negative examples from other Arctic tourism destinations – where foreign operators are running tours labelled as traditional ethnic experiences – have also made an impression. DMOs are therefore highly aware of the need to be sharp and clear on what kind of tourism they want to have in the future, as argued by this destination manager: 'We are the ethnic majority in Greenland, but we are also not naïve. In the future when we have the longer [airport] runway, we do not want e.g. Filipinos doing the drum dance at tourist arrivals'.

So far, and unlike in many other destinations in the Nordic Arctic, it seems that a fully-fledged 'ethnic' tourism development path is challenged in Greenland, much in line with Greenland's lack of interest on the Indigenous political stage, outlined above. Rather, many new tourism entrepreneurs develop products which display the Arctic everyday life and innovative use and reinterpretation of cultural practices, places and materials. Next to this development, we find a plethora of cultural products drawing on Indigenous heritage, such as the publication on Greenlandic myths and creatures, '*Bestiarium Groenlandica*', the fashion brand 'Inuit', and Inuit hand tattoo-like jewellery by PanikPanik. Examples such as these cultural products generously blend Inuit heritage with innovative materials and hybrid symbolism.

The exhibits of hybrid cultural product in and around tourism leave Smith's (1982) vision on reinforced ethnic identity unfulfilled. Instead, they unravel an 'un-exotifying' or transformed hybrid representation and interpretation of modern Arctic everyday life and culture. This hybrid form might very well be able to reduce the level of internal disputes over Indigenous identity and culture and their use in destination promotion. While such activities point towards an un-dramatic merger of modern and ethnic logics in tourism development, something that we also see within political discourse in Greenland, other concerns also emerge in the growing field of tourism.

12.4 Discussion: re-colonising the destination?

In December 2019, Greenland's capital city of Nuuk played host to the first Greenland Science Week. A key component of this major event was a broad range of public activities sharing and discussing the state of affairs in research areas of public interest – from climate research to Arctic constructions. One of these activities was an open tourism workshop, 'Tourism Today and Tomorrow – Learning from the Arctic'. Based on presentations by tourism scholars from Denmark, Iceland, Norway and Greenland, the aim of the workshop was, as stated in the invitation, to 'consider how [tourism] development can become socially and environmentally sustainable'.

The event was held at the University of Greenland and attended by a range of mostly North Atlantic and Nordic guests. For this reason, the English language was used. Different from other official gatherings, where discussions about a meeting's working language can become sensitive and difficult, the employment of English 'short-cuts' such debates by offering a 'neutral' third space between Greenlandic and Danish in which to meet and talk. The workshop was well attended by students and educators, government and municipal stakeholders, DMOs and private companies. In line with our presentation of tourism logics, we might think of the event as a tourism activity inscribed with a logic of modernity, discussing, in a knowledge-informed and comparative manner in relation to other Arctic destinations, how Greenland as a desti-nation should proceed in its development endeavours.

During the workshop, participants were encouraged to share views about the current tourism situation and to discuss its future. At one point, the owner of a Greenlandic small boat chartering company stood up and offered his views on tourism development. He described a situation where competing foreign companies arrived in the early summer to reap profits during the high season, only to leave again in early autumn without having made any significant contribution to the local community. In his words, what Greenland was witnessing was 'tourism colonialism'.

The statement left the atmosphere in the room notably changed. Not only because it pointed to the need to take seriously the challenges facing Greenlandic communities and businesses as tourism develops, a very common global concern in developing tourism. But also because it pointed to a historical experience and the lingering presence of colonial ties most often unaddressed in an industry that had seemingly been able to rewire usual post-colonial discussions, for instance around language and in between the two logics. By calling out and naming (re-)emerging colonial power relations, the idea of tourism colonialism points to how tourism development in the future might not only follow one or both logics, but might also – at least potentially – carry with it post-colonial conflicts.

This suggestion points to how tourism is not only, as argued by Ren and Chimirri (2018) 'a fish-like resource to be "trawled" or mined' but 'a powerful societal force to be carefully managed'. Public discussion is needed on where tourism development in Greenland is currently heading, how it could be planned and developed in the future, and how it might also be connected to other societal challenges as 'more than an industry'. This entails recognising tourism as a social force with the power to reignite ongoing controversies – or to create new ones.

12.5 Concluding remarks: bridging conflict through hybridity or new tourism colonialism?

This contribution has identified a link between tourism, (desti-)nation-building and future independence in Greenland and has pointed to the existence of a vibrant and hybrid cultural production to challenge the nature of 'ethnic tourism'. However, the territory's current post-colonial and quasi-independent status remains open to controversy over access and land ownership issues, and suggests a possibly turbulent development path for tourism in times to come, challenging the idea of Greenlandic exceptionalism as a modern, Indigenous desti-nation state. These controversies indicate the need for the development of culturally sensitive tourism – not to protect a fragile culture, which, as indicated, is far from the case, but to facilitate and support a solid foundation for respectful and mutually enriching tourism encounters.

While the need for such considerations is clear, it remains to be seen what cultural sensitivity in the context of Greenland's tourism will entail: whether it will involve new forms of governing, or take inspiration from Sámí tourism and turn to certification (de Bernardi, 2019). This may support local tourism entrepreneurs, but might also spark new logics and new controversies into being, concerning the 'authenticity' and legitimacy of tourism actors, experiences and services.

Acknowledgement

The authors would like to acknowledge the EU-Northern Periphery Arctic programme for financing the project 'ARCTISEN: Culturally Sensitive Tourism in the Arctic' and enabling the research behind this contribution.

Notes

[1] For example, during presentations at the 'Towards More Tourism' seminar in Nuuk, 2018 and the 'Nuuk City Development – A Project for Billions' seminar organised by the Confederation of Danish Industry and Sermersooq Municipality in Copenhagen, 2019.

Part 3:

Conclusions

13

Summary and Conclusions

13.1 Summary

The context, nature and role of tourism in Greenland was set within an overlapping geopolitical frame of:

- heightening climate crisis;
- Greenland's trajectory towards political independence from Denmark;
- its concept of economic 'self-sustainability' in supporting this trajectory;
- a contemporary culture exhibiting hybridity; and
- growing international interest in, and competition for, Greenland's natural resources and infrastructure development, reflecting improving land and sea accessibility afforded by climate change.

Such dynamic processes were seen to both challenge and encourage Greenland's concepts of sustainable development, within which tourism plays an ambivalent role.

Following a brief summary, this final chapter attempts to draw some of the intertwined strands of the foregoing chapters together in looking to Greenland's post-Covid future and tourism's role within it.

Part 1 sought to establish an Arctic context for later chapters. In Chapter 1, 'Framing the Arctic', 'imaginaries' of the Circumpolar North, most having been constructed by outsiders, were counter-balanced with an evaluation of debates on indigeneity and intersectionality in the Arctic.

Chapter 2, 'The changing role of the Arctic: transforming peripherality', offered an overview of Arctic dynamics while acknowledging the dangers of generalisation. The chapter emphasised the significance of the climate crisis in the Arctic and inter-relationships between climate change and the geopolitics of development. In an assessment of the role of the Arctic Council in mediating between these dynamic forces, a number of major challenges were recognised.

In Chapter 3, the Arctic context for tourism and the interplay of sustainability and identity were examined. Acknowledging changing emphases in Arctic tourism research literature, it was concluded that there remains much potential to explore, and various data comparability shortcomings remain challenging for meaningful analysis.

In Part 2, Chapter 4 addressed the nature and evolution of Greenland as an inhabited Arctic territory. It evaluated the island's colonial, neo-colonial and dependency roles, and highlighted substantial contemporary geopolitical tensions surrounding its pursuit of political independence while being squeezed between superpower Arctic aspirations.

Chapter 5, 'Greenland's 'self-sustainability'', focused on the 'hesitant transition' of Greenland's recent economic development, its inter-relationship with aspirations for political independence and the philosophy of economic 'self-sustainability' that has emerged in the face of growing global climate crisis pressures.

DOI: 10.1079/9781789246728.0013

Chapter 6, 'Greenlandic identity and culture within development processes', addressed the nature of Greenlandic traditional and contemporary culture, and debates over identity, language, indigeneity and hybridity. Greenlandic culture has largely escaped commodification and inertia, while a sharp decline in the numbers of sled dogs was seen as a barometer of changing cultural circumstances, connecting climate change, traditional pursuits and tourism.

Chapter 7 examined tourism as one of the three central economic planks for Greenland's future, through the evolving interrelationships between the country's political trajectory and its tourism administration and development. The pre-Covid basis for projecting Greenland's tourism future was closely examined.

Chapter 8, 'Peripherality, tourism and geopolitical dimensions of accessibility in Greenland' examined in both conceptual and logistical terms the international and domestic dimensions of Greenland's multiple peripheralities – not least in relation to tourism development – and the geopolitical complications arising from attempts to address some of these.

In Chapter 9, 'Fulcrum of climate change?' the nature and ambivalent implications of climate and climate change for tourism development in Greenland were emphasised. In evaluating Greenland's role – if not responsibility – towards climate crisis amelioration and adaptation, two dimensions of tourism's role were exemplified: as climate education, and as climate experience.

Chapter 10, 'Tourism and imagery: soft power, branding and cultural disconnection', examined the nature of tourism's soft power and Greenland's branding and tourism imagery. The chapter addressed the difficult area of cultural interpretation of imagery and exemplified this through apparent mutual misunderstandings in relation to the marketing and development of adventure tourism for Chinese visitors, reflecting a 'softening' of adventure tourism noted elsewhere in the Arctic.

Chapter 11, 'Paradoxes of cruise tourism to Greenland' discussed a number of paradoxes that have surrounded the growing numbers of passengers gazing at, and visiting, Greenland from those oceanic juggernauts of globalisation. Questions to arise include the ability of Greenland to construct sustainable frameworks within which to withstand the power inequalities that cruise ship corporations exert, and the extent to which 'gazing at the ice' from a distance reinforces the false imagery of an unpeopled land.

Chapter 12 was the invited contribution from Carina Ren and Mette Simonsen Abildgaard: 'Greenlandic Independence and Tourism Futures – Exploring *Modern* and *Ethnic* Logics'. It interwove debates concerning the future of Greenland as a likely independent state with that of tourism, and the role of a hybrid Greenlandic people in that future. This contribution provided an appropriate culmination to the major part of this volume, raising important questions for future debate.

13.2 Tourism and travel articulating, and located within, Greenland's development trajectory

13.2.1 Environmental context

At the Arctic-wide scale, with increasing and/or revived superpower interest in the Circumpolar North, there is the danger that the environmentally-related work the Arctic Council and its agencies have achieved will be diverted by increasing geopolitical concerns over Arctic resource exploitation competition. Heightened geopolitical tensions could act to deter, or at least constrain, such a broadly risk-averse activity as tourism.

As Bjørst (2019) has forcefully shown, the Greenland elite's conception of sustainable development, of necessity, turns a blind eye to the climate crisis and explicit environmental considerations, to conceive it in specific terms, focused at the national scale as the socio-economic sustainability of Kalaallit Nunaat for the purposes of achieving political independence. Both the means and perceived objective of this process remain, at best, uncertain. Again, the risk-averse nature of the tourism sector would appear to reduce its ability to flourish within an environment of uncertainty.

13.2.2 Tourism and economic development

How far are there parallels between, on the one hand, bucket list/last chance tourists (and exogenous tourism companies?) visiting Greenland and the Arctic (Chapter 3), moving in (briefly) before moving on, and, on the other hand, explorers and exploiters of mineral resources moving in, often with some degree of local disruption, exhausting reserves and then moving on (Chapters 4 and 5)? Is there also a comparable precariousness between global tourism markets and world prices of raw materials, a precariousness which has contributed to Greenland's economic 'hesitant transition' (Chapter 5)?

Again, there would appear to be comparable issues between (a) evaluating the gains and losses between, for example, land-based visitors and cruise ship tourists, juggling with their numerical weight, demographic profiles and likely economic impact, to arrive at some judgement on priorities for strategic planning and management (Chapter 7), compared to (b) the relative economic, political and strategic value of Greenland's range of raw materials, each weighed against the human, environmental and economic costs of their extraction. Both reflect sets of unequal power relationships. In particular, the logistically difficult questions of measuring and managing tourism carrying capacities would appear to parallel questions concerning an optimum level of mining activity in the pursuit of 'self-sustainability'. Issues of scale, location and logistics would appear critical for both.

Given Greenland's location on the globe and the nature of its internal geography, the issue of accessibility is critical in social, economic and political terms. The constraints of this dimension add considerable cost and time for visiting tourists just as they do for the movement of commodities in, out and within Greenland. On the one hand, while new/enhanced airports in Ilulissat, Nuuk and Qaqortoq (Chapter 8) are likely to intensify the spatial concentration of tourism activity in Ilulissat (Chapter 9) and exacerbate the pressures of cruise tourism activity in all three (Chapter 11), internal migration patterns are also emphasising an increasing concentration of the domestic population in Greenland's five largest settlements (Chapter 6), with greatest concentration in the capital Nuuk, placing further pressures on infrastructure and related facilities, while exacerbating core and 'peripheries' issues.

The geopolitical games over involvement in the airports development process between Greenland, Denmark, the USA and China (Chapter 8), only act to remind us of the reality that tourism and travel are embedded within wider socio-cultural, economic, environmental and (geo)political dynamic processes. Such processes, in this case, ascend from the local scale of replacing a helipad with a regional airport, to Greenland's global strategic role and security value vis-à-vis superpower muscle-flexing across the Arctic region.

In Chapter 9 tourism was, in part, idealised as a potential means of climate education as well as a vehicle for direct climate experience. Yet such a notional, if indirect, climate mitigation role could be neutralised by the superficial 'success' of increasing numbers of, and pressure from, last chance/bucket list tourists drawn to such iconic climate symbols as Ilulissat Icefjord. In this context, the paradox of tourism's climate crisis 'responsibility' being hosted within Greenland's economic 'self-sustainability' ideology is manifest.

As Martin and Ren (2020) have revealed, there appears to have been an almost schizophrenic approach taken by the Greenland tourism sector towards 'Other' tourism markets such as visitors from Asia. The good intentions of Visit Greenland in segmenting tourists along psychographic dimensions was seen to be confounded by practitioners referring specifically to 'Chinese' tourists and stereotyping their assumed preferences. Such bifurcated approaches were shown to be exacerbated with an apparent mutual misunderstanding between Greenland tourism marketers and Chinese tourists as to the meaning and nature of 'adventure tourism' (Chapter 10). Are there comparable misunderstandings and stereotyping to be found in Copenhagen's apparent sinophobic paranoia – allegedly prompted by security-driven US interests (Chapters 4 and 8) – in contrast to Greenland's apparently open attitude towards Chinese involvement in resource and infrastructure development?

The potential danger of Greenland moving from political/economic dependence on Denmark to an economic/geopolitical dependence upon resource development corporations (both private and state run), can be compared to the neo-liberal power of multinational tourism and travel companies, and especially cruise ship corporations. For a physically large but demographically small country such as Greenland, pre-Covid trends were suggesting that the latter companies were deploying larger ships calling at fewer ports and thereby geographically polarising the impacts of cruise tourism onshore, exacerbating potential congestion and over-tourism at key locations within a relatively short visiting season, confounding Visit Greenland objectives (Chapter 11). Such increasing spatial-temporal concentration effects mirror the political/economic concentration of power in the hands of a small number of global cruise line corporations and major tour companies.

Greenland's staycation response to Covid restrictions (Chapters 8 and 12) presented an interesting, if short-term example of what can be achieved by adaptation to the often precarious nature of Arctic development processes. Visit Greenland's 'flattening the curve' context for post-Covid recovery (Box 9.4) is notable for extending the range of potential visitor activities, emphasising younger age groups than hitherto – those less vulnerable to pandemics – and in particular extolling Greenland's vast dramatic open spaces with their (usual) absence of crowding, to overcome lingering risk perceptions (Por, 2020).

13.3 'After' Covid-19

13.3.1 Tourism

Within the global tourism and travel sector, there is likely to be a hesitant return to a modified version of previous 'normality', despite calls for a 'reset' (Renaud, 2020), 'degrowth' (Galvani, Lew and Sotelo Perez, 2020), a fork away from 'path-dependent institutional mindsets' (Ioannides and Gyimóthy, 2020: 630) and a move away from the neo-liberal tourism development model. Realistically, any change is likely to be incremental, and concern for responding to the climate crisis needs to be clearly prioritised (Prideaux, Thompson and Pabel, 2020), even if this appears not to be consonant with Greenland's current 'self-sustainability' priorities.

Particularly in the cruise tourism sector, Covid-19 and its variants are unlikely to be wholly eliminated. The medium-term global biosecurity environment is likely to be, at best, characterised by occasional isolated viral outbreaks resulting in short-term unforeseen disruption in virtually any travel and tourism sub-sector at any location.

If the Covid-19 pandemic revealed global tourism to be 'fragile and unreliable' (Higgins-Desbiolles, 2020: 618), Greenland's limited range of source markets, relatively low numbers and balance between land-based and cruise-ship tourism is especially so. This would certainly be exacerbated if 'mainland' Danish visitors were significantly reduced in number as a result of a too hasty independence process.[1]

If completion of the new/upgraded airports coincides with a degree of post-Covid euphoria, there is likely to be a short- to medium-term surge in tourist numbers – notionally responding to Visit Greenland's 'flattening the curve' imagery of uncrowded places – which, ironically, could pose problems of congestion and carrying capacity at the country's few honeypots. This may assist the extension of tourism seasons with a spill-over effect away from the most congested periods. But completion of the airports will inevitably require other comparable new/upgraded infrastructure.

13.3.2 Elephants in the room (i)

The continuing and likely intensifying climate crisis will have ambivalent impacts with feedback loops providing both local and regional causes for concern while possibly encouraging more last chance/bucket list tourism. When local livelihoods are further impacted by changes in the ice (and a diminution in accessible mammals), adaptation to participate in the tourism sector could offer for increasing numbers of Greenlanders an alternative, if likely seasonal, possible livelihood (although some winter tourism sports activities will struggle to continue). But equally, some degree of social unrest could be a consequence. Further glacier retreat and increased iceberg calving will provide more offshore hazards for both hydrocarbon exploration and exploitation, and cruise ship tourism. By contrast, the further opening of Arctic sea routes with the diminution of pack ice will have additional deleterious consequences for Arctic marine mammals.

If (recurring) local or regional outbreaks of Covid-19 that interfere with economic activities coincide with local and regional climate crisis events, unpredictable social and political consequences could follow at either or both ends of the tourism demand–supply chain.

13.3.3 Elephants in the room (ii)

Greenland's ideology of economic 'self-sustainability' is based on uncertain ground. If an estimated 20 (sustained) major mineral exploitation projects are required for Greenland's political independence to appear financially viable (Taagholt and Brooks, 2016), then Nuuk needs to find a way to proceed way beyond the continuing state of hesitant transition in its economic development. None of the three main planks are ideally placed: while the notion that climate warming is improving the accessibility of Greenland's mineral resources has tended to be talked up, and the post-Covid role of tourism remains somewhat ambivalent, the current relatively healthy state of fish stocks could diminish if, for example, marine terminating glaciers retreat to become land terminating and thereby deny fjords and coastal waters their critically important (for fish stocks) freshwater upwellings (Box 9.1).

Meanwhile, as repeated in this volume, economic 'self-sustainability' appears to brush aside the causes and negative consequences of the climate crisis.

13.3.4 Political independence

Political independence for a population of 56,000 scattered across a land the size of Western Europe located in a zone of (in)security and strategic significance for major global powers, appears, superficially, hard to envision.[2] Perpetuating the status quo would appear to be in the current best interests of both Greenland and Denmark. Denmark needs Greenland in order to be able to maintain a credible foothold in NATO and the Arctic Council, while Greenland will continue to require Denmark's annual financial subsidy, its Nordic cachet and its, albeit not always convenient, political cushion to buffer the various pressures that the major powers may wish to exert on Nuuk in an increasingly complicated post-colonial world.

13.3.5 And finally

Greenland is experiencing a series of hesitant transitions – in environmental, socio-cultural, economic and (geo)political terms, in the short-term exacerbated by Covid-19 – that any society and political system might find difficult to steer through. The hydra-like impacts of climate change are unlikely to recede any time soon – quite the reverse – despite the country's attempt to downplay them while pursuing (economic) 'self-sustainability'. As a notionally 'less unsustainable' means (compared to mining and hydrocarbons extraction) of deriving income to support a trajectory towards independence, tourism has a currently uncertain footing, in terms of tourist markets, availability of infrastructure and trained personnel, and the wider philosophy upon which an expansion of tourism within fragile ecosystems is based.

To the extent that it was able to safely close down and isolate at a number of different scales, Greenland had a 'good pandemic'. A post-Covid, climate crisis world (of recurrent pandemics?) could become more challenging.

Notes

[1] In 2019, 79 per cent of all 'foreign' visitors staying in paid accommodation were Danish (Martin, 2020).

[2] While Iceland is often looked to as a model, it currently has 6.5 times the population of Greenland within a territory less than one twenty-third of the size.

References

ACIA (Arctic Climate Impact assessment) (2004) *Impacts of a Warming Arctic*. Cambridge, UK: Cambridge University Press.

ACIA (Arctic Climate Impact assessment) (2005) *Arctic Climate Impact Assessment*. Cambridge, UK: Cambridge University Press.

Ackrén, M. and Jacobsen, U. (2015) Greenland as a self-governing sub-national territory in international relations: past, current and future perspectives. *Polar Record* 51(4), 404–412.

Adler, N., Ülkü, T. and Yazhemsky, E. (2013) Small regional airport sustainability: lessons from benchmarking. *Journal of Air Transport Management* 33, 22–31.

Adler-Nissen, R. and Gad, U.P. (2017) In the post-colonial waiting room. How overseas countries and territories play games with the norm of sovereignty. In: Epstein, C. (ed.) *Against International Relations Norms, Postcolonial Perspectives*. Abingdon, UK: Routledge, pp. 175–192.

Adve, N. (2019) A hesitant transition. Renewable energy growth and carbon emissions. *The India Forum* 2 August. https://www.theindiaforum.in/article/hesitant-transition-future-energy-use-and-carbon-emissions (accessed 3 December 2020).

AEPS (Arctic Environment Protection Strategy) (1991) *Arctic Environmental Protection Strategy. Meeting in June 1991 in Rovaniemi, Finland*. Helsinki: Finnish Ministry of the Environment. http://library.arcticportal.org/1542/1/artic_environment.pdf (accessed 7 February 2020).

AFP/The Local (2020) Danish ice cream maker drops 'Eskimo' name. *The Local* (Copenhagen) 16 July. https://www.thelocal.dk/20200716/danish-ice-cream-maker-drops-eskimo-name (accessed 27 September 2020).

Ahren, M. (2016) *Indigenous Peoples' Status in the International Legal System*. Oxford, UK: Oxford University Press.

Air Greenland (2020) Route map. *Suluk* 06, 84.

Aksenov, Y., Popova, E., Yool, A., Nurser, A.J.G. *et al.* (2017) On the future navigability of Arctic sea routes: high resolution projections of the Arctic Ocean and sea ice. *Marine Policy* 75, 300–317.

AMAP (Arctic Monitoring and Assessment Programme) (1997) *Arctic Pollution Issues: A State of the Arctic Environment Report.* Oslo: Arctic Monitoring and Assessment Programme. https://www.amap.no/documents/doc/arctic-pollutionissues-a-state-of-the-arctic-environment-report/67 (accessed 7 February 2020).

AMAP (Arctic Monitoring and Assessment Programme) (2017) *Adaptation Actions for a Changing Arctic (AACA) – Barents Area Overview Report*. Oslo: Arctic Monitoring and Assessment Programme. https://www.amap.no/documents/download/2885/inline (accessed 7 February 2020).

AMAP (Arctic Monitoring and Assessment Programme) (2018) *Adaptation Actions for a Changing Arctic: Perspectives from the Baffin Bay/Davis Strait Region*. Oslo: Arctic Monitoring and Assessment Programme. https://www.amap.no/documents/download/3015/inline (accessed 7 February 2020).

AMAP (Arctic Monitoring and Assessment Programme) (2019) *Arctic Climate Change Update 2019: An Update to Key Findings of Snow, Water, Ice and Permafrost in the Arctic (SWIPA) 2017*. Tromsø, Norway: Arctic Council Secretariat. https://oaarchive.arcticcouncil.org/bitstream/handle/11374/2353/ccupdate18.pdf?sequence=1&isAllowed=y (accessed 9 July 2020).

Ambrose, J. (2020) Denmark ends oil and gas exploration in North Sea under green energy plan. *The Guardian* 5 December, pp. 50–51.

Amelung, B. and Lamers, M. (2007) Estimating the greenhouse gas emissions from Antarctic tourism. *Tourism in Marine Environments* 4(2-3), 121–134.

Amos, J. (2019) Climate change: Greenland ice melt 'is accelerating'. *BBC News* 10 December. https://www.bbc.co.uk/news/science-environment-48387030 (accessed 11 December 2019).

Andersen, A., Jensen, L. and Hvenegård-Lassen, K. (2016) Qullissat: historicising and localising the Danish scramble for the Arctic. In: Huggan, G. and Jensen, L. (eds) *Postcolonial Perspectives on the European High North*. London: Palgrave, pp. 93–116.

Andersen, A.N. (2020) Personal communication. Aarhus, Denmark: School of Culture and Society, Aarhus University, 15 September.

Andersson, P., Zeuthen, J.W. and Kalvig, P. (2018) Chinese Mining in Greenland: Arctic Access or Access to Minerals? In: Bislev, A., Gad, U.P. and Zeuthen, J.W. (eds) (2018) *Arctic Yearbook 2018. Special Section: China & the Arctic*. Akureyri, Iceland: Northern Research Forum, pp. 102–117,

Archer, C. (1988) The US defence areas in Greenland. *Cooperation and Conflict* 23(3), 123–144.

Archer, C. (2003) Greenland, US bases and missile defense: new two-level negotiations? *Cooperation and Conflict* 38(2), 125–147.

Arctic Council (2002) *Inari Declaration on the Occasion of the Third Ministerial Meeting of the Arctic Council*. Inari, Finland: The Arctic Council.

Arctic Council (2009) *Arctic Marine Shipping Assessment 2009 Report*. Tromsø, Norway: Arctic Council.

Arctic Council (2013) *Rules of Procedure*. Tromsø, Norway: Arctic Council Secretariat.

Arctic Council (2015) *New ACAP Project on Climate Mitigation.*Tromsø, Norway: Arctic Council Secretariat. https://arctic-council.org/en/news/new-acap-project-on-climate-mitigation/ (accessed 9 July 2020).

Arctic Council (2017) *Telecommunications Infrastructure in the Arctic. A Circumpolar Assessment*. Tromsø, Norway: Arctic Council Secretariat.

Arctic Council (2018) *Observers*. http://www.arctic-council.org/index.php/en/about-us/arctic-council/observers (accessed 19 December 2019).

Arctic Council (2019) *Statement by the Chair*. Rovaniemi, Finland: Arctic Council Secretariat. https://oaarchive.arctic-council.org/handle/11374/2343 (accessed 7 February 2020).

Arctic Council Secretariat (2013) *Kiruna Declaration*. Kiruna, Sweden: Arctic Council. https://oaarchive.arctic-council.org/handle/11374/93 (accessed 7 February 2020).

Arctic Council Secretariat (2015) Iqaluit Declaration. Tromsø, Norway: Arctic Council Secretariat. https://oaarchive.arctic-council.org/handle/11374/662 (accessed 7 February 2020).

Arctic Economic Council (2014) Fostering circumpolar business partnerships. Tromsø, Norway: Arctic Economic Council. www.arctic-council.org/index.php/en/news/arctic-economic-council (accessed 13 September 2020).

Arctic Science Ministerial (2018) Joint Statement of Ministers on the Occasion of the Second Arctic Science Ministerial. Berlin: Federal Republic of Germany, 26 October. https://asm-2-joint-statement.pdf (accessed 30 October 2020).

Arctic Science Ministerial (2020) About ASM3. https://asm3.org/about-asm3/ (accessed 30 October 2020).

Árnadóttir, R.E. (2019) *Nordic Tourism Policy Analysis*. Copenhagen: Norden/Nordic Council of Ministers. http://norden.diva-portal.org/smash/get/diva2:1299201/FULLTEXT01.pdf (accessed 6 February 2020).

Arneborg, J. (2000) Nordboerne. In: Jakobsen, B.H., Bøchner, J., Nielsen, N., Guttersen, R., Humlun, O. and Jensen, E. (eds) *Topografisk Atlas Grønland*. Copenhagen: Det Kongelige Danske Geografiske Selskab og Kort & Matrikelstyrelsen, pp. 54–55.

Arneborg, J. (2003) Norse Greenland: reflections on settlement and depopulation. In: Barrett, J.H. (ed.) *Contact, Continuity and Collapse: The Norse Colonization of the North Atlantic*. Turnhout, Belgium: Prepols, pp.163–181.

Arneborg, J., Lynnerup, N. and Heinemeier, J. (2012) Human diet and subsistence patterns in Norse Greenland AD c980 – AD c1450: archaeological interpretations. *Journal of the North Atlantic* 3, 119–133.

Asian Development Bank (2011) *Asia 2050: Realizing the Asian Century*. Manila: Asian Development Bank. https://www.adb.org/publications/asia-2050-realizing-asian-century (accessed 9 December 2020).

Assef, N.A. (2018) China's Polar Silk Road: Overview, Challenges and Opportunities. *SSRN* (Elsevier) 18 May. https://papers.ssrn.com/sol3/papers.cfm?abstract_id=3179218 (accessed 30 November 2020).

Åtland, K. (2008) Mikhail Gorbachev, the Murmansk Initiative, and the desecuritization of interstate relations in the Arctic. *Cooperation and Conflict* 43(3), 289–311.

ATTA (Adventure Travel Trade Association) (2013) *Adventure Tourism Market Sizing*. Monroe, WA: ATTA.

Aurora Zone, The (2020) About us and our trips. Stannington, UK: The Aurora Zone. https://www.theaurorazone.com/search (accessed 27 August 2020).

Axelsson, R. (2020) *Arctic Heroes: A Tribute to the Sled Dogs of Greenland*. Heidelberg, Germany: Kehrer Verlag.

Baldacchino, G. (1993) Bursting the bubble: the pseudo-development strategies of microstates. *Development and Change* 24(1), 29–52.

Baldacchino, G. (2006) Warm versus cold water island tourism: a review of policy implications. *Island Studies Journal* 1, 183–200.

Bamber, J., Van den Broeke, M., Ettema, J., Lenaerts, J. and Rignot, E. (2012) Recent large increases in freshwater fluxes from Greenland into the North Atlantic. *Geophysical Research Letters* 39, 8–11.

Bankes, N. and Kuivorova, T. (eds) (2013) *The Proposed Nordic Saami Convention: National and International Dimensions of Indigenous Property Rights*. Oxford, UK: Hart Publishing.

Bannikin Travel and Tourism (2019) *China Adventure Tourism Market Study*. Toronto, ON, and Hong Kong: Bannikin Travel and Tourism.

BARK Rådgivning (2016) *Ilulissat Icefjord Centre: Communication Concept*. Copenhagen: Government of Greenland, Qaasuitsup Municipality and Realdania.

Barnes, J. (2019) Sustainable development and notions of security. In: Heininen, L., Exner-Pirot, H. and Barnes, J. (eds) *Redefining Arctic Security: Arctic Yearbook 2019.* Akureyri, Iceland: Arctic Portal, pp. 197–209.

Barr, S. (2015) The effects of climate change on cultural heritage in the polar regions. In: Petzet, M. and Ziesemer, J. (eds) *Heritage at Risk 2006-2007*. Paris: ICOMOS, pp. 203–205.

Barry, R.G. (2017) The Arctic cryosphere in the twenty-first century. *Geographical Review* 107(1), 69–88.

Barth, F. (ed.) (1969) *Ethnic Groups and Boundaries: The Social Organization of Culture Differences*. Oslo: Universitetsforlaget.

Bauer, B., Watson, D. and Gylling, A.C. (2018) *Sustainable Consumption and Production. An Analysis of Nordic Progress Towards SDG12, and the Way Ahead*. Copenhagen: Norden/Nordic Council of Ministers.

BBC (2019) Donald Trump and Greenland: why would he want to buy it? *BBC News* 21 August. https://www.bbc.co.uk/news/newsbeat-49422832 (accessed 21 August 2019).

Becken, S. and Hays, J.E. (2007) *Tourism and Climate Change: Risks and Opportunities*. Clevedon, UK: Channel View.

Becken, S. and Hays, J.E. (2012) *Climate Change and Tourism: From Policy to Practice*. Abingdon, UK: Earthscan.

Bendtsen, J., Mortensen, J., Lennert, K. and Rysgaard, S. (2015) Heat sources for glacial ice melt in a West Greenland tidewater outlet glacier fjord: the role of subglacial freshwater discharge. *Geophysical Research Letters* 42, 4089–4095.

Bendtsen, J., Mortensen, J. and Rysgaard, S. (2015) Modelling subglacial discharge and its influence on ocean heat transport in Arctic fjords. *Ocean Dynamics* 65, 1535–1546.

Benn, D.I., Cowton, T., Todd, J. and Luckman, A. (2017) Glacier calving in Greenland. *Current Climate Change Reports* 3(4): 282–290.

Bennike, O., Mikkelson, N., Pedersen, H.K. and Weidick, A. (2004) *Ilulissat Icefjord: A World Heritage Site*. Copenhagen: Geological Survey of Denmark and Greenland (GEUS), Ministry of the Environment.

Berbeka, J. (2018) The value of remote Arctic destinations for backcountry skiers. *Scandinavian Journal of Hospitality and Tourism* 18(4), 393–418.

Berger, A.A. (2010) Selling the adventure of a lifetime: an ethnographic report on cruising in the Antarctic. In: Lück, M., Maher, P.T. and Stewart, E.J. (eds) *Cruise Tourism in the Polar Regions: Promoting Environmental and Social Sustainability?* London: Earthscan, 43–54.

Bertelsen, R.G. and Gallucci, V.F (2016) The Return of China, Post-Cold War Russia and the Arctic: Changes on Land and at Sea. *Marine Policy* 72, 240–245.

Bertelsen, R.G. and Justinussen, J.C.S. (2017) Knowledge and Natural Resources: a Crucial Connection for Local Benefits and Sustainable Arctic Development. In: Conde, E. and Sánchez, S.I. (eds) *Global Challenges in the Arctic Region: Sovereignty, Environment and Geopolitical Balance*. Abingdon, UK: Routledge, pp. 281–305.

Berthelsen, A. (2020) This happened last night ... *Twitter* 21 June. https://twitter.com/uyarakq/status/1274689761127600128 (accessed 22 June 2020).

Berthelsen, T. (2014) *Coastal Fisheries in Greenland*. Nuuk: KNAPK (Kalaallit Nunaanni Aalisartut Piniartullu Kattuffiat).

Bevis, M., Harig, C., Khan, S.A., Brown, S., Simons, F.J. *et al.* (2019) Accelerating changes in ice mass within Greenland, and the ice sheet's sensitivity to atmospheric forcing. *Proceedings of the National Academy of Sciences* 116, 1934–1939.

Bintanja, R. and Andry, O. (2017) Towards a rain-dominated Arctic. *Nature Climate Change* 7, 263–267.

Birkhold, M.H. (2019) A brief history of the indignities heaped upon Greenland. *New York Times* 22 August. https://www.nytimes.com/2019/08/22/opinion/trump-greenland-denmark.html (accessed 22 June 2020).

Biskaborn, B.K., Smith, S.L., Noetzli, J., Matthes, H., Vieira, G. *et al.* (2019) Permafrost is warming at a global scale. *Nature Communications* 10(1), 264.

Bislev, A., Gad, U.P and Zeuthen, J.W. (2018) China seeking Arctic resources – the Arctic seeking resources in China. In: Bislev, A., Gad, U.P. and Zeuthen, J.W. (eds) (2018) *Arctic Yearbook 2018. Special Section: China & the Arctic*. Akureyri, Iceland: Northern Research Forum, pp. 2–5.

Bislev, A. and Smed, K. (2018) When will the iceberg melt? Narrating the Arctic among Chinese and Danish tourists aboard a cruise ship in Greenland. In: Bislev, A., Gad, U.P. and Zeuthen, J.W. (eds) (2018) *Arctic Yearbook 2018. Special Section: China & the Arctic*. Akureyri, Iceland: Northern Research Forum, pp. 29–42.

Bjerregard, P. and Larsen, C.V.L. (2015) Time trend by region of suicides and suicidal thoughts among Greenland Inuit. *International Journal of Circumpolar Health* 74(1), 26053–26058.

Bjørst, L.R. (2008) *En Anden Verden: Fordomme og Stereotyper om Grønland og Arktis*. Copenhagen: BIOS.

Bjørst, L.R. (2011) Arktiske Diskurser og klimaforandringer i Grønland. Fire (post) humanistiske klimastudier. Unpublished doctoral dissertation. Copenhagen: Det Humanistiske Fakultet, Københavns Universitet.

Bjørst, L.R. (2012) Climate testimonies and climate crises narratives: Inuit delegated to talk on behalf of the climate. *Acta Borelia* 29(1), 98–113.

Bjørst, L.R. (2016) Saving or destroying the local community? Conflicting spatial storylines in the Greenlandic debate on uranium. *The Extractive Industries and Society* 3(1), 34–40.

Bjørst, L.R. (2019) The right to 'sustainable development' and Greenland's lack of a climate policy. In: Gad, U.P. and Strandsbjerg, J. (eds) *The Politics of Sustainability in the Arctic: Reconfiguring Identity, Space, and Time*. Abingdon, UK: Routledge, pp. 121–135.

Bjørst, L.R. and Ren, C. (2015) Steaming up or staying cool? Tourism development and Greenlandic futures in the light of climate change. *Arctic Anthropology* 52(1), 91–101.

Blanchette, J. and Hillman, J.E. (2020) China's Digital Silk Road after the Coronavirus. *Commentary* (Center for Strategic and International Studies, Washington DC) 13 April. https://www.csis.org/analysis/chinas-digital-silk-road-after-coronavirus (accessed 14 August 2020).

Blaxekjær, L.Ø., Lanteigne, M. and Shi, M. (2018) The Polar Silk Road & The West Nordic Region. In: Heininen, L. and Exner-Pirot, H. (eds) *Arctic Yearbook 2018: Arctic Development in Theory and Practice*. Akureyri, Iceland: Northern Research Forum, pp. 437–455.

Blichfeldt, B.S. (2005) Unmanageable place brands? *Place Branding* 1(4), 388–401.

Bloomfield, C., Harreveld, B. and Fisher, R. (2020) Hesitant hopes: how a comprehensive approach to learning impacts on the transition hopes of marginalised young people in an alternative learning programme in regional Australia. *British Educational Research Journal* 46(1), 75–91.

Boas, F. ([1888]1964) *The Central Eskimo*. Lincoln, NE: University of Nebraska Press.

Bogoyavlenskiy, D. and Siggner, A. (2004) Arctic demography. In: Einarsson, N., Larsen, J.N., Nilsson, A. and Young, O.R. (eds) (2004) *Arctic Human Development Report (AHDR)*. Akureyri, Iceland: Stefansson Arctic Institute, pp. 27–41.

Bosangit, C., Hibbert, S. and McCabe, S. (2015) 'If I was going to die I should at least be having fun': travel blogs, meaning and tourist experience. *Annals of Tourism Research* 55, 1–14.

Bowitz, E. and Ibenholt, K. (2009) Economic impacts of cultural heritage: research and perspectives. *Journal of Cultural Heritage* 10, 1–8.

Brady, A.M. (2017) *China as a Polar Great Power*. Cambridge, UK: Cambridge University Press.

Breum, M. (2016) Analysis: did the Danish PM prevent a Chinese acquisition on Greenland? *High North News* 20 December. https://www.highnorthnews.com/en/analysis-did-danish-pm-prevent-chinese-acquisition-greenland (accessed 26 February 2020).

Breum, M. (2020) New row over language divides Greenland. *High North News* 13 November. https://www.highnorthnews.com/en/new-row-over-language-divides-greenland (accessed 15 November 2020).

Brigham, L. (2018) China's polar ships and future operations. In: Heininen, L. and Exner-Pirot, H. (eds) *Arctic Yearbook 2018: Arctic Development in Theory and Practice.* Akureyri, Iceland: Northern Research Forum, pp. 487–489.

Brigham, L., Exner-Pirot, H., Heininen, L. and Plouffe, J. (2016) Introduction. In: Heininen, L., Exner-Pirot, H. and Plouffe, J. (eds) *Arctic Yearbook 2016*. Akureyri, Iceland: Northern Research Forum, pp. 9–15.

Britton, S.G. (1982) The political economy of tourism in the Third World. *Annals of Tourism Research* 9, 331–358.

Broadbent, N.D. and Lantto, P. (2016) Terms of engagement: an Arctic perspective on the narratives and politics of global climate change. In: Crate, S.A. and Nuttall, M. (eds) *Anthropology and Climate Change: from Encounters to Actions*. Abingdon, UK: Routledge, pp. 341–355.

Broeke, M. van den, Bamber, J., Ettema, J., Rignot, E., Schrama, E. *et al.* (2009) Partitioning recent Greenland mass loss. *Science* 326, 984–986.

Brøsted, J. (1986) Territorial rights in Greenland: some preliminary notes. *Arctic Anthropology* 23(1/2), 325–338.

Brouder, P. (2014a) Evolutionary economic geography: a new path for tourism studies? *Tourism Geographies* 16(1), 2–7.

Brouder, P. (2014b) Evolutionary economic geography and tourism studies: extant studies and future research directions. *Tourism Geographies* 16(4), 540–545.

Brouder, P., Clavé, S.A., Gill, A. and Ioannides, D. (2016) *Dynamic Destinations: Evolutionary Change in Tourism Areas.* Abingdon, UK: Routledge.

Brown, F. and Hall, D. (2000) Introduction: the paradox of peripherality. In: Brown, F. and Hall, D. (eds) *Tourism in Peripheral Areas*. Clevedon, UK: Channel View, pp.1–6.

Buckley, R. (2005) In search of the narwhal: ethical dilemmas in ecotourism. *Journal of Ecotourism* 4(2), 129–134.

Buckley, R., McDonald, K., Duan, L., Sun, L. and Chen, L.X. (2014) Chinese model for mass adventure tourism. *Tourism Management* 44, 5–13.

Burnett, M. (2001) Melting golf-courses and polar bears. *BBC Sport Online* 22 February. http://news.bbc.co.uk/sport1/hi/funny_old_game/1184073.stm (accessed 27 January 2020).

Büsser, S., Stucki, M. and Jungbluth, N. (2010) *Umweltbelastungen Verschiedener Ferienszenarien.* Uster, Switzerland: ESU Services GmbH for WWF Switzerland. http://esu-services.ch/publications/ (accessed 21 August 2020).

Bye, H-G. (2020) Russia to appoint honorary consul in Greenland. *High North News* 14 October. https://www.highnorthnews.com/en/russia-appoint-honorary-consul-greenland (accessed 4 December 2020).

Bykova, A. (2021) Two United States – Greenland collaborations get US State Department funding. *High North News* 18 January. https://www.highnorthnews.com/en/two-united-states-greenland-collaborations-get-us-state-department-funding (accessed 22 January 2021).

CAFF (Conservation of Arctic Flora and Fauna) (2001) *Arctic Flora and Fauna: Status and Conservation.* Helsinki: Edita.

CAFF (Conservation of Arctic Fauna and Flora) (2010) *Arctic Biodiversity Trends 2010: Selected Indicators of Change.* Akureyri, Iceland: CAFF Secretariat.

CAFF (Conservation of Arctic Fauna and Flora) (2019) *Conservation of Arctic Fauna and Flora.* Reykjavík: CAFF. https://caff.is/ (accessed 29 September 2019).

Cai, L.A., Li, M. and Knutson, B.J. (2008). Research on China outbound market: a meta-review. *Journal of Hospitality and Leisure Marketing* 16(1–2), 5–20.

Callaghan, T.V., Johansson, M., Brown, R.D., Groisman, P.Y., Labba, N. *et al.* (2011) Multiple effects of changes in Arctic snow cover. *Ambio* 40(1), 32–45.

Campbell, N. (2018) *The Library of Ice: Readings From a Cold Climate*. London: Simon & Schuster.

Capoccia, G. and Keleman, R.D. (2007) The study of critical junctures: theory, narrative and counterfactuals in historical institutionalism. *World Politics* 59(3), 341–369.

Cassinger, C., Lucarelli, A. and Gyimóthy, S. (2019) The Nordic wave in place branding: moving back and forth in time and space. In: Cassinger, C., Lucarelli, A. and Gyimóthy, S. (eds) *The Nordic Wave in Place Branding: Poetics, Practices, Politics*. Cheltenham, UK: Edward Elgar, pp. 1–9.

Castañeda, Q. (2012) The neoliberal imperative of tourism: rights and legitimization in the UNWTO Global Code of Ethics for Tourism. *Practicing Anthropology* 34(3), 47–51.

Castree, N. (2001) Socializing nature: theory, practice, and politics. In: Castree, N. and Braun, B. (eds) *Social Nature: Theory, Practice, and Politics*. Oxford, UK: Blackwell Publishing, pp. 1–21.

Cavell, J. (2008) *Tracing the Connected Narrative: Arctic Exploration in British Print Culture, 1818–1860*. Toronto, ON: University of Toronto Press.

Chambers, D. and Buzinde, C. (2015) Tourism and decolonization: locating research and self. *Annals of Tourism Research* 51, 1–16.

Chater, A. (2016) Explaining non-Arctic states in the Arctic Council. *Strategic Analysis* 40(3), 173–184.

Chater, A., Greaves, W. and Sarson, L. (2020) Assessing security governance in the Arctic. In: Hoogensen Gjørv, G., Lanteigne, M. and Sam-Aggrey, H. (eds) *Routledge Handbook of Arctic Security*. Abingdon, UK: Routledge, pp. 43–56.

Chen, N., Masiero, L. and Hsu, C.H. (2019) Chinese outbound tourist preferences for all-inclusive group package tours: a latent class choice model. *Journal of Travel Research* 58(6), 916–931.

Chiappa, G. del, Lorenzo-Romero, C. and Gallarza, M. (2018) Host community perceptions of cruise tourism in a homeport: a cluster analysis. *Journal of Destination Marketing Management* 7, 170–181.

Chimirri, D. (2020) Collaborative configurations of tourism development: a Greenlandic example. *Journal of Tourism Futures* 6(1), 24–39.

China Daily (2017) Chinese visitors to Iceland up by 660 percent. *China Daily* 13 June. http://www.chinadaily.com.cn/world/2017-06/13/content_29728471.htm (accessed 12 January 2020).

Chinese Tourist Agency (2017) Chinese tourists in Canada on the rise. *GMA* (Shanghai) 3 November. https://www.chinesetouristagency.com/chinese-tourists-canada-rise/ (accessed 8 December 2020).

Choi, M., Chung, H., Yamaguchi, H. and Nakagawa, K. (2015) Arctic sea route path planning based on an uncertain ice prediction model. *Cold Regions Science and Technology* 109, 61–69.

Chow, I. and Murphy, P. (2008) Travel activity preferences of Chinese outbound tourists for overseas destinations. *Journal of Hospitality and Leisure Marketing* 16(1–2), 61–80.

Christensen, L., Nielsen, O.A., Rich, J. and Knudsen, M. (2020) Optimizing airport infrastructure for a country: the case of Greenland. *Research in Transportation Economics* 79, doi: 10.1016/j.retrec.2019.100773.

Christiansen, M. (2020) Greenland was Henrik Kauffmann's trump card. Nuuk: Visit Greenland. https://visitgreenland.com/articles/greenland-was-henrik-kauffmanns-triumph/?utm_source=B2C+newsletter+from+Visit+Greenland&utm_campaign=ebcac794bf-EMAIL_CAMPAIGN_2020_03_11_10_34_COPY_01&utm_medium=email&utm_term=0_d3ac890d57-ebcac794bf-146225322 (accessed 27 August 2020).

Christensen, S.A. (2009) *The Marshal's Baton. There Is No Bomb, There Was No Bomb, They Were Not Looking For A Bomb.* Copenhagen: Danish Institute for International Studies, DIIS Report 2009: 18.

Christensen, T.R. (1990) Tourism in Polar Environments with Special Reference to Greenland and Antarctica. Cambridge, UK: University of Cambridge unpublished M.Phil thesis.

Christensen, T.R. (1992) Greenland wants tourism. *Polar Record* 28(164), 62–63.

CIA (Central Intelligence Agency) (2020) *The World Factbook*. Washington DC: CIA.

CLIA (Cruise Lines International Association) (2020) *State of the Cruise 2020, Industry Outlook 2020*. Washington DC: CLIA. https://cruising.org/-/media/research-updates/research/state-of-the-cruise-industry.pdf (accessed 4 September 2020).

ClimateGreenland.com (2014) *A Greenlandic Reality*. Nuuk: ClimateGreenland.com. www.silap-pissusaa.gl/om-climategreenland.aspx

Coalition Agreement (2014) *Siumut, Demokraatit and Atassut. Coalition Agreemant, Election term 2014-2018 'Fellowship-Security-Development'*. Nuuk: Naalakkersuisut. http://naalakkersuisut.gl/~/media/nnooq/Files/Attched%20Files/Naalakkersuisut/DK/Koalitionsaftaler/Koalitionsaftale%202014-2018%20engelsk.pdf

Coates, K.S. and Broderstad, E.G. (2020) Indigenous peoples of the Arctic: re-taking control of the Far North. In: Coates, K.S. and Holroyd, C. (eds) *The Palgrave Handbook of Arctic Policy and Politics*. Cham, Switzerland: Palgrave Macmillan, pp. 9–25.

Coburn, C. (2019) Why industry is going green on the quiet. *The Guardian* 8 September. https://www.theguardian.com/science/2019/sep/08/producers-keep-sustainable-practices-secret (accessed 3 December 2020).

Coldwell, W. (2017) First Venice and Barcelona: now anti-tourism marches spread across Europe. *Guardian* 10 August.

Cole, S. (2007) Beyond authenticity and commodification. *Annals of Tourism Research* 34(4), 943–960.

Colgan, W., Machguth, H., MacFerrin, M., Colgan, J.D., As, D. *et al.* (2016) The abandoned ice sheet base at Camp Century, Greenland, in a warming climate. *Geophysical Research Letters* 43(15), 8091–8096.

Collis, S.M. (1890) *A Woman's Trip to Alaska, Being an Account of a Voyage Through the Inland Seas of the Sitkan Archipelago in 1890*. New York, NY: Cassell.

Committee for Greenlandic Mineral Resources to the Benefit of Society (2014) *To the Benefit of Greenland*. Nuuk and Copenhagen: Ilisimatusarfik and University of Copenhagen.

Connolly, K. (2019) Polar cruise boom harming the Arctic, explorer warns. *The Guardian* 13 August. https://www.theguardian.com/world/2019/aug/13/polar-cruise-increase-harming-the-arctic-explorer-arved-fuchs-warns (accessed 22 August 2019).

Conway, W.M. (1897) *The First Crossing of Spitsbergen*. London: Dent.

Copus, A. (ed.) (2004) *Aspatial Peripherality, Innovation and the Rural Economy (AsPIRE): Final Report to the European Commission, EU Fifth Framework Programme*. Aberdeen, UK: Scottish Agricultural College.

Corcoran, K. (2017) Cruise ships are being banned from sailing through Venice after locals got sick of them dwarfing their city. *Business Insider Nederland* 9 November. https://www.businessinsider.nl/author/kieran-corcoran/4/ (accessed 29 September 2020).

Crawford, A. (2013) When an iceberg melts, who owns the riches beneath the ocean? *Smithsonian Magazine* April. www.smithsonianmag.com/innovation/when-an-iceberg-melts-who-owns-the-riches-beneath-the-ocean-4800218/?no-ist (accessed 3 July 2020).

Creed, G.W. (2010) *Domesticating Revolution: From Socialist Reform to Ambivalent Transition in a Bulgarian Village*. University Park, PA: Pennsylvania State University Press.

Crenshaw, K. (1991) Mapping the margins: intersectionality, identity politics, and violence against women of color. *Stanford Law Review* 43(6), 1241–1299.

CRT (2013) *Turisme i Ilulissat. Fakta og Potentialer*. Naxø, Denmark: Centre for Regional and Tourism Research.

Cruise Adviser (2020) Cruise & Maritime Voyages goes into administration. *Cruise Adviser* 3 August. https://cruise-adviser.com/cruise-maritime-goes-administration/ (accessed 10 August 2020).

Cui, F. and Shen, B. (2019) Progress and prospects of China in implementing the Goal 14 of the 2030 Agenda for Sustainable Development. *Chinese Journal of Urban and Environmental Studies* 7(1), doi: 10.1142/S2345748119400062.

Curson, P. (2009) When fantasy cruises run aground on reality. *New Zealand Herald* 3 February.

Cussler, C. and Cussler, D. (2009) *Arctic Drift*. London: Penguin.

Dahl, J. (2013) The United Nations and the indigenous space. In: Dahl, I.J. and Fihl, E. (eds) *A Comparative Ethnography of Alternative Spaces*. Basingstoke, UK: Palgrave-Macmillan, pp. 19–40.

Dai, A., Luo, D., Song, M. and Liu, J. (2019) Arctic amplificationis caused by sea-ice loss under increasing CO_2. *Nature Communications* 10(1), 121.

Dai, B., Jiang, Y., Yang, L. and Ma, Y. (2017) China's outbound tourism – stages, policies and choices. *Tourism Management* 58, 253–258.

Dalby, S. (2013) Global Environmental Security. In: Falkner, R. (ed.) *The Handbook of Global Climate and Environmental Policy*. Chichester, UK: John Wiley & Sons, pp. 163–178.

Dale, B., Bay-Larsen, I. and Skorstad, B. (2018) *The Will to Drill: Mining in Arctic Communities*. Cham, Switzerland: Springer.

Dangerfield, K. (2018) 'Year of Canada-China Tourism' ends with travel agencies hitting pause amid diplomatic rift. *Global News* 21 December. https://globalnews.ca/news/4780777/justin-trudeau-canada-china-tourism/ (accessed 8 December 2020).

Danish Ecological Council, The (2013) *Recent Debates in the Uranium Debate in Greenland*. Copenhagen: Danish Ecological Council.

David, R.G. (2000) *The Arctic in the British Imagination 1818–1914*. Manchester, UK: Manchester University Press.

Davies, S. and Michie, R. (2011) *Peripheral Regions: A Marginal Concern?* Glasgow, UK: European Policies Research Centre, University of Strathclyde.

Davis Cross, M.K. and Melissen, J. (2013) *European Public Diplomacy: Soft Power at Work*. Basingstoke, UK: Palgrave Macmillan.

Dawson, J., Johnston, M.J., Stewart, E.J., Lemieux, C.J., Lemelin, H. *et al.* (2011) Ethical considerations of last chance tourism. *Journal of Ecotourism* 10(3), 250–265.

Dawson, J., Stewart, E.J., Lemelin, H. and Scott, D. (2010) The carbon cost of polar bear viewing tourism in Churchill, Manitoba. *Journal of Sustainable Tourism* 18(3), 319–336.

de Bernardi, C. (2019) A critical realist appraisal of authenticity in tourism: the case of the Sámi. *Journal of Critical Realism* 18(4), 437–452.

De Historiske, Den Norske Turistforening, Fellesforbundet, HANEN, Innovasjon Norge *et al.* (2017) *Towards Sustainable Travel and Tourism in Norway. A Roadmap*. Oslo: Innovasjon Norge.

de la Barre, S. (2013) Wilderness and cultural tour guides: place identity and sustainable tourism in remote areas. *Journal of Sustainable Tourism* 21(6), 825–844.

de la Barre, S. and Brouder, P. (2013) Consuming stories: placing food in the Arctic tourism experience. *Journal of Heritage Tourism* 8(2–3), 213–223.

Deboick, S. (2021) Nuuk power. *The New European* 18 March, pp. 40–41.

Demiroglu, O.C., Dannevig, H. and Aall, C. (2018) Climate change acknowledgement and responses of summer (glacier) ski visitors in Norway. *Scandinavian Journal of Hospitality and Tourism* 18(4), 419–438.

Demiroglu, O.C., Lundmark, L., Saarinen, J. and Müller, D.K. (2019) The last resort? Ski tourism and climate change in Arctic Sweden. *Journal of Tourism Futures* 6(1), 91–101.

Denchak, M. (2018) *Ocean Pollution: the Dirty Facts*. New York, NY: Natural Resources Defense Council, 23 January. https://www.nrdc.org/stories/ocean-pollution-dirty-facts (accessed 19 October 2020).

Deng, B. (2018) Shipping matters: the role of Arctic shipping in shaping China's engagement in Arctic resource development. In: Bislev, A., Gad, U.P. and Zeuthen, J.W. (eds) (2018) *Arctic Yearbook 2018. Special Section: China & the Arctic*. Akureyri, Iceland: Northern Research Forum, pp. 59–70.

Dent, C.T. (1892) *The Badmington Library of Sports and Pastimes – Mountaineering*. London: Longmans, Green, and Co.

Department of Defense, US (2011) *Report to Congress on Arctic Operations and the Northwest Passage*. Washington, DC: US Department of Defense.

Destination Canada (2020) *Year-end Review, 2019*. Ottawa: Destination Canada. https://www150.statcan.gc.ca/n1/daily-quotidien/200221/dq200221b-eng.htm (accessed 8 October 2020).

Dillon, P. (2019) *Trekking in Greenland: The Arctic Circle Trail*. Kendal, UK: Cicerone, 2nd. edn.

Dodds, K., and Nuttall, M. (2016) *The Scramble for the Poles: The Geopolitics of the Arctic and Antarctic*. Cambridge, UK: Polity Press.

Dodds, K., and Nuttall, M. (2018) Materialising Greenland within a critical Arctic geopolitics. In: Kristensen, K.S. and Rahbek-Clemmensen, J. (eds) *Greenland and the International Politics of a Changing Arctic: Postcolonial Paradiplomacy Between High and Low Politics*. Abingdon, UK: Routledge, pp. 139–154.

Doerr, L., Dorn, F., Gaebler, S. and Potrafke, N. (2020) How new airport infrastructure promotes tourism: evidence from a synthetic control approach in German regions. *Regional Studies* 54(10), 1402–1412.

Domenèch, A. and Gutiérrez, A. (2019) Determinants of cruise tourists' expenditure visiting port cities. *Tourism Review* doi: 75(5), 793–808.

Dorte Mandrup Arkitekter (2018) *The Icefjord Centre*. Copenhagen: Dorte Mandrup Arkitekter, 26 November. https://www.youtube.com/watch?v=K9CSTICsEtM (accessed 26 January 2020).

Dubois, G. and Ceron, J-P. (2006) Tourism and climate change: proposals for a research agenda. *Journal of Sustainable Tourism* 14(4), 399–415.

Dubois, G., Ceron, J-P. and Peeters, P. (2013) *Tourism Sensitivity to Climate Change Mitigation Policies: Lessons from Recent Surveys*. Abingdon, UK: Routledge.

Dubois, K. and Gagaridis, A. (2018) The security implications of China-Greenland relations. *The Polar Connection* 10 July. http://polarconnection.org/security-china-greenland-relations/ (accessed 4 January 2020).

Due, R. and Ingerslev, T. (eds) (2000) *Naturbeskyttelse i Grønland*. Nuuk: Grønlands Naturinstitut.

Dugmore, A.J., Keller, C. and McGovern, T.H. (2007) Norse Greenland settlement: reflections on climate change, trade, and the contrasting fates of human settlement on the North Atlantic islands. *Arctic Anthropology* 44(1), 12–36.

Duhaime, G. (2004) Economic systems. In: Einarsson, N., Larsen, J.N., Nilsson, A. and Young, O.R. (eds) (2004) *Arctic Human Development Report (AHDR)*. Akureyri, Iceland: Stefansson Arctic Institute, pp. 69–84.

Durfee, M. and Johnstone, R.L. (2019) *Arctic Governance in a Changing World*. Lanham, MD: Rowman and Littlefield.

Eckerstorfer, M., Christiansen, H.H. and Neumann, U. (2009) Avalanches and snowmobile traffic around Longyearbyen, Svalbad. International Snow and Ice Workshop, Davos, Switzerland, pp. 44–47.

Edensor, T. (2001) Performing tourism, staging tourism. (Re)producing tourist space and practice. *Tourist Studies* 1, 59–81.

Edensor, T. (2010) Aurora landscapes: affective atmospheres of light and dark. In: Benediktsson, K. and Lund, K. (eds) *Conversations with Landscape.* Farnham, UK: Ashgate, pp. 227–240.

EFLA Verkfræðistofa (2018) *Álagsmat umhverfis, innviða og samfélags gagnvart fjölda ferðamanna á Íslandi: 1. áfangi: Þróun vísa fyrir álagsmat.* Reykjavik: EFLA. https://www.stjornstodin.is/calendar-event/alagsmat-throun-alagsvisa-1-afangi (accessed 14 October 2020).

Egede, A.B. (2017) *The SDGs from the Greenland perspective*. Speech for the High Level Conference about the SDG 2030 agenda, Minister of Finance and Internal Revenue, Greenland, 30 November.

Egri, C.P. and Ralston, D.A. (2004) Generational cohorts and personal values: a comparison of China and the United States. *Organizational Science* 15(2), 210–220.

Eijgelaar, E., Thaper, C. and Peeters, P. (2010) Antarctic cruise tourism: the paradoxes of ambassadorship, 'last chance tourism' and greenhouse gas emissions. *Journal of Sustainable Tourism* 18(3), 337–354.

Einarsson, N., Larsen, J.N., Nilsson, A. and Young, O.R. (eds) (2004) *Arctic Human Development Report (AHDR)*. Akureyri, Iceland: Stefansson Arctic Institute.

Eiterjord, T.A. (2018) The growing institutionalization of China's Polar Silk Road. *The Diplomat* 7 October. https://thediplomat.com/2018/10/the-growing-institutionalization-of-chinas-polar-silk-road (accessed 13 January 2020).

Emmerson, C. (2010) *The Future History of the Arctic*. London: The Bodley Head.

Enderlin, E.M., Howat, I.M., Jeong, S., Noh, M-J., Angelen, J.H. van *et al.* (2014) An improved mass budget for the Greenland ice sheet. *Geophysical Research Letters* 41, 866–872.

Enge, M. (2020) 'No coloniser deserves to be on top of a mountain'. *Kunstkritikk* 22 June https://kunstkritikk.com/no-coloniser-deserves-to-be-on-top-of-a-mountain/ (accessed 22 June 2020).

English, J. (2013) *Ice and Water: Politics, Peoples, and the Arctic Council*. Toronto, ON: Allen Lane.

Epler Wood, M. (2017) The cruise industry. Empire of the seas. In: Epler Wood, M. (ed.) Sustainable Tourism on a Finite Planet. Abingdon, UK: Routledge, pp. 225–260.

Erokhin, V., Gao, T. and Zhang, X. (2018) Arctic blue economic corridor: China's role in the development of a new connectivity paradigm in the north. In: Heininen, L. and Exner-Pirot, H. (eds) *Arctic Yearbook 2018: Arctic Development in Theory and Practice.* Akureyri, Iceland: Northern Research Forum, pp. 456–474.

European Commission (EC) (1997) *Cohesion Report*. Brussels: EC.

European Parliament (2018) *China's Arctic Policy. How China Aligns Rights and Interests. Brussels: European Parliament*. https://EPRS_BRI(2018)620231_EN.pdf (accessed 4 September 2020).

European Union (2019) *Agreement to Prevent Unregulated High Seas Fisheries in the Central Arctic Ocean.* Brussels: EU, Decision (EU) 2019/407, 3 October.

Exner-Pirot, H. (2020) The Arctic in international affairs. In: Coates, K.S. and Holroyd, C. (eds) *The Palgrave Handbook of Arctic Policy and Politics*. Cham, Switzerland: Palgrave Macmillan, pp. 307–318.

Falardeau, M. and Bennett, E.M. (2020) Towards integrated knowledge of climate change in Arctic marine systems: a systematic literature review of multidisciplinary research. *Arctic Science* 6, 1–23.

Fan, Y. (2006) Branding the nation: what is being branded? *Journal of Vacation Marketing* 12(1), 5–14.

Fan, Y. (2008) Soft power: power of attraction or confusion? *Place Branding and Public Diplomacy* 4(2), 147–158.

Fan, Y. (2010) Branding the nation: towards a better understanding. *Place Branding and Public Diplomacy* 6(2), 97–103.

Fay, G. and Karlsdóttir, A. (2011) Social indicators for Arctic tourism: observing trends and assessing data. *Polar Geography* 34(1–2), 63–86.

Fazzini, M., Fratianni, S., Biancotti, A. and Billi, P. (2004) Skiability conditions in several skiing complexes on Piedmontese and Dolomitic Alps. *Meteorologische Zeitschrift* 13(3), 253–258.

Feng, A. and Saha, S. (2018) China's Arctic ambitions in Alaska. *The Diplomat* 20 April. https://thediplomat.com/2018/04/chinas-arctic-ambitions-in-alaska/ (accessed 3 July 2020).

Feng, J., Fan, W., Wang, B. and Zhang, T. (2016) Research on the innovation ability evaluation of traditional enterprise's business model for internet transition with hesitant fuzzy information. *Journal of Intelligent & Fuzzy Systems* 31(1), 91–97.

Fettweis, X., Franco, B., Tedesco, M., van Angelen, J.H., Lenaerts, J.T.M. *et al.* (2013) Estimating the Greenland ice sheet surface mass balance contribution to future sea level rise using the regional atmospheric climate model MAR. *The Cryosphere* 7, 469–489.

Fettweis, X., Hanna, E., Lang, C., Belleflamme, A., Erpicum, M. *et al.* (2013) Important role of the mid-tropospheric atmospheric circulation in the recent surface melt increase over the Greenland ice sheet. *The Cryosphere* 7, 241–248.

Finnish Chairmanship (2017) *Finland's Chairmanship Program for the Arctic Council 2017-2019*. Helsinki: Ministry for Foreign Affairs of Finland.

Fischer, H., Meissner, K.J., Mix, A.C., Abram, N.J., Austermann, J. *et al.* (2018) Palaeoclimate constraints on the impact of 2°C anthropogenic warming and beyond. *Nature Geoscience* 11, 474–485.

Fleischer, J. (2003) *Grønlands Historie - Kort Fortalt*. Copenhagen: Aschehoug.

Flora, J. (2011) 'I don't know why he did it; it happened by itself': causality and suicide in Northwest Greenland. In: High, C., Kelly, A. and Mair, J. (eds) *The Anthropology of Ignorance: An Ethnographic Approach*. New York, NY: Palgrave, pp. 137–161.

Flora, J. (2019) *Wandering Spirits: Loneliness and Longing in Greenland*. Chicago, IL: University of Chicago Press.

Foley, K. (2018) The politics of economic security: Denmark, Greenland and Chinese mining investment. In: Kristensen, K.S. and Rahbek-Clemmensen, J. (eds) *Greenland and the International Politics of a Changing Arctic: Postcolonial Paradiplomacy Between High and Low Politics*. Abingdon, UK: Routledge, pp. 98–112.

Fondahl, G., Espiritu, A.A. and Ivanova, A. (2020) Russia's Arctic regions and policies. In: Coates, K. and Holroyd, C. (eds) *The Palgrave Handbook of Arctic Policy and Politics*. Cham, Switzerland: Palgrave Macmillan, pp. 195–216.

Ford, J. and Smit, B. (2004) A framework for assessing the vulnerability of communities in the Canadian Arctic to risks associated with climate change. *Arctic* 57(4), 389–400.

Foster, G.M. (1998) South Seas cruise: a case study of a short-lived society. *Annals of Tourism Research* 13(2), 215–238.

Franklin, A. (2003) *Tourism: An Introduction*. London: Sage.

Franklin, A. (2004) Tourism as an ordering: towards a new ontology of tourism. *Tourist Studies* 4(3), 277–301.

Franklin, A.S. and Crang, M. (2001) The trouble with tourism and travel theory? *Tourist Studies* 1(1), 5–22.

Friedman, J. (2008) Indigeneity: anthropological notes on a historical variable. In: Minde, H. (ed.) *Indigenous Peoples: Self-Determination, Knowledge, Indigeneity*. Delft, Netherlands: Eburon Academic Publishers, pp. 29–37.

Fritz, M., Vonk, J.E. and Lantuit, H. (2017) Collapsing Arctic coastlines. *Nature Climate Change* 7, 6–7.

Fu, Y. (2012) Chinese tourists early adopters the Arctic Circule [sic], Antartic [sic] and Arctic tourism starting price 50,000 RMB.*China News* 16 October.

Gad, F. (1967) *Grønlands Historie. Vol. 1, Indtil 1700*. Copenhagen: Nyt Nordisk Forlag Arnold Busck.

Gad, F. (1969) *Grønlands Historie. Vol. 2, 1700 til 1782*. Copenhagen: Nyt Nordisk Forlag Arnold Busck.

Gad, F. (1984) *Grønlands Politikens Danmarkshistorie.* Copenhagen: Politikens Forlag.

Gad, U.P. (2014) Greenland: a post-Danish sovereign nation state in the making. *Cooperation and Conflict* 49(1), 98–118.

Gad, U.P. (2016) *National Identity Politics and Postcolonial Sovereignty Games: Greenland, Denmark and the EU*. Copenhagen: Museum Tusculanum Press.

Gad, U.P. (2017a) A reverse Greenland? In: Riddoch, L. and Bort, E. (eds) *McSmörgåsbord: What Post-Brexit Scotland can Learn from the Nordics*. Edinburgh, UK: Luath Press, pp. 87–98.

Gad, U.P. (2017b) Sex, løgn og landlingsbaner; Thule set relationelt. *Grønland* 65(3), 216–241.

Gad, U.P. (2019) Grønlandsk identitet og udvikling – Danske trusler og muligheder: sprogdebatten under hjemme- og selvstyre. In: Høiris, O., Marquardt, O. and Reimer, G.A. (eds) *Grønlændernes syn på Danmark: Historiske, Kulturelle og Sproglige Perspektiver.* Århus, Denmark: Aarhus Universitetsforlag, pp. 481–512.

Gad, U.P. (2020) Greenland, the Faroe Islands, and Denmark: unity or community? In: Munk Christiansen, P., Elklit, J. and Nedergaard, P. (eds) *The Oxford Handbook of Danish Politics.* Oxford, UK: Oxford University Press, pp. 28–45.

Gad, U.P., Graugaard, N.D., Holgersen, A., Jacobsen, M., Lave, N. *et al.* (2018) Imagining China on Greenland's road to independence. In: Bislev, A., Gad, U.P. and Zeuthen, J.W. (eds) *Arctic Yearbook 2018. Special Section: China & the Arctic*. Akureyri, Iceland: Northern Research Forum, pp. 6–28.

Gad, U.P., Jacobsen, M. and Strandsbjerg, J. (2019) Introduction: sustainability as a political concept in the Arctic. In: Gad, U.P. and Strandsbjerg, J. (eds) *The Politics of Sustainability in the Arctic: Reconfiguring Identity, Space, and Time*. Abingdon, UK: Routledge, pp. 1–18.

Galvani, A., Lew, A.A. and Sotelo Perez, M. (2020) COVID-19 is expanding global consciousness and the sustainability of travel and tourism. *Tourism Geographies* 22(3), 567–576.

Gao, F., Wang, J.P. and Tang, T.B. (2009) 世界主要海洋国家海洋发展战略分析 [Analysis of the ocean development strategies of the world's major ocean countries]. 世界科技研究与发展 [*World Sci-Tech R&D*] 31(5), 973–976.

Gardiner, S. and Kwek, A. (2017) Chinese participation in adventure tourism: a study of generation Y international students' perceptions. *Journal of Travel Research* 56(4), 496–506.

Gauthier, F. (2017) The economic development of Arctic navigation: a systematic literature review. In: Heininen, L., Exner-Pirot, H. and Plouffe, J. (eds) *Arctic Yearbook 2017: Change and Innovation in the Arctic: Policy, Society and Environment.* Akureyri, Iceland: Northern Research Forum, pp. 352–367.

Gautier, D.L., Bird, K.J., Charpentier, R.R., Grantz, A., Houseknecht, D.W. *et al.* (2009) Assessment of undiscovered oil and gas in the Arctic. *Science* 324(5931), 1175–1179.

Gavron, S. and Katznelson, D. (2012) *Village at the End of the World*. Copenhagen: MET Film, 78 minutes.

Gelbman, A. and Timothy, D.J. (2010) From hostile boundaries to tourist attractions. *Current Issues in Tourism* 13(3), 239–259.

Gentile, M. and Marcińczak, S. (2014) Housing inequalities in Bucharest: shallow changes in hesitant transition. *GeoJournal* 79 (4), 449–465.

Gerhardt, H. (2011) The Inuit and sovereignty: the case of the Inuit Circumpolar Conference and Greenland. *Tiddskriftet Politik* 14(1), 6–14.

Gerhardt, H. (2018) The divergent scalar strategies of the Greenlandic government and the Inuit Circumpolar Council. In: Kristensen, K.S. and Rahbek-Clemmensen, J. (eds.) *Greenland and the International Politics of a Changing Arctic: Postcolonial Paradiplomacy between High and Low Politics*. Abingdon, UK: Routledge, pp. 113–124.

Germany, L. (2018) *Exploring Bluie East Two – an Abandoned WWII Air Base in East Greenland*. Nuuk: guidetogreenland.com. https://guidetogreenland.com/travel-blogs/lisa-germany/exploring-bluie-east-two-an-abandoned-wwii-air-base-in-east-greenland/ (accessed 26 August 2020).

Gilchrist, H., Povey, R., Dickinson, A. and Povey, R. (1995) The Sensation Seeking Scale: its use in a study of the characteristics of people choosing 'adventure holidays'. *Personality and Individual Differences* 19(4), 513–516.

Gorbachev, M.S. (1987) *The Speech in Murmansk At the Ceremonial Meeting on the Occasion of the Presentation of the Order of Lenin and the Gold Star Medal to the City of Murmansk, 1 October 1987*. Moscow: Novosti Press Agency. https://www.barentsinfo.fi/docs/Gorbachev_speech.pdf (accessed 12 December 2019).

Gössling, S., Hall, C.M., Peeters, P. and Scott, D. (2010) The future of tourism: can tourism growth and climate policy be reconciled? A mitigation perspective. *Tourism Recreation Research* 35(2), 119–130.

Gössling, S., Scott, D. and Hall, C.M. (2021) Pandemics, tourism and global change: a rapid assessment of COVID-19. *Journal of Sustainable Tourism* 29(1), 1–20.

Gössling, S., Scott, D., Hall, C.M., Ceron, J-P. and Dubois, G. (2012) Consumer behaviour and demand response of tourists to climate change. *Annals of Tourism Research* 39(1), 36–58.

Gover, K. (2015) Settler-state political theory, CANZUS, and the UN Declaration on the Rights of Indigenous Peoples. *European Journal of International Law* 26, 345–373.

Government of Canada (2018) Canada's Arctic and Northern Policy Framework. Ottawa, ON: Government of Canada. https://www.rcaanc-cirnac.gc.ca/eng/1560523306861/1560523330587 (accessed 1 November 2020).

Government of Greenland (2012) *National Turismetrategi 2012-2015*. Nuuk: Government of Greenland.

Government of Greenland/Naalakkersuisut (2014) *Greenland Oil and Mineral Strategy 2014–2018*. Nuuk: Naalakkersuisut http://naalakkersuisut.gl/da/Naalakkersuisut/Departementer/Erhverv-og-Energi/Naalakkersuisuts-strategier

Government of Greenland/Naalakkersuisut (2015) *Turismeudvikling i Grønland Hvad Skal der Til?* Nuuk: Naalakkersuisut. http://naalakkersuisut.gl/~/media/Nanoq/Files/Hearings/2015/Turismestrategi/Documents/Turismestrategi%202016-2020%20FINAL%20DK.pdf (accessed 6 September 2019).

Government of Greenland/Naalakkersuisut (2017) *Speech at Future Greenland. Speech delivered by Minister for Fishery and Hunting Mr. Karl-Kristian Kruse on behalf of The Premier of Greenland Mr. Kim Kielsen*. Nuuk: Naalakkersuisut. http://naalakkersuisut.gl/en/Naalakkersuisut/News/2017/05/090517_KKK_Future-Greenland (accessed 13 September 2020).

Graham, R.M., Itkin, P., Meyer, A., Sundfjord, A., Spreen, G. *et al.* (2019) Winter storms accelerate the demise of sea ice in the Atlantic sector of the Arctic Ocean. *Scientific Reports* 9: 9222.

Greaves, W. (2016) Environment, identity, autonomy: Inuit perspectives on Arctic security. In: Hossain, K. and Petrétei, A. (eds) *Understanding the Many Faces of Human Security: Perspectives of Northern Indigenous Peoples*. Leiden, Netherlands: Brill, pp. 35–55.

Greaves, W. (2019) Arctic break up: climate change, geopolitics, and the fragmenting Arctic security region. In: In: Heininen, L., Exner-Pirot, H. and Barnes, J. (eds) *Redefining Arctic Security: Arctic Yearbook 2019.* Akureyri, Iceland: Arctic Portal, pp.156–172.

Green, J. (ed.) (2007) *Making Space for Indigenous Feminism*. Halifax, NS: Fernwood.

Greene, C.H. and Monger, B.C. (2012) An Arctic wild card in the weather. *Oceanography* 25 (2 June), 7–9.

Greenland Tourism (1995) *Oplæg til Turismestrategi 1996-2005.* Nuuk: Greenland Tourism.

Greenland Tourism (2001a) *Årsberetning 2000 – Fremtiden i Sigte*. Nuuk: Greenland Tourism.

Greenland Tourism (2001b) *Estimates of Number of Tourists to Greenland 1993-2000*. Nuuk: Greenland Tourism.

Greenland Tourism (2003) *Tre Vigtig År – Oplæg til Strategi og Handling for Turismen i Grønland 2003-2005*. Nuuk: Greenland Tourism.

Greenland Tourism Board (2005) *Tourism Economic Development Strategy*. Nuuk: Greenland Tourism Board. http://greenland.com (accessed 19 December 2019).

Gren, M. and Huijbens, E.H. (2014) Tourism and the anthropocene. *Scandinavian Journal of Hospitality and Tourism* 14(1), 6–22.

Grenier, A.A. and Müller, D. (eds) (2011) *Polar Tourism: a Tool for Regional Development*. Québec City, QC: Presses de l'Université du Québec.

Grimwood, B.S.R. (2015) Advancing tourism's moral morphology: relational metaphors for just and sustainable Arctic tourism. *Tourist Studies* 15(1), 3–26.

Grix, J. and Brannagan, P.M. (2016) Of mechanisms and myths: conceptualising states''soft power' strategies through sports mega-events. *Diplomacy and Statecraft* 27(2), 251–272.

Gronholt-Pedersen, J. (2018) Mining firms from Canada to China watch as Greenland holds election. *Reuters* 20 April. https://www.reuters.com/article/greenland-election/rpt-mining-firms-from-china-to-canada-watch-as-greenland-holds-election-idUKL8N1RX5S4?edition-redirect=uk (accessed 11 January 2020).

Gronholt-Pedersen, J. and Pamuk, H. (2020) US extends economic aid to Greenland to counter China, Russia in Arctic. *Reuters* 23 April. https://www.reuters.com/article/uk-usa-trump-greenland/u-s-economic-aid-to-greenland-draws-criticism-in-denmark-idUKKCN225202?edition-redirect=uk (accessed 4 June 2020).

Gronholt-Pedersen, J., Qiu, S. and Heavens, L. (2019) China withdraws bid for Greenland airport projects: *Sermitsiaq* newspaper. *Reuters* 4 June. https://www.reuters.com/article/us-china-silkroad-greenland-idUSKCN1T5191 (accessed 2 April 2020).

Grønlands Landsstyre (1998) *Turismeredegørelse*. Nuuk: Grønlands Landsstyre.

Grønlands Selvstyre (2009) *Redegørelse for Virkemidler til Reduktion af Udledningen af Drivhusgasser*. Nuuk : Grønlands Selvstyre EM2009/127.

Grønow, B. (2000) Palæo-eskimoerne: de Første Mennesker I Grønland. In: Jakobsen, B.H., Bøchner, J., Nielsen, N., Guttersen, R., Humlun, O. and Jensen, E. (eds) *Topografisk Atlas Grønland*. Copenhagen: Det Kongelige Danske Geografiske Selskab og Kort & Matrikelstyrelsen, pp. 46–51.

Groulx, M., Boluk, K., Lemieux, C. and Dawson, J. (2019) Place stewardship among last chance tourists. *Annals of Tourism Research* 75, 202–212.

Groulx, M., Lemieux, C., Dawson, J., Stewart, E. and Yudina, O. (2016) Motivations to engage in last chance tourism in the Churchill wildlife management area and Wapusk National Park: the role of place identity and nature relatedness. *Journal of Sustainable Tourism* 24(11), 1523–1540.

Grydehøj, A. (2014) Constructing a centre on the periphery: urbanization and urban design in the island city of Nuuk, Greenland. *Island Studies Journal* 9(2), 205–222.

Grydehøj, A. (2018a) Decolonising the economy in micropolitics: rents, government spending and infrastructure development in Kalaallit Nunaat (Greenland). *Small States and Territories* 1, 69–94.

Grydehøj, A. (2018b) Hearing voices: colonialism, outsider perspectives, island and Indigenous issues, and publishing ethics. *Island Studies Journal* 13, 3–12.

Grydehøj, A. (2020a) Government, policies, and priorities in Kalaallit Nunaat (Greenland): roads to independence. In: Coates, K.S. and Holroyd, C. (eds) *The Palgrave Handbook of Arctic Policy and Politics*. Cham, Switzerland: Palgrave Macmillan, pp. 217–231.

Grydehøj, A. (2020b) Unravelling economic dependence and independence in relation to island sovereignty: the case of Kalaallit Nunaat (Greenland). *Island Studies Journal* 15(1), 89–112.

Grydehøj, A. and Kelman, I. (2016) The eco-island trap: climate change mitigation and conspicuous sustainability. *Area* 49(1), 106–113.

Grydehøj, A., Kelman, I. and Su, P. (2020) Island geographies of separation and cohesion: the Coronavirus (COVID-19) pandemic and the geopolitics of Kalaallit Nunaat (Greenland). *Tijdschrift voor Economische en Sociale Geografie* 111(3), 288–301.

guidetogreenland.com (2020) *Staycation in Greenland*. Nuuk: guidetogreenland.com. https://guidetogreenland.com/book-trips-holiday/staycation-in-greenland (accessed 8 June 2020).

Guissard, I. and Lee, Y-S. (2017) Experiencing the Arctic in the past: French visitors to Finnmark in the late 1700s and early 1800s. In: Lee, Y-S., Weaver, D.B. and Prebensen, N.K. (eds) *Arctic Tourism Experiences: Production, Consumption and Sustainability*. Wallingford, UK: CABI, pp. 9–18.

Guldager, O. (2016) *The History of Narsarsuaq*. Narsarsuaq, Greenland: Arctic Sun.

Gulløv, H.C. (2000) Thulekulturen. In: Jakobsen, B.H., Bøchner, J., Nielsen, N., Guttersen, R., Humlun, O. and Jensen, E. (eds) *Topografisk Atlas Grønland*. Copenhagen: Det Kongelige Danske Geografiske Selskab og Kort & Matrikelstyrelsen, pp. 52–53.

Gunnarsson, B. (2021) Recent ship traffic and developing shipping trends on the Northern Sea Route – policy implications for future arctic shipping. *Marine Policy* 124, 104369.

Guo, F., Lane, J., Duan, Y., Stoltman, J.P., Khlebosolova, O. *et al.* (2018) Sustainable development in geography education for middle school in China. *Sustainability* 10(11), doi: 10.3390/su10113896.

Guo, L-F. and Zhang, S. (2008) A study on outbound tourism ethics of Chinese citizens and 'soft power' upgrade. *Tourism Tribune* 12. www.en.cnki.com.cn (accessed 10 June 2015).

Guttorm, H., Kantonen, L. and Kramvig, B. (2019) Pluriversal stories with indigenous wor(l)ds creating paths behind the next mountain. *Dutkansearvi* (Helsinki) 3(2), 148–172.

Gyimóthy, S. and Mykletun, R.J. (2004) Play in adventure tourism: the case of Arctic trekking. *Annals of Tourism Research* 31(4), 855–878.

Halkier, H. and Therkelsen, E. (2013) Exploring tourism destination path plasticity: the case of coastal tourism in North Jutland, Denmark. *Zeitschrift für Wirtschaftsgeographie* 57(1–2), 39–51.

Hall, C.M. (2007) North-South perspectives on tourism, regional development and peripheral areas. In: Müller, D.K. and Jansson, B. (eds) *Tourism in Peripheries: Perspectives From the Far North and South*. Wallingford, UK: CAB International, pp. 19–37.

Hall, C.M. (2013) Framing tourism geography: notes from the underground. *Annals of Tourism Research* 43, 601–623.

Hall, C.M. (2014a) Will climate change kill Santa Claus? Climate change and high latitude Christmas place branding. *Scandinavian Journal of Hospitality and Tourism* 14(1), 23–40.

Hall, C.M. (2014b) You can check out any time you like but you can never leave: can ethical consumption in tourism ever be sustainable? In: Weeden, C. and Boluk, K. (eds) *Managing Ethical Consumption in Tourism:Compromise and Tension*. Abingdon, UK: Routledge, pp. 32–56.

Hall, C.M. and Higham, J. (2005) *Tourism, Recreation and Climate Change*. Clevedon, UK: Channel View.

Hall, C.M. and Johnston, M.E. (1995) *Polar Tourism: Tourism in the Arctic and Antarctic Regions*. Chichester, UK: John Wiley & Sons.

Hall, C.M. and Saarinen, J. (2021) 20 years of Nordic climate change crisis and tourism research: a review and future research agenda. *Scandinavian Journal of Hospitality and Tourism* 21(1), 102–110.

Hall, D. (1999a) Conceptualising tourism transport: inequality and externality issues. *Journal of Transport Geography* 7(3), 181–188.

Hall, D. (1999b) Destination branding, niche marketing and national image projection in Central and Eastern Europe. *Journal of Vacation Marketing* 5(3), 227–237.

Hall, D. (2002) Brand development, tourism and national identity: the re-imaging of former Yugoslavia. *Brand Management* 9(4-5), 323–334.

Hall, D. (ed.) (2004) *Tourism and Transition: Governance, Transformation and Development*. Wallingford, UK: CABI Publishing.

Hall, D. (2017a) Bringing geopolitics to tourism. In: Hall, D. (ed.) *Tourism and Geopolitics: Issues and Concepts from Central and Eastern Europe*. Wallingford, UK: CABI, pp. 3–14.

Hall, D. (ed.) (2017b) *Tourism and Geopolitics: Issues and Concepts from Central and Eastern Europe*. Wallingford, UK: CABI.

Hall, D. (2020) *Brexit and Tourism: Process, Impacts and Non-Policy*. Bristol, UK: Channel View.

Hall, D. and Brown, F. (2017) Interrogating tourism's relevance: mediating between polarities in Kosovo. In: Hall, D. (ed.) *Tourism and Geopolitics: Issues and Concepts from Central and Eastern Europe*. Wallingford, UK: CABI, pp. 250–263.

Halpern, N. and Graham, A. (2015) Airport route development: a survey of current practice. *Tourism Management* 46, 213–221.

Hamilton, A.K., Laval, B.E., Mueller, D.R., Vincent, W.F. and Copland, L. (2017) Dynamic response of an Arctic epishelf lake to seasonal and long term forcing: implications for ice shelf thickness. *The Cryosphere* 11(5), 2189–2211.

Hansen, A.M. and Johnstone, R.L. (2020) Arctic voices: strategies for community engagement. In: Johnstone, R.L. and Hansen, A.M. (eds) *Regulation of Extractive Industries: Community Engagement in the Arctic*. London: Routledge, pp. 286–297.

Hansen, A.M., Larsen, S.V. and Noble, B. (2018) Social and environmental impact assessments of the Arctic. In: Nuttall, M., Christensen, T.R. and Siegert, M. (eds) *The Routledge Handbook of the Polar Regions*. Abingdon, UK: Routledge, pp. 380–390.

Hansen, A.M. and Ren, C. (eds) (2020) *Collaborative Methods in Arctic Research: Experiences from Greenland*. Aalborg, Denmark: Aalborg Universitetsforlag.

Hansen, A.M. and Tejsner, P. (2016) Identifying challenges and opportunities for residents in Upernavik as oil companies are making the first entrance into Baffin Bay. *Arctic Anthropology* 23(1), 84–94.

Hansen, A.M., Vanclay, F., Croal, P. and Skjervedal, A.S. (2016) Managing the social impacts of the rapidly expanding extractive industries in Greenland. *The Extractive Industries and Society* 3(1), 25–33.

Hansen, K. (2001) Til den bitre ende. *Polarfronten* 2, 15.

Hansson, H. (2012) An Arctic Eden: Alexander Hutchinson's 'Try Lapland and the Hospitable North'. *Northern Review* (Whitehorse, YT) 35, 147–166.

Hansson, H. (2018) The Arctic in literature and the popular imagination. In: Nuttall, M., Christensen, T.R. and Siegert, M. (eds) *The Routledge Handbook of the Polar Regions*. Abingdon, UK: Routledge, pp. 45–56.

Harris, P.G. (2017) China's Paris pledge on climate change: inadequate and irresponsible. *Journal of Environmental Studies and Sciences* 7(1), 102–107.

Harrison, R., Newholm, T. and Shaw, D. (2005) *The Ethical Consumer*. London: Sage.

Harvey, F. (2019) Greenland's ice sheet melting seven times faster than in 1990s. *The Guardian* 10 December. https://www.theguardian.com/environment/2019/dec/10/greenland-ice-sheet-melting-seven-times-faster-than-in-1990s (accessed 11 December 2019).

Hauser, D.D.W., Laidre, K.L. and Stern (2018) Vulnerability of Arctic marine mammals to vessel traffic in the increasingly ice-free Northwest Passage and Northern Sea Route. *Proceedings of the National Academy of Sciences of the United States* 115(20), 7617–7622.

Haycox, S. (2020) Arctic policy of the United States: an historical survey. In: Coates, K.S. and Holroyd, C. (eds) *The Palgrave Handbook of Arctic Policy and Politics*. Cham, Switzerland: Palgrave Macmillan, pp. 233–250.

Hazbun, W. (2004) Globalisation, re-teritorialisation and the political economy of tourism development in the Middle East. *Geopolitics* 9(2), 30–341.

Hegelund, L.E. (2009) *Tikeraaq. Tourism in Greenland*. Nuuk: Milik.

Heikkinen, M., Pihkala, S., Pääsky, L. and Harmoinen, S. (2020) Intersectional gender-responsibility in STEM: co-creating sustainable Arctic knowledge. In: Heininen, L., Exner-Pirot, H. and Barnes, J.

(eds) *Climate Change and the Arctic: Global Origins, Regional Responsibilities? Arctic Yearbook 2020.* Akureyri, Iceland: Arctic Portal, pp. 175–188.

Heimtun, B. and Lovelock, B. (2017) Communicating paradox: uncertainty and the Northern Lights. *Tourism Management* 61, 63–69.

Hein, C. (2016) Port cityscapes: conference and research contributions on port cities. *Planning Perspectives* 31(2), 313–326.

Heinämäki, L. (2015) Global context – Arctic importance: Free, Prior and Informed Consent, a new paradigm in international law related to Indigenous peoples. In: Herrmann, T.M. and Martin, T. (eds) *Indigenous Peoples Governance of Land and Protected Territories in the Arctic*. Cham, Switzerland: Springer, pp. 209–243.

Heininen, L., Exner-Pirot, H. and Barnes, J. (2020a) *Climate Change and the Arctic: Global Origins, Regional Responsibilities?* In: Heininen, L., Exner-Pirot, H. and Barnes, J. (eds) *Climate Change and the Arctic: Global Origins, Regional Responsibilities? Arctic Yearbook 2020.* Akureyri, Iceland: Arctic Portal, pp. 7–11.

Heininen, L., Exner-Pirot, H. and Barnes, J. (eds) (2020b) *Climate Change and the Arctic: Global Origins, Regional Responsibilities: Arctic Yearbook 2020.* Akureyri, Iceland: Arctic Portal.

Helm, T. and McKie, R. (2020) Follow Danes and halt North Sea exploration, Britain told. *The Observer* 6 December, p. 24.

Henley, J. (2016) Greenland's receding icecap to expose top-secret US nuclear project. *Guardian* 27 September. https://www.theguardian.com/world/2016/sep/27/receding-icecap-top-secret-us-nuclear-project-greenland-camp-century-project-iceworm (accessed 2 September 2020).

Herald (Glasgow) (2020) Denmark pledges to end its oil and gas offshore activities in the North Sea. *Herald*, 5 December, p. 14.

Hicks, J. (2007) The social determinants of elevated rates of suicide among Inuit youth. *IWGIA Indigenous Affairs* 4, 30–37.

Higgins-Desbiolles, F. (2020) Socialising tourism for social and ecological justice after COVID-19. *Tourism Geographies* 22(3), 610–623.

Hillgrén, A., Bröckl, M., Descombes, L., Kontiokari, V. and Halonen, M. (2016) *Nordic Best Practices Relevant for UNEP 10YFP on Sustainable Tourism and Consumer Information*. Copenhagen: Norden/Nordic Council of Ministers/TemaNord 546.

Hinch, T. (1998) Ecotourists and indigenous hosts: diverging views on their relationships with nature. *Current Issues in Tourism* 1(1), 120–124.

Hindley, A. and Font, X. (2018) Values and motivations in tourist perceptions of last-chance tourism. *Tourism and Hospitality Research* 18(1), 3–14.

Hjort, J., Karjalainen, O., Aalto, J., Westermann, S., Romanovsky, V.E. *et al.* (2018) Degrading permafrost puts Arctic infrastructure at risk by mid-century. *Nature Communications* 9(1), 5147.

HLN (2019) 'Feestschepen hebben hier geen plaats': cruises richting noordpoolgebied winnen aan populariteit. *HLN* (Brussels) 13 August. https://www.hln.be/reizen/-feestschepen-hebben-hier-geen-plaats-cruises-richting-noordpoolgebied-winnen-aan-populariteit~adf6708f/?referer=https%3A%2F%2Fwww.google.com%2F (accessed 22 August 2019).

Høeg, P. (1992) *Frøken Smillas Fornemmelse for Sne*. Copenhagen: Rosinante/Munksgaard. (1993) *Miss Smilla's Feeling for Snow*. (Trans. F. David) London: Harvill.

Holland, D. M., Thomas, R. H., de Young, B., Ribergaard, M. H. and Lyberth, B. (2008) Acceleration of Jakobshavn Isbrae triggered by warm subsurface ocean waters. *Nature Geoscience* 1(10), 659–664.

Home Rule Authority (1992) *Tourism Development Plan for Greenland 1991-2005 – Summary*. Nuuk: Greenland Home Rule Authority.

Hønneland, G. (2010) East-West collaboration in the European north. *International Journal* 65(4), 837–850.

Hoogensen Gjørv, G. (2017) Finding gender in the Arctic: a call to intersectionality and diverse methods. In: Latola, K. and Savela, H. (eds) *The Interconnected Arctic – UArctic Congress 2016*. Cham, Switzerland: Springer Open, pp. 293–303.

Hook, L. (2019) Boom town: Greenland's climate change gold rush. *FT Magazine Life & Arts* 20 September. https://www.ft.com/content/65dbd11a-d9a4-11e9-8f9b-77216ebe1f17 (accessed 3 November 2019).

Horejsova, T. and Paris, C.M. (2013) Tourism and the challenge of Arctic governance. *International Journal of Tourism Policy* 5(1-2), 113–127.

Hossain, K. and Maruyama, H. (2016) Japan's admission to the Arctic Council and commitment to the rights of its Indigenous Ainu people. *Polar Journal* 6(1), 169–187.

Huang, K. (2017) Will the Arctic be the next stop on China's new Silk Road? *South China Morning Post* 21 May. http://www.scmp.com/news/china/diplomacy-defence/article/2095078/will-arctic-be-next-stop-chinas-new-silk-road (accessed 10 January 2020).

Huang, M-F, Tang, C. and Weaver, D.B. (2017) The Arctic tourism experiences from an evolving Chinese perspective. In: Lee, Y-S., Weaver, D.B. and Prebensen, N.K. (eds) *Arctic Tourism Experiences: Production, Consumption and Sustainability*. Wallingford, UK: CABI, pp. 89–99.

Huang, R. (2020) Implications of Brexit for Chinese educational tourists in the UK. In: Hall, D. *Brexit and Tourism: Process, Impacts and Non-Policy*. Bristol, UK: Channel View, pp. 199–206.

Huang, R. and Xu, T. (2018) Opportunities and challenges brought by the China outbound travel market. *Worldwide Hospitality and Tourism Themes* 10(6), 642–651.

Huang, S., Keating, B.W., Kriz, A. and Heung, V. (2015) Chinese outbound tourism: an epilogue. *Journal of Travel and Tourism Marketing* 32(1–2), 153–159.

Huebert, R., Exner-Pirot, H., Lajeunesse, A. and Gulledge, J. (2012) *Climate Change and International Security: The Arctic as a Bellwether*. Arlington, VA: Center for Climate and Energy Solutions.

Huijbens, E.H. and Alessio, D. (2015) Arctic 'concessions' and icebreaker diplomacy? Chinese tourism development in Iceland. *Current Issues in Tourism* 18(5), 433–449.

Huijbens, E. and Lamers, M. (2017) Sustainable tourism and natural resource conservation in the polar regions: an editorial. *Resources* 6(3), 45–52.

Humpert, M. (2020) Greenland and U.S. agree on improved cooperation at Thule Air Base. *High North News* 30 October. https://www.highnorthnews.com/en/greenland-and-us-agree-imprioved-cooperation-thule-air-base (accessed 4 December 2020).

Huntington, H.P., Carey, M., Apok, C., Forbes, B.C., Fox, S. *et al.* (2019) Climate change in context: putting people first in the Arctic. *Regional Environmental Change* 19(4), 1217–1223.

Huskey, L., Mäenpää, I. and Pelyasov, A. (2014) Economic systems. In: Larsen, J.N. and Fondahl, G. (eds) *Arctic Human Development Report (AHDR-II)*. Copenhagen: Norden/Tema Nord/Nordic Council of Ministers.

Hussin, H. (2018) Gastronomy, tourism, and the soft power of Malaysia. *SAGE Open* October-December, 1-11, doi: 10.1177/2158244018809211.

Hyldal, C. (2018) Danmark betaler 180 millioner for at rydde op efter USA's militær i Grønland. *DR* (Danish Broadcasting Corporation) 11 January. https://www.dr.dk/nyheder/indland/danmark-betaler-180-millioner-rydde-op-efter-usas-militaer-i-groenland (accessed 1 September 2020).

Hynne, A. (2002) *Kultur & Turisme i Grønland: En Analyse og Diskussion af Turismens Kulturelle Aspekt med Udgangspunkt i en Diskursanalyse af Greenland Tourisms Brochurer*. Copenhagen: Institut for Eskimologi, Det Humanistiske Fakultet, Københavens Universitet.

IAATO (International Association of Antarctic Tour Operators) (1992–2019) *Antarctic Tourism Statistics*. http://www.iaato.org/ (accessed 21 August 2020).

IASC (International Arctic Science Committee) (2020a) International Arctic Science Committee. Akureyri, Iceland: IASC. https://www.iasc.info/ (accessed 29 October 2020).

IASC (International Arctic Science Committee) (2020b) T-MOSAiC. Akureyri, Iceland: IASC. https://www.t-mosaic.com/ (accessed 29 October 2020).

ICC (Inuit Circumpolar Council) (2011) *A Circumpolar Inuit Declaration on Resource Development Principles in Inuit Nunaat*. Ottawa, ON: ICC. https://iccalaska.org/wp-icc/wp-content/uploads/2016/01/Declaration-on-Resource-Development-A4-folder-FINAL.pdf (accessed 19 October 2020).

ICES (International Council for the Exploration of the Sea) (2015) *Report of the North-Western Working Group (NWWG): 28 April-5 May 2015*. Copenhagen: ICES.

Igloo Mountain ApS (2017) Solbakken Ski Area and Tøsebakken Mini-lift. Sisimiut, Greenland: Igloo Mountain ApS. http://ski.gl (accessed 21 August 2020).

ILO (International Labour Organization) (1989) *Convention Concerning Indigenous and Tribal Peoples in Independent Countries*. Paris: ILO.

IMO (International Maritime Organization) (2017) *International Code for Ships Operating in Polar Waters* [The Polar Code]. London: IMO. www.imo.org/en/MediaCentre/HotTopics/polar/pages/default.aspx (accessed 2 September 2019).

Ingeman-Nielsen, T. and Foged, N.N. (2016) The construction of Kangerlussuaq Airport - a case story from West Greenland. In: Günther, F. and Morgenstern, A. (eds) *Proceedings of the XI. International Conference on Permafrost: Exploring Permafrost in a Future Earth*. Potsdam, Germany: Bibliothek Wissenschaftspark Albert Einstein, pp. 1131–1133.

Innovasjon Norge (2017) *Standard for Bærekraftig Reisemål – Kriterier og Indikatorer*. Oslo: Innovasjon Norge.

Instituttet for Fremtidsforskning (2013) *Fremtidsscenarier for Grønland: Scenariebeskrivelser for Grønland*. Copenhagen: Institute for Future Studies.

Ioannides, D. (2019) Greenland's tourism policy making and the risk of overtourism. In: Milano, C., Cheer, J.M. and Novelli, M. (eds) *Overtourism: Excesses, Discontents and Measures in Travel and Tourism*. Wallingford, UK: CABI, pp. 209–223.

Ioannides, D. and Gyimóthy, S. (2020) The COVID-19 crisis as an opportunity for escaping the unsustainable tourism path. *Tourism Geographies* 22(3), 624–632.

IPCC (Intergovernmental Panel on Climate Change) (1997) *The Regional Impacts of Climate Change: An Assessment of Vulnerability. Summary for Policymakers. A Special Report of Intergovernmental Panel on Climate Change Working Group II*. Geneva, Switzerland: IPCC. https://www.ipcc.ch/site/assets/uploads/2018/03/regionen-2.pdf (accessed 9 February 2020).

IPCC (Intergovernmental Panel on Climate Change) (2018) *Special Report on Global Warming of 1.5°C (SR15)*. Geneva, Switzerland: IPCC and World Meterological Organization. https://www.ipcc.ch/sr15/ (accessed 15 January 2020).

IPCC (Intergovernmental Panel on Climate Change) (2020) *AR6 Synthesis Report: Climate Change 2022*. Geneva, Switzerland: IPCC. https://www.ipcc.ch/report/sixth-assessment-report-cycle/ (accessed 5 June 2020).

Islam, M.R. and Khan, N.A. (2018) Threats, vulnerability, resilience and displacement among the climate change and natural disaster-affected people in South-East Asia: an overview. *Journal of the Asia Pacific Economy* 23(2), 297–323.

Jaakson, R. (2004) Beyond the tourist bubble? Cruiseship passengers in port. *Annals of Tourism Research* 31(1), 44–60.

Jacobsen, M. (2019) Greenland's Arctic advantage: articulations, acts and appearances of sovereignty games. *Cooperation and Conflict* 55(2), 170–192.

James, L., Olsen, L.S. and Karlsdóttir, A. (2020) Sustainability and cruise tourism in the arctic: stakeholder perspectives from Ísafjörður, Iceland and Qaqortoq, Greenland. *Journal of Sustainable Tourism* 28(9), 1425–1441.

Jensen, O.G. (2007) *The Culture of Greenland in Glimpses*. Nuuk: Milik.

Jex, C. (2016) Arctic soils: a ticking climate time bomb. High North News (Tromsø, Norway) 1 December. https://www.highnorthnews.com/en/arctic-soils-a-ticking-climate-time-bomb (accessed 18 June 2020).

Jin, X. and Sparks, B. (2017) Barriers to offering special interest tour products to the Chinese outbound group market. *Tourism Management* 59, 205–215.

Jin, X.and Wang, Y. (2016) Chinese outbound tourism research: a review. *Journal of Travel Research* 55(4), 440–453.

Jin, X., Xiang, Y., Weber, K. and Liu, Y. (2019) Motivation and involvement in adventure tourism activities: a Chinese tourists' perspective. *Asia Pacific Journal of Tourism Research* 24(11), 1066–1078.

Jóhannesson G.T. (2016) A fish called tourism: emergent realities of tourism policy in Iceland. In: van der Duim, R.T., Ren, C. and Jóhannesson, G.T. (eds) *Tourism Encounters and Controversies: Ontological Politics of Tourism Development*. Farnham, UK: Ashgate, pp.181–200.

Jóhannesson, G.T., Huijbens, E. and Sharpley, R. (2010) Icelandic tourism: past directions – future challenges. *Tourism Geographies* 12(2), 278–301.

Jóhannesson, G.T. and Lund, K.A. (2017) Aurora Borealis: choreographies of darkness and light. *Annals of Tourism Research* 63, 183–190.

Johnson, D. (2002) Environmentally sustainable cruise tourism: a reality check. *Marine Policy* 26, 261–270.

Johnson, N. (2020) Extractive energy and Arctic communities. In: Coates, K.S. and Holroyd, C. (eds) *The Palgrave Handbook of Arctic Policy and Politics*. Cham, Switzerland: Palgrave Macmillan, pp. 97–116.

Johnston, A. I. (2013) How new and assertive is China's new assertiveness? *International Security* 37(4), 7–48.

Johnston, A. M. (2006) *Is the Sacred for Sale? Tourism and Indigenous Peoples*. London: Earthscan.

Johnston, M.E. (1995) Patterns and issues in Arctic and sub-Arctic tourism. In: Hall, C.M. and Johnston, M.E. (eds) *Polar Tourism: Tourism in Arctic and Antarctic Regions*. Chichester, UK: John Wiley & Sons, pp. 27–42.

Johnston, M.E. (2011) Arctic tourism introduction. In: Maher, P.T. Stewart, E.J. and Lück, M. (eds) *Polar Tourism: Human, Environmental and Governance Dimensions*. Putnam Valley, NY: Cognizant Communication Corporation, pp. 17–32.

Johnston, M.E. and Viken, A. (1997) Tourism development in Greenland. *Annals of Tourism Research* 24(4), 978–982.

Johnston, M., Viken, A. and Dawson, J. (2012) Firsts and lasts in Arctic tourism: last chance tourism and the dialectics of change. In: Lemelin, R.H., Dawson, J. and Stewart, E.J. (eds) *Last Chance Tourism: Adapting Tourism Opportunities in a Changing World*. Abingdon, UK: Routledge, pp. 10–24.

Johnston, V.K. (1933) Canada's title to the Arctic islands. *Canadian Historical Review* 14, 24–41.

Johnstone, R.L. and Hansen, A.M. (2020a) Introduction. In: Johnstone, R.L. and Hansen, A.M. (eds) *Regulation of Extractive Industries: Community Engagement in the Arctic*. London: Routledge, pp. 1–8.

Johnstone, R.L. and Hansen, A.M. (eds) (2020b) *Regulation of Extractive Industries: Community Engagement in the Arctic*. London: Routledge.

Jolly, J. (2020) China among 15 nations to sign historic free-trade deal. *The Guardian* 16 November, p. 26.

Jonassen, T. (2020a) 'For every centimeter the sea level rises, one million more people will have to evacuate. *High North News* 18 September. https://www.highnorthnews.com/en/uk-for-every-centimeter-the-sea-level-rises-one-million-more-people-will-have-to-evacuate (accessed 19 September 2020).

Jonassen, T. (2020b) UK Parliament establishes cross-party group dedicated to Greenland. *High North News* 19 November. https://www.highnorthnews.com/en/uk-parliament-establishes-cross-party-group-dedicated-greenland (accessed 4 December 2020).

Jones, B. (2011) 'White Heat' by MJ McGrath: review. *Daily Telegraph* 13 March. www.telegraph.co.uk/culture/books/8371383/White-Heat-by-MJ-McGrath-review.html (accessed 7 July 2019).

Jonson, L. (1990) What's in store for northern Europe in the Gorbachev era? In: Harle, V. and Iivonen, J. (eds) *Gorbachev and Europe*. London: Pinter, pp. 130–152.

Jørgensen, M.T. and Bertelsen, R.G. (2020) Chinese tourism in the Nordic Arctic – opportunities beyond the economic. *Scandinavian Journal of Hospitality and Tourism* 20(2) 166–177.

Jørgensen, M.T., Law, R. and King, B.E. (2017) Understanding the past, anticipating the future – a critical assessment of China outbound tourism research. *Journal of Travel and Tourism Marketing* 34(7), 880–891.

Jørgensen, M.T. and Ren, C. (2015) Getting 'China ready'. Challenging static and practice based configurations of the Chinese tourist. *Aei Insights* 1(1), 19–35.

Joyce, L. and Burns, M. (eds) (2013) *Greenland & North Pole*. Richmond, BC: ITM Publishing, 3rd. edition.

Jussila, H. (1998) Marginality in regional policy research: a view from the Nordic countries. In: Jussila, H., Leimgruber, W. and Majoral, R. (eds) *Perception of Marginality: Theoretical Issues and Regional Perceptions of Marginality*. Aldershot, UK: Ashgate, pp. 217–235.

Kaae, B.C. (2002) Nature and tourism in Greenland. In: Watson, A.E., Alessa, L. and Sproull, J. (eds) *Wilderness in the Circumpolar North: Searching for Compatibility in Ecological, Traditional, and Ecotourism Values*. Ogden, UT: US Department of Agriculture, Forest Service, Rocky Mountain Research Station, pp. 43–53.

Kaae, B.C. (2006) Greenland/Kalaallit Nunaat. In: Baldacchino, G. (ed.) *Extreme Tourism: lessons From the World's Cold Water Islands*. Oxford, UK: Elsevier, pp. 100–112.

Kaae, B.C. (2007) Tourism research in Greenland. In: Müller, D.K. and Jansson, B. (eds) *Tourism in Peripheries: Perspectives From the Far North and South*. Wallingford, UK: CAB International, pp. 205–219.

Kaján, E. (2013) An integrated methodological framework: engaging local communities in Arctic tourism development and community-based adaptation. *Current Issues in Tourism* 16(3), 286–301.

Kaján, E. (2014a) Arctic tourism and sustainable adaptation: community perspectives to vulnerability and climate change. *Scandinavian Journal of Hospitality and Tourism* 14(1), 60–79.

Kaján, E. (2014b) Community perceptions to place attachment and tourism development in Finnish Lapland. *Tourism Geographies* 16(3), 490–511.

Kankaanpää, P. and Young, O.R. (2012) The effectiveness of the Arctic Council. *Polar Research* 31(1), 10–14.

Käpylä, J. and Mikkola, H. (2015) *On Arctic Exceptionalism: Critical Reflections in the Light of the Arctic Sunrise Case and the Crisis in Ukraine*. Helsinki: The Finnish Institute of International Affairs, FIIA Working Paper 85.

Katz, C. (2018) Alien waters: neighboring seas are flowing into a warming Arctic Ocean. *Yale Environment 360,* 10 May. https://e360.yale.edu/features/alien-waters-neighboring-seas-are-flowing-into-a-warming-arctic-ocean (accessed 7 March 2020).

Kazda, A., Hromádka, M. and Mrekaj, B-R. (2017) Small regional airports operation: unnecessary burdens or key to regional development? *Transportation Research Procedia* 28, 59–68.

Keating, B., Huang, S., Kriz, A. and Heung, V. (2015) A systematic review of the Chinese outbound tourism literature: 1983–2012. *Journal of Travel and Tourism Marketing* 32(1-2), 2–17.

Keating, B. and Kriz, A. (2008) Outbound tourism from China: literature review and research agenda. *Journal of Hospitality and Tourism Management* 15(1), 32–41

Keohane, R. and Nye, J. (1998) Power and interdependence in the information age. *Foreign Affairs* 77(5), 81–94.

Keskitalo, E.C.H. (2017) Images of the Northern and 'Arctic' in tourism and regional literature. In: Viken, A. and Müller, D.K. (eds) *Tourism and Indigeneity in the Arctic*. Bristol, UK: Channel View, pp. 33–49.

Keskitalo, E.C.H. and Schilar, H. (2017) Co-constructing 'northern' tourism representations among tourism companies, DMOs and tourists. An example from Jukkasjärvi, Sweden. *Scandinavian Journal of Hospitality and Tourism* 17(4), 406–422.

Keskitalo, E., Schilar, H., Heldt Cassel, S. and Pashkevich, A. (2019) Deconstructing the indigenous in tourism: the production of indigeneity in tourism-oriented labelling and handicraft/souvenir development in Northern Europe. *Current Issues in Tourism* 24(1), 16–32.

Khadaroo, J. and Seetanah, B. (2007) Transport infrastructure and tourism development. *Annals of Tourism Research* 34(4), 1021–1032.

Khazendar, A., Fenty, I.G., Carroll, D., Garner, A., Lee, C.M. *et al.* (2019) Interruption of two decades of Jakobshavn Isbrae acceleration and thinning as regional ocean cools. *Nature Geoscience* 12, 277–283.

Kikkert, P. and Lackenbauer, P.W. (2019) Canada's Arctic and Northern Policy Framework: a roadmap for the future? In: Heininen, L., Exner-Pirot, H. and Barnes, J. (eds) *Redefining Arctic Security: Arctic Yearbook 2019.* Akureyri, Iceland: Arctic Portal, pp. 322–331.

Kim, D., Lee, C.K. and Sirgy, M.J. (2016) Examining the differential impact of human crowding versus spatial crowding on visitor satisfaction at a festival. *Journal of Travel and Tourism Marketing* 33, 293–312.

Kingdom of Denmark (2008) *Ilulissat Declaration on the Arctic Ocean, Adopted 28 May 2008*. Ilulissat, Greenland. www.oceanlaw.org/downloads/arctic/Ilulissat_Declaration.pdf (accessed 15 April 2020).

Kintisch, E. (2017) Meltdown. *Science* 355(6327), 788–791.

Kirchner, S. (2018) Greening Arctic cruise shipping through law and technology: a role for China? In: Bislev, A., Gad, U.P. and Zeuthen, J.W. (eds) (2018) *Arctic Yearbook 2018. Special Section: China & the Arctic*. Akureyri, Iceland: Northern Research Forum, pp.71–83.

Kirk, L. (2018) Greenland votes with eye on independence. *EU Observer* 24 April. https://euobserver.com/nordic/141674 (accessed 10 January 2020).

Kirkegaard, J-E. and Kielsen, K. (2014) *Vores råstoffer skal skabe velstand: Grønlands olie- og mineralstrategi 2014–2018. Sammenfatning*. Nuuk: Naalakkersuisut/Government of Greenland.

Kirshenblatt-Gimblett, B. (1998) *Destination Culture: Tourism, Museums and Heritage*. Berkeley, CA: University of California Press.

Kivelä, T. (2014) Actor-Networks of Northern Lights Tourism in Iceland, Norway and Finland. Rovaniemi, Finland: University of Lapland, unpublished thesis.

Klare, M.T. (2013) Rushing for the Arctic's riches. *New York Times* 7 December. www.nytimes.com/2013/12/08/opinion/sunday/rushing-for-the-arctics-riches.html?pagewanted=all (accessed 15 July 2020).

Klarskov, K. (2018) Løkke advarer Grønland mod at lade kinesere bygge lufthavne. *Politiken* 24 May.

Klein, R.A. (2010) Cruises and bruises: safety, security and social issues on Polar cruises. In: Lück, M., Maher, P.T. and Stewart, E.J. (eds) *Cruise Tourism in the Polar Regions: Promoting Environmental and Social Sustainability?* London: Earthscan, pp. 57–74.

Kleivan, I. (1992) The Arctic Peoples' Conference in Copenhagen, November 22–25, 1973. *Études/Inuit/Studies* 16(1/2), 227–236.

Kobzeva, M. (2019) China's Arctic policy: present and future. *The Polar Journal* 9(1), 94–112.

Kohut, R. and Prior, T. (2016) Overlooking a regional crux of vulnerability: missing women in the Arctic. In: Heininen, L., Exner-Pirot, H. and Plouffe, J. (eds) *Arctic Yearbook 2016. The Arctic Council: 20 Years of Regional Cooperation and Policy-Shaping*. Akureyri, Iceland: Northern Research Forum, pp. 336–340.

Koivurova, T. (2010) Limits and possibilities of the Arctic Council in a rapidly changing scene of Arctic governance. *Polar Record* 46(2), 146–156.

Koivurova, T. and Heinämäki, L. (2006) The participation of Indigenous peoples in international norm-making in the Arctic. *Polar Record* 42(2), 101–109.

Kopra, S. and Nojonen, M. (2020) The elusive norm of climate responsibility: the Belt and Road Initiative and COVID-19. In: Heininen, L., Exner-Pirot, H. and Barnes, J. (eds) *Climate Change and the Arctic: Global Origins, Regional Responsibilities? Arctic Yearbook 2020.* Akureyri, Iceland: Arctic Portal, pp. 13–25.

Korneliussen, N. (2014) *HOMO Sapienne*. Nuuk: Milik. (2018) *Crimson*. [Trans. A. Halager] London: Virago.

Kramvig, B. (2019) Foreward: towards a Nordic manifesto. In: Cassinger, C., Lucarelli, A. and Gyimóthy, S. (eds) *The Nordic Wave in Place Branding: Poetics, Practices, Politics*. Cheltenham, UK: Edward Elgar, pp. xix–xxii.

Kramvig, B. and Førde, A. (2020) Stories of reconciliation enacted in the everyday lives of Sámi tourism entrepreneurs. *Acta Borealia* 37(1–2), 27–42.

Kreutzmann, M. and Engell, U. (2020) *5 Mythical Creatures from Greenland: Mother of the Sea, Tupilak, Qivittoq and More*. Nuuk: Visit Greenland. https://www.youtube.com/watch?v=qc7Cy9NxHmg (accessed 10 November 2020).

Kristensen, K.S. and Rahbek-Clemmensen, J. (2018) Introduction: Greenland and the international politics of a changing Arctic. In: Kristensen, K.S. and Rahbek-Clemmensen, J. (eds) *Greenland and the International Politics of a Changing Arctic: Postcolonial Paradiplomacy Between High and Low Politics*. Abingdon, UK: Routledge, pp. 1–10.

Kruse, K-K. (2017) *Speech at Future Greenland 9 May*. Nuuk: Naalakkersuisut – Government of Greenland. https://naalakkersuisut.gl//en/Naalakkersuisut/News/2017/05/090517_KKK_Future-Greenland (accessed 10 December 2019).

Kühn, M. (2015) Peripheralization. Theoretical concepts explaining socio-spatial inequalities. *European Planning Studies* 23(2), 367–378.

Kuokkanen, R. (2008) Globalization as racialized, sexualized violence: the case of Indigenous women. *International Feminist Journal of Politics* 10(2), 216–233.

Kuokkanen, R. (2009) Indigenous women in traditional economies: the case of Sámi reindeer herding. *Signs* 34(3), 499–504.

Kuokkanen, R. (2014) Gendered violence and politics in Indigenous communities. *International Feminist Journal of Politics* 17(2), 271–288.

Kuokkanen, R. (2017) The pursuit of Inuit sovereignty in Greenland. *Northern Public Affairs* July, 46–49.

Kuusama, T. (2020) Evolution of the Arctic Council agenda: from environmental protection to the effects of climate change. In: Heininen, L., Exner-Pirot, H. and Barnes, J. (eds) *Climate Change and the Arctic: Global Origins, Regional Responsibilities? Arctic Yearbook 2020.* Akureyri, Iceland: Arctic Portal, pp. 478–494.

Kyhn, L.A., Sveegaard, S. and Tougaard, J. (2014) Underwater noise emissions from a drillship in the Arctic. *Marine Pollution Bulletin* 86(1-2), 424–433.

Lackenbauer, P.E., Lajeunesse, A., Manicom, J. and Lasserre, F. (2018) *China's Arctic Ambitions and What They Mean for Canada*. Calgary, AB: University of Calgary Press.

Lahey, K., Svensson, E-M. and Gunnarsson, Å. (2014) Gender challenges and human capital in the Arctic. In: Heininen, L., Exner-Pirot, H. and Plouffe, J. (eds) *Arctic Yearbook 2014: Human Capital in the North*. Akureyri, Iceland: Northern Research Forum, pp. 183–199.

Lamers, M. and Amelung, B. (2010) Climate change and its implications for cruise tourism in the Polar regions. In: Lück, M., Maher, P.T. and Stewart, E.J. (eds) *Cruise Tourism in the Polar Regions: Promoting Environmental and Social Sustainability?* London: Earthscan, pp. 147–163.

Landauer, M., Pröbstl, U. and Haider, W. (2012) Managing cross-country skiing destinations under the conditions of climate change – scenarios for destinations in Austria and Finland. *Tourism Management* 33, 741–751.

Lang, H. (2017) See the abandoned WWII base on Greenland leaking toxic waste. *National Geographic* 25 July (updated 11 January 2018), https://www.nationalgeographic.co.uk/travel-and-adventure/2018/01/see-abandoned-wwii-base-greenland-leaking-toxic-waste (accessed 26 August 2020).

Langgård, K. (2014) Greenlandic attitudes towards Norwegians and Danes from Nansen's icecap crossing to the 1933 World Court verdict in The Hague. *Études/Inuit/Studies* 38(1–2), 53–1.

Lanteigne, M. (2020) The Internet in the Arctic: crucial connections. *Over The Circle* 5 April. https://overthecircle.com/2020/04/05/the-internet-in-the-arctic-crucial-connections/ (accessed 25 June 2020).

Lantto, P. and Mörkenstam, U. (2008) Sami rights and Sami challenges: the modernization process and the Swedish Sami movement, 1886-2006. *Scandinavian Journal of History* 33(1), 26–51.

Larsen, J.N. and Fondahl, G. (eds) (2014) *Arctic Human Development Report: Regional Processes and Global Linkages (AHDR-II)*. Copenhagen: Norden/Tema Nord/Nordic Council of Ministers.

Larsen, J.N. and Petrov, A.N. (2020) The economy of the Arctic. In: Coates, K.S. and Holroyd, C. (eds) *The Palgrave Handbook of Arctic Policy and Politics*. Cham, Switzerland: Palgrave Macmillan, pp. 79–95.

Larsen, M. and Alslund-Lantén, A. (2017) *Bumps on the Road to 2030*. Copenhagen: Norden/Nordic Council of Ministers.

Larsson, N. (2015) *Mind the Gender Gap*. Toronto, ON: The Gordon Foundation.

Lasserre, F. (2018) Arctic shipping routes: from the Panama myth to reality. *International Journal* 66(4), 793–808.

Lat Multilingual (2018) Canada-China marketing tourism 2018 – attracting Chinese visitors to Canada. *lat Multilingual* (Vancouver and Montreal) 16 April. https://www.latmultilingual.com/canada-china-tourism-attracting-chinese-visitors/ (accessed 8 December 2020).

Law, R., Sun, S., Fong, D.K.C., Fong, L.H.N. and Fu, H. (2016) A systematic review of China's outbound tourism research. *International Journal of Contemporary Hospitality Management* 28(12), 2654–2674.

Lebrun, M., Vancoppenolle, M., Madec, G. and Massonnet, F. (2019) Arctic Sea-ice-free season projected to extend into autumn. *The Cryosphere* 13(1), 79–96.

Lee, Y-S. (2020) Asia and Arctic tourism. *Scandinavian Journal of Hospitality and Tourism* 20(2) 105–109.

Lee, Y-S., Weaver, D.B. and Prebensen, N.K. (2017a) Arctic destinations and attractions as evolving peripheral settings for the production and consumption of peak tourism experiences. In: Lee, Y-S., Weaver, D.B. and Prebensen, N.K. (eds) *Arctic Tourism Experiences: Production, Consumption and Sustainability*. Wallingford, UK: CABI, pp. 1–8.

Lee, Y-S., Weaver, D.B. and Prebensen, N.K. (2017b) Arctic tourism experiences: opportunities, challenges and future research directions for a changing periphery. In: Lee, Y-S., Weaver, D.B. and Prebensen, N.K. (eds) *Arctic Tourism Experiences: Production, Consumption and Sustainability*. Wallingford, UK: CABI, pp. 191–197.

Leichenko, R.M. and O'Brien, K. (2008) *Environmental Change and Globalization: Double Exposure*. Oxford, UK: Oxford University Press.

Leineweber, M., Bjerregaard, P., Baerweldt, P., Voestermans, C. and Voestermans, P. (2001) Suicide in a society in transition. *International Journal of Circumpolar Health* 60(2), 280–287.

Lemelin, H. and Whipp, P. (2019) Last chance tourism: a decade in review. In: Timothy, D.J. (ed.) *Handbook of Globalisation and Tourism*. Cheltenham, UK: Edward Elgar, pp. 316–322.

Lemelin, R.H., Dawson, J., Stewart, E.J., Maher, P. and Lück, M. (2010) Last chance tourism: the boom, doom, and gloom of visiting vanishing destinations. *Current Issues in Tourism* 13(5), 477–493.

Lemelin, R.H., Dawson, J. and Stewart, E.J. (eds) (2012) *Last Chance Tourism: Adapting Tourism Opportunities in a Changing World*. Abingdon, UK: Routledge.

Lemieux, C.J., Groulx, M., Halpenny, E., Stager, H., Dawson, J. *et al.* (2017) 'The end of the ice age?' Disappearing world heritage and the climate change communication imperative. *Environmental Communication* 12(5), 653–671.

Lenzen, M., Sun, Y.Y., Faturay, F., Ting, Y.P., Geschke, A. *et al.* (2018) The carbon footprint of global tourism. *Nature Climate Change* 8, 522–528.

Leu, T.C. and Müller, D.K. (2016) Maintaining inherited occupations in changing times. The role of tourism among reindeer herders in Northern Sweden. *Polar Geography* 39(1), 40–57.

Levinson-King, R. (2019) Huawei heats up the battle for internet in Canada's north. *BBC News* 9 September. https://www.bbc.com/news/world-us-canada-49415867 (accessed 13 January 2020).

Lew, A.A. (2018) Why travel? – Travel, tourism and global consciousness. *Tourism Geographies* 20(4), 742–749.

Li, W. and Huang, Z. (2017) Andersen awards highlight Danish writer's appeal in China. *China Daily* 31 May. https://www.chinadaily.com.cn/culture/2017-05/31/content_29552856.htm#:~:text=Chinese%20versions%20of%20Andersen's%20fairy,appeared%20in%20the%20early%201930s (accessed 2 December 2020).

Lidegaard, M. (1991) *Grønlands Historie*. Copenhagen: Nyt Nordisk Forlag Arnold Busck.

Lim, K.S. (2018) China's Arctic policy and the Polar Silk Road vision. In: Heininen, L. and Exner-Pirot, H. (eds) *Arctic Yearbook 2018: Arctic Development in Theory and Practice*. Akureyri, Iceland: Northern Research Forum, pp. 420–436.

Lincoln, A. (2020) The first Arctic culture in North America. In: Lincoln, A., Cooper, J. and Laurens Loovers, J.P. (eds) *Arctic Culture and Climate*. London: Thames & Hudson and The British Museum, pp. 199–206.

Lindquist, L.A. and Golub, R.M. (2004) Cruise ship care: a proposed alternative to assisted living facilities. *Journal of the American Geriatrics Society* 52(11), 1951–1954.

Lindstrøm, M. (2020) Kim Kielsen: drop hundeslæde og snescooter til nabobyer. *Sermitsiaq AG* (Nuuk) 17 March. https://sermitsiaq.ag/node/220171 (accessed 9 July 2020).

Linehan, D., Clark, I.D. and Xie, P.F. (eds) (2020) *Colonialism, Tourism and Place: Global Transformations in Tourist Destinations.* Cheltenham, UK: Edward Elgar.

Lino, M.R (2020) *Understanding China's Arctic Activities*. London: International Institute of Strategic Studies (IISS). https://www.iiss.org/blogs/analysis/2020/02/china-arctic (accessed 30 November 2020).

Liu, N. (2019) China's Arctic policy and Belt and Road Initiative: synergy or conflict? *The Yearbook of Polar Law Online* 10(1), 431–434.

Löfgren, O. (1991) The nationalization of culture: constructing Swedishness. *Studia Ethnologica* 3, 101–116.

Lopez, B. (1986) *Arctic Dreams*. New York, NY: Charles Scribner's Sons; (1999) London: The Harvill Press.

Lov om Grønands Selvystre (2009) 12 June. http://www.stm.dk/_p_5490.html.

Lowenthal, D. (1997) *The Heritage Crusade*. London: Viking.

Lück, M. (2010) Environmental impacts of Polar cruises. In: Lück, M., Maher, P.T. and Stewart, E.J. (eds) *Cruise Tourism in the Polar Regions: Promoting Environmental and Social Sustainability?* London: Earthscan, 109–131.

Lukovich, J. and McBean, G. (2009) Addressing human security in the Arctic in the context of climate change through science and technology. *Mitigation and Adaptation Strategies for Global Change* 14(8), 697–710.

Lulu J. (2017) *China, Greenland and Competition for the Arctic*. Nottingham, UK: University of Nottingham, Asia Research Institute. https://theasiadialogue.com/2017/01/02/china-greenland-and-competition-for-the-arctic/ (accessed 10 January 2020).

Lumholt, M. (2020) *Tourism Statistics Report 2019 Greenland*. Nuuk: Visit Greenland.

Lund, K.A. (2016) Chasing the Lights: darkness, tourism and the Northern Lights. In: Abram S. and Lund K. (eds) *Green Ice: Tourism Ecologies in the European High North*. Basingstoke, UK: Palgrave Macmillan, pp. 49–71.

Lunde, L., Yang, J. and Stensdal, I. (2015) *Asian Countries and the Arctic Future.* Singapore: World Scientific.

Lundqvist, A., Nilstun, T. and Dykes, A-K. (2002) Experiencing neonatal death: an ambivalent transition into motherhood. *Pediatric Nursing* 28(6), 621–625.

Lutz, S., Anesio, A., Jorge Villar, S. and Benning, L. (2014) Variations of algal communities cause darkening of a Greenland glacier. *FEMS Microbiology Ecology*. 89, 402–414.

Lutz, S., McCutcheon, J., McQuaid, J.B. and Benning, L.G. (2018) The diversity of ice algal communities on the Greenland Ice Sheet as revealed by oligotyping. *Microbial Genomics* 4(3), e000159.

Lycke, L. (ed.) (1998) *Turismestrategi og Udvikling i Grønland*. Copenhagen: Nordic Press.

Lynge, I. (1994) Suicide in Greenland. *Arctic Medical Research* 53(S2), 551–554.

MacCannell, D. (1992) *Empty Meeting Grounds: The Tourist Papers*. London: Routledge.

Machleit, K.A., Eroglu, S.A. and Mantel, S.P. (2000) Perceived retail crowding and shopping satisfaction: what modifies this relationship? *Journal of Consumer Psychology* 9, 29–42.

Maher, P.T. (2007) Arctic tourism: a complex system of visitors, communities, and environments. *Polar Geography* 30(1-2), 1–5.

Maher, P.T. (2017) Tourism futures in the Arctic. In: Latola, K. and Savela, H. (eds) *The Interconnected Arctic – UArctic Congress 2016*. Cham, Switzerland: Springer Open, pp. 213–220.

Maher, P.T., Gelter, H., Hillmer-Pegram, K., Hovgaard, G., Hull, J. *et al.* (2014) Arctic tourism: realities and possibilities. In: Heininen, L. (ed.) *Arctic Yearbook*. Akureyri, Iceland: Northern Research Forum, pp. 290–306.

Maher, P.T., Stewart, E.J. and Lück, M. (2011) *Polar Tourism: Human, Environmental and Governance Dimensions*. Putnam Valley, NY: Cognizant Communication Corporation.

Mairs, J. (2016) Dorte Mandrup unveils plans for tent-like climate research centre in Greenland. *Dezeen* (Copenhagen) 30 June. https://www.dezeen.com/2016/06/30/dorte-mandrup-unveils-plans-for-tent-like-climate-research-centre-in-greenland/ (accessed 26 January 2020).

Máliková, L., Farrell, M. and McDonagh, J. (2016) Perception of marginality and peripherality in an Irish rural context. *Quaestiones Geographicae* (Poznań) 35(4), 93–105.

Mallory, T.G (2015) Preparing for the Ocean Century: China's changing political institutions for ocean governance and maritime development. *Issues and Studies* 51(2), 111–138.

Mancini Billson, J. and Mancini, K. (2007) *Inuit Women: Their Powerful Spirit in a Century of Change*. Lanham, MD: Rowman & Littlefield.

Mangano, S. and Ugolini, G.M. (2020) New opportunities for cruise tourism: the case of Italian historic towns. *Sustainability* 12, 4646, doi: 10:3390/su12114616.

Marcińczak, S., Gentile, M., Rufat, S. and Chelcea, L. (2014) Urban geographies of hesitant transition: tracing socioeconomic segregation in post-Ceaușescu Bucharest. *International Journal of Urban and Regional Research* 38 (4), 1399–1417.

Marfelt, M.M. (2016) Grounded intersectionality: key tensions, a methodological framework, and implications for diversity research. *Equality, Diversity and Inclusion* 35(1), 31–47.

Mariani, M.M., Buhalis, D., Longhi, C. and Vitouladiti, O. (2014) Managing change in tourism destinations: key issues and current trends. *Journal of Destination Marketing and Management* 2, 269–272.

Markussen, U. (2017) Towards an Arctic awakening: neo-colonialism, sustainable development, emancipatory research, collective action, and Arctic regional policymaking. In: Latola, K. and Savela, H. (eds) *The Interconnected Arctic – UArctic Congress 2016*. Cham, Switzerland: Springer Open, pp. 305–311.

Martin, N. (2020) *Preparing for Post-Covid Travel to Greenland*. Nuuk: Visit Greenland. https://traveltrade.visitgreenland.com/latest-news/webinar-summary-of-preparing-for-post-covid-travel-to-greenland/ (accessed 2 July 2020).

Martin, N. and Ren, C. (2020) Adventurous Arctic encounters? Exploring Chinese adventure tourism. *Scandinavian Journal of Hospitality and Tourism* 20(2), 126–143.

Mason, O.K. (2020) The Punuk/Birnirk interregnum of the 9th to 11th centuries. In: Lincoln, A., Cooper, J. and Laurens Loovers, J.P. (eds) *Arctic Culture and Climate*. London: Thames & Hudson and The British Museum, pp. 212–219.

Mason, P. (1994) A visitor code for the Arctic. *Tourism Management* 15(2), 93–97.

Mason, P. (1997) Tourism codes of conduct in the Arctic and sub-Arctic region. *Journal of Sustainable Tourism* 5(2), 151–165.

Mason, P., Johnston, M.E. and Twynam, G.D. (2000) The World Wide Fund for Nature Arctic tourism project. *Journal of Sustainable Tourism* 8(4), 305–323.

Mason, P. and Legg, S.J. (1999) Antarctic tourism: activities, impacts, management issues and a proposed research agenda. *Pacific Tourism Review* 3, 71–84.

Mathisen, S.R. (2010) Indigenous spirituality in the touristic borderzone: virtual performances of Sámi Shamanism in Sápmi Park. *Temenos* 46, 53–72.

Mathisen, S.R. (2017) Northern Lights experiences in the Arctic dark: old imaginaries and new tourism narratives. In: Lee, Y-S., Weaver, D.B. and Prebensen, N.K. (eds) *Arctic Tourism Experiences: Production, Consumption and Sustainability*. Wallingford, UK: CABI, pp. 67–78.

Matzen, E. (2017) Denmark spurned Chinese offer for Greenland base over security: sources. *Reuters* 6 April. https://www.reuters.com/article/us-denmark-china-greenland-base-idUSKBN1782EE (accessed 1 September 2020).

Matzen, E. and Daly, T. (2018) Greenland's courting of China for airport projects worries Denmark. *Reuters* 22 March. https://uk.reuters.com/article/uk-china-arctic-greenland/greenlands-courting-of-china-for-airport-projects-worries-denmark-idUKKBN1GY2LK (accessed 15 August 2019).

Mauss, M. ([1906] 1979) *Seasonal Variations of the Eskimo: A Study in Social Morphology*. London: Routledge and Kegan Paul.

McBoyle, G., Scott, D. and Jones, B. (2007) Climate change and the future of snowmobiling in non-mountainous regions of Canada. *Managing Leisure* 12, 237–250.

McClory, J. (2019) *The Soft Power 30: A Global Ranking of Soft Power 2019*. Los Angeles, CA: Portland PR and USC (University of Southern California) Center on Public Diplomacy.

McCormick, E. and Greenfield, P. (2020) Revealed: 100,000 crew never made it off cruise ships amid coronavirus crisis. *The Guardian* 30 April. https://www.theguardian.com/environment/2020/apr/30/no-end-in-sight-100000-crew-on-cruise-ships-stranded-at-sea-coronavirus (accessed 1 May 2020).

McDowell, G. and Ford, J.D. (2014) The socio-ecological dimensions of hydrocarbon development in the Disko Bay region of Greenland: opportunities, risks, and tradeoffs. *Applied Geography* 46, 98–110.

McEwan, I. (2010) *Solar*. London: Jonathan Cape.

McGrath, M.J. (2011) *White Heat*. London: Pan Macmillan.

McGuire, A.D., Lawrence, D.M., Koven, C., Clein, J.S., Burke, E. *et al*. (2018) Dependence of the evolution of carbon dynamics in the Northern Permafrost Region on the trajectory of climate change. *Proceedings of the National Academy of Sciences* 115(15), 3882–3887.

McGwin, K. (2020a) A US agency is ready to fund a major Greenland zinc mine. *Arctic Today* 2 November. https://www.arctictoday.com/a-us-agency-is-ready-to-fund-a-major-greenland-zinc-mine/ (accessed 22 November 2020).

McGwin, K. (2020b) Greenland looks to grow tourism by tapping adventure, winter travel trends. *Arctic Today* 27 February. https://www.arctictoday.com/greenland-looks-to-grow-tourism-by-taping-adventure-winter-travel-trends/ (accessed 27 July 2020).

McGwin, K. (2020c) In an online poll, Nuuk residents lean towards keeping Egede statue in place. *Arctic Today* 17 July. https://www.arctictoday.com/in-an-online-poll-nuuk-residents-lean-towards-keeping-egede-statue-in-place/ (accessed 20 July 2020).

Meire, L., Mortensen, J., Meire, P., Juul-Pedersen, T., Sejr, M.K. *et al*. (2017) Marine-terminating glaciers sustain high productivity in Greenland fjords. *Global Change Biology* 23, 5344–5357.

Mered, M. (2018) Groenland: de l'autonomie danoise à la tutelle chinoise? *La Revue Parlementaire* 25 May. https://www.larevueparlementaire.fr/index.php?option=com_k2&view=item&id=1465:groenland-de-l-autonomie-danoise-a-la-tutelle-chinoise&Itemid=1524 (accessed 11 January 2020).

Merlan, F. (2009) Indigeneity, global and local. *Current Anthropology* 50, 303–333.

Meskell, L. (2015) Transacting UNESCO World Heritage: gifts and exchanges on a global stage. *Social Anthropology* 23, 3–21.

Miles, T. (2019) Europe's record heatwave threatens Greenland ice sheet. *Reuters* 26 July. https://www.reuters.com/article/us-europe-weather-greenland/europes-record-heatwave-threatens-greenland-ice-sheet-idUSKCN1UL15C (accessed 10 December 2019).

Mileski, J., Gharehgozli, A., Ghoram, L. and Swaney, R. (2018) Cooperation in developing a disaster prevention and response plan for Arctic shipping. *Marine Policy* 92, 131–137.

Milne, R. (2019) Greenland's cherished ambitions tied to its mineral bounty. *Financial Times* 29 August. https://www.ft.com/content/b71354a0-c990-11e9-a1f4-3669401ba76f (accessed 2 February 2020).

Ministeriet for Grønland (1973) *Turisme i Grønland*. Copenhagen: Ministeriet for Grønland. https://www.foxylex.dk/media/betaenkninger/Turisme_i_Groenland_naitsumik_kalaatdlisut_ nailisagartalik.pdf (accessed 1 November 2020).

Ministry for Foreign Affairs, Iceland (2015) *Gender Equality in the Arctic: Current Realities Future Challenges (2014 Conference Report)*. Akureyri, Iceland: Ministry for Foreign Affairs.

Ministry of Foreign Affairs, Denmark (2017a) *China-Denmark tourism year 2017 launched in Beijing*. Copenhagen: Ministry of Foreign Affairs. https://kina.um.dk/en/about-us/news/newsdisplaypage/?newsid=b816bd15-506d-4f7c-91cf-fc495b083fca (accessed 2 August 2020).

Ministry of Foreign Affairs, Denmark (2017b) *SDGs in the Arctic: Local and Global Perspectives – Key Insights*. Copenhagen: Ministry of Foreign Affairs. http://um.dk/en/foreign-policy/the-arctic/the-sdgs-in-the-arctic/key-insights (accessed 13 September 2020).

Mofengwo (2020) About us. Beijing: Mofengwo.http://www.mofengwo.cn/s/about.html (accessed 30 November 2020).

Møller-Sørensen, L. (2018) *Sermersooq Business Councils Erhvervsplan 2018*. Nuuk: Kommuneqarfik Sermersooq (accessed 2 September 2020).

Moon, T., Joughin, I., Smith, B. and Howat, I. (2012) 21st-century evolution of Greenland outlet glacier velocities. *Science* 336, 576–578.

Moriarty, L.F., Plucinski, M.M., Marston, B.J., Kurbatova, E.V., Knust, B. *et al*. (2020) Public health responses to COVID-19 outbreaks on cruise ships - worldwide, February–March 2020. *Morbidity and Mortality Weekly Report* 69(12), 347–352.

Moscato, D. (2018) The Polar Silk Road in the popular press: global media framing of China's 2018 Arctic policy white paper. In: Heininen, L. and Exner-Pirot, H. (eds) *Arctic Yearbook 2018: Arctic Development in Theory and Practice*. Akureyri, Iceland: Northern Research Forum, pp. 475–486.

Moss, S. (2009) *Cold Earth*. London: Granta.

Mulford, C.L. and Rogers, D.L. (1982) Definitions and models. In: Rogers, D.L. and Whetten, D.A. (eds) *Interorganizational Coordination: Theory, Research and Implementation*. Ames, IA: Iowa State University Press, pp. 9–31.

Müller, D.K. (2015) Issues in Arctic tourism. In: Evengård, B., Larsen, J.N. and Paasche, Ö. (eds) *The New Arctic*. Heidelberg, Germany: Springer, pp. 147–158.

Müller, D.K. and Kuoljok Huuva, S. (2009) Limits to Sami tourism development: the case of Jokkmokk, Sweden. *Journal of Ecotourism* 8(2), 115–127.

Müller, D.K. and Pettersson, R. (2006) Sami heritage at the winter festival in Jokkmokk, Sweden. *Scandinavian Journal of Hospitality and Tourism* 6(1), 54–69.

Müller, D.K. and Viken, A. (2017a) Indigenous tourism in the Arctic. In: Viken, A. and Müller, D.K. (eds) *Tourism and Indigeneity in the Arctic*. Bristol, UK: Channel View, pp. 3–15.

Müller, D.K. and Viken, A. (2017b) Toward a de-essentializing of indigenous tourism? In: Viken, A. and Müller, D.K. (eds) *Tourism and Indigeneity in the Arctic*. Bristol, UK: Channel View, pp. 281–289.

Mykletun, R.J. (2018) Adventure tourism in the North – six illustrative cases. *Scandinavian Journal of Hospitality and Tourism* 18(4), 319–329.

Naatsorsueqqissaartarfik (Statistics Greenland) (2018) *Cruise Ship Statistics*. Nuuk: Naatsorsueqqissaartarfik.

Naatsorsueqqissaartarfik (Statistics Greenland) (2019) *Cruise Ship Statistics 2018*. Nuuk: Naatsorsueqqissaartarfik.

Naatsorsueqqissaartarfik (Statistics Greenland) (2020) *Cruise Ship Statistics 2019*. Nuuk: Naatsorsueqqissaartarfik.

NATA (North Atlantic Tourism Association) (2015) *Strategy 2015-2025*. Tórshaven: NATA. https://www.northatlantic-islands.com/about-nata/strategy.html (accessed 11 October 2020).

NATA (North Atlantic Tourism Association) (2020) *About NATA*. Tórshaven: NATA. http://www.northatlantic-islands.com/about-nata.html (accessed 11 October 2020).

National Bureau of Statistics (2020) *China Statistical Yearbook 2020*. Beijing: National Bureau of Statistics. http://www.stats.gov.cn/tjsj/ndsj/2020/indexeh.htm (accessed 12 January 2021).

NDRC and SOA (2017) *Vision for Maritime Cooperation under the Belt and Road Initiative*. Beijing: Chinese National Development and Reform Commission and the State Oceanic Administration. http://www.xinhuanet.com/english/2017-06/20/c_136380414.htm (accessed 16 October 2019).

Neue Osnabrücker Zeitung (2019) Arctic explorer Arved Fuchs: party ships have no business in the Arctic. *Neue Osnabrücker Zeitung* 12 August. https://www.presseportal.de/pm/58964/4345937 (accessed 22 August 2019).

Neumann, R.B., Moorberg, C.J., Lundquist, J.D., Turner, J.C., Waldrop, M.P. *et al.* (2019) Warming effects of spring rainfall increase methane emissions from thawing permafrost. *Geophysical Research Letters* 46(3), 1393–1401.

Neuvonen, M., Sievänen, T., Fronzek, S., Lathinen, I., Veijalainen, N. *et al.* (2015) Vulnerability of cross-country skiing to climate change in Finland - an interactive mapping tool. *Journal of Outdoor Recreation and Tourism* 11, 64–79.

Newman, D. (2020) International Indigenous rights law and contextualized decolonization of the Arctic. In: Coates, K.S. and Holroyd, C. (eds) *The Palgrave Handbook of Arctic Policy and Politics*. Cham, Switzerland: Palgrave Macmillan, pp. 427–437.

NHO Reiseliv (2017) *Norsk Reiseliv Har Større Tro på Asia Enn Noe Annet Marked*. Oslo: NHO Reiseliv. https://www.nhoreiseliv.no/bransjerog-fakta/fakta-reiselivsnaringen/nyhet/2017/norsk-reiseliv-harstorre-tro-pa-asia-enn-noe-annet-marked/ (accessed 2 August 2020).

NHO Reiseliv (2020) *This is the Norwegian Hospitality Association*. Oslo: NHO Reiseliv. https://www.nhoreiseliv.no/tall-og-fakta/ (accessed 11 October 2020).

Nielsen, H.P. (2020) Being (un)stuck in Qaqortoq: attachment, ambivalence, and affect in contemporary Greenland. In: Gaini, F. and Nielsen, H.P. (eds) *Gender and Island Communities*. Abingdon, UK: Routledge, pp. 46–63.

Nielsen, K.H., Nielsen, H. and Martin-Nielsen, J. (2014) City under the ice: the closed world of Camp Century in Cold War culture. *Science as Culture* 23(4), 443–464.

Noël, B., van de Berg, W.J., Lhermitte, S., Wouters, B., Machguth, H. *et al.* (2017) A tipping point in refreezing accelerates mass loss of Greenland's glaciers and ice caps. *Nature Communications* 8, 14730.

Noël, B., van de Berg, W.J., Lhermitte, S., Wouters, B., Schaffer, N. *et al.* (2018) Six decades of glacial mass loss in the Canadian Arctic archipelago. *Journal of Geophysical Research: Earth Surface* 123(6), 1430–1449.

Nomura (2017) *Asia Economic Outlook 2018*. Tokyo: Nomura. https://nomuraconnects.com/focused-thinking-posts/asia-economic-outlook-2018/ (accessed 9 December 2020).

Nordbo, M.P. (2017) *Pigen Uden Hud*. Copenhagen: Politikens Forlag. (2018) *The Girl Without Skin*. [Trans. C. Barsland] Melbourne, Australia: Text Publishing.

Nordbo, M.P. (2018) *Kold Angst*. Copenhagen: Politikens Forlag. (2019) *Cold Fear*. [Trans. C. Barsland] Melbourne, Australia: Text Publishing.
Nordic Council of Ministers (2001) *Towards a Sustainable Nordic Tourism*. Copenhagen: Norden/Tema Nord/ Nordic Council of Ministers.
Nordic Council of Ministers (2013) *A Good Life in a Sustainable Nordic Region Nordic Strategy for Sustainable Development*. Copenhagen: Norden/Nordic Council of Ministers.
Nordic Council of Ministers (2017) *Generation 2030*. Copenhagen: Norden/ Nordic Council of Ministers.
Notzke, C. (1999) Indigenous tourism development in the Arctic. *Annals of Tourism Research* 26(1), 55–76.
NSIDC (National Snow and Ice Data Center) (2020) *Climate Change in the Arctic*. Boulder, CO: NSIDC, 4 May. https://nsidc.org/cryosphere/arctic-meteorology/climate_change.html (accessed 9 July 2020).
Nuttall, M. (2012) Imagining and governing the Greenlandic resource frontier. *The Polar Journal* 2(1), 113–124.
Nuttall, M. (2013) Zero-tolerance, uranium and Greenland's mining future. *The Polar Journal* 3(2), 368–383.
Nuttall, M. (2016) Living in a world of movement: human resilience to environmental stability in Greenland. In: Crate, S.A. and Nuttall, M. (eds) *Anthropology and Climate Change: from Encounters to Actions*. Abingdon, UK: Routledge, pp. 292–310.
Nuttall, M. (2017) *Under the Ice: Climate, Society and Subsurface Politics in Greenland*. Abingdon, UK: Routledge.
Nye, J.S. (1990) Soft power. *Foreign Policy* 80(3), 153–171.
Nye, J. (2004) Soft power and American foreign policy. *Political Science Quarterly* 119(2), 255–270.
Nye, J.S. (2008) *The Powers to Lead*. Oxford, UK: Oxford University Press.
Nye, J.S. (2009) Get smart: combining hard and soft power. *Foreign Affairs* 88(4), 160–163.
O'Neill, J., Moffatt, M.E.K., Tate, R.B. and Young, T.K. (1994) Suicidal behaviour among Inuit in the Keewatin Region. *Arctic Medical Research* 53(S2), 558–561.
O'Neill, P.H. (1996) Hungary's hesitant transition. *Current History* 95(199), 135–139.
Oddsdóttir, E.E., Ágústsson, H.Ó., Svensson, E-M., Retter, G-B., Hayfield, E.A. *et al*. (eds) (2021) *Gender and Diversity in the Arctic*. Akureyri, Iceland: Icelandic Arctic Cooperation Network.
Oddsdóttir, E., Sigurdsson, A.M. and Svandal, S. (eds) (2015) *Gender Equality in the Arctic: Current Realities, Future Challenges: Conference Report*. Reykjavik: Ministry for Foreign Affairs.
OECD (Oganisation for Economic Cooperation and Development) (2020) *OECD Tourism Trends and Policies 2020*. Paris: OECD Publishing.
OHCHR (Office of the United Nations High Commissioner for Human Rights) (2006) *Convention on the Elimination of All Forms of Discrimination against Women New York, 18 December 1979*. Geneva, Switzerland: OHCHR.
OHCHR (Office of the United Nations High Commissioner for Human Rights) (2020) *Human Rights and Climate Change*. Geneva, Switzerland: OHCHR.
Øian, H., Fredman, P., Sandell, K., Sæþórsdóttir, A.D., Tyrväinen, L. *et al*. (2018) *Tourism, Nature and Sustainability. A Review of Policy Instruments in the Nordic Countries*. Copenhagen: Norden/Nordic Council of Ministers/TemaNord 534.
Olafsdóttir, G. (2013) On nature-based tourism. *Tourist Studies* 13(2), 127–138.
Olsen, K. (2017) World Heritage List = tourism attractiveness? In: Lee, Y-S., Weaver, D.B. and Prebensen, N.K. (eds) *Arctic Tourism Experiences: Production, Consumption and Sustainability*. Wallingford, UK: CABI, pp. 48–55.
Olsen, L. (2018) Løkke vil bruge 700 millioner på lufthavne i Grønland: 'En historisk dag'. *DR* (Copenhagen) 10 September. https://www.dr.dk/nyheder/politik/loekke-vil-bruge-700-millioner-paa-lufthavne-i-groenland-en-historisk-dag (accessed 18 October 2020).
Olsen, L.S. (2016) Sami tourism in destination development: conflict and collaboration. *Polar Geography* 39(3), 179–195.
Olsen, N. (2010) Kalaalimernit – an intangible cultural heritage. *The Journal Greenland* 3, 218.
Oltmanns, M., Straneo, F. and Tedesco, M. (2019) Increased Greenland melt triggered by large-scale, year-round precipitation events. *The Cryosphere* 13(3), 815–825.
Olwig, K.R. and Lowenthal, D. (eds) (2006) *The Nature of Cultural Heritage and the Culture of Natural Heritage. Northern Perspectives on a Contested Patrimony*. London: Routledge.
Ooi, C. (2016). Soft power, tourism. In: Jafari, J. and Xiao, H. (eds) *Encyclopedia of Tourism*. Cham, Switzerland: Springer, pp. 1–2.
Ooi, C.S. (2019) Asian tourists and cultural complexity: implications for practice and the Asianisation of tourism scholarship. *Tourism Management Perspectives* 31, 14–23.

Overland, J., Dunlea, E., Box, J.E., Corell, R., Forsius, M. *et al.* (2018) The urgency of Arctic change. *Polar Science* 21, 6–13.

Pattyn, F., Ritz, C., Hanna, E., Asay-Davis, X., DeConto, R. *et al.* (2018) The Greenland and Antarctic ice sheets under 1.5°C global warming. *Nature Climate Change* 8, 1053–1061.

Pearce, P.L. and Lee, U. (2005) Developing the travel career approach to tourist motivation. *Journal of Travel Research* 43, 226–237.

Pedersen, A. (2002) *World Heritage Manuals 1. Managing Tourism at World Heritage Sites: A Practical Manual for World Heritage Site Managers*. Paris: UNESCO World Heritage Centre.

Peeters, P. (2007) *Tourism and Climate Change Mitigation: Methods, Greenhouse Gas Reductions and Policies*. Breda, Netherlands: Strichting NHTV Breda.

Pelaudeix, C. (2015) China's interests in the Arctic and the EU's Arctic policy: towards a proactive EU foreign policy. *Yearbook of Polar Law* 7(1), 128–150.

Pelaudeix, C. (2017) EU-Greenland relations and sustainable development in the Arctic. In: Conde, E. and Sánchez, S.I. (eds) *Global Challenges in the Arctic Region: Sovereignty, Environment and Geopolitical Balance*. Abingdon, UK: Routledge, pp. 306–324.

Pelaudeix, C. (2018) *Along the Road: China in the Arctic*. Paris: European Union Institute for Security Studies (EUISS). https://www.jstor.org/stable/pdf/resrep21468.pdf (accessed 30 November 2020).

Persily, L. (2020) Energy numbers in US – China trade deal don't add up. *Alaska Journal of Commerce* (Anchorage) 29 January. https://www.alaskajournal.com/2020-01-29/energy-numbers-us-china-trade-deal-dont-add-up (accessed 4 February 2020).

Petersen, M.L.D. (1986) The impact of planning on ethnic culture: aspects of Danish resettlement policies in Greenland after World War II. *Arctic Anthropology* 23(1-2), 271–280.

Petersen, N. (2008) The iceman that never came: 'Project Iceworm', the search for a NATO deterrent, and Denmark, 1960-1962. *Scandinavian Journal of History* 33(1), 75–98.

Petersen, N. (2011) Greenland in the U.S. Polar Strategy. *Journal of Cold War Studies* 13(2), 90–115.

Petersen, N. (2013) The politics of US military research in Greenland in the early Cold War. *Centaurus* 55(3), 294–318.

Petersen, R. (1995) Colonialism as seen from a former colonized area. *Arctic Anthropology* 32(2), 118–126.

Pettersson, R. and Viken, A. (2007) Sami perspectives on indigenous tourism in northern Europe: commerce or cultural development? In: Butler, R. and Hinch, T. (eds) *Tourism and Indigenous Peoples: Issues and Implications.* Oxford, UK: Elsevier, pp. 176–187.

Petrick, J. (2005) Segmenting cruise passengers with price sensitivity. *Tourism Management* 26(5), 753–762.

Petridou, E., Olausson, P.M. and Ioannides, D. (2019) Nascent island tourism policy in Greenland: a network perspective. *Island Studies Journal* 14(2), 227–244.

Petrov, A. (2016) Exploring the Arctic's other economies: knowledge, creativity and the new frontier. *The Polar Journal* 6(1), 30–50.

Petrov, A.N., BurnSilver, S., Chapin III, F.S., Fondahl, G., Graybill, J.K. *et al.* (2017) *Arctic Sustainability Research: Past, Present and Future*. London: Routledge.

Piggott-McKellar, A.E. and McNamara, K.E. (2017) Last chance tourism and the Great Barrier Reef. *Journal of Sustainable Tourism* 25(3), 397–415.

Poppel, B. (2006) Interdependency of subsistence and market economies in the Arctic. In: Glomsrød, S. and Alaksen, I. (eds) *The Economy of the North*. Oslo-Kongsvinger: Statistics Norway, pp. 65–80.

Por, T. (2020) *Future of Travel in Greenland: Flattening the Curve of the Coronavirus*. Nuuk: Visit Greenland, 25 November. https://www.youtube.com/watch?v=E8hYhii51qQ&feature=youtu.be&utm_source=B2C+newsletter+from+Visit+Greenland&utm_campaign=eadde07291- (accessed 27 November 2020).

Poto, M.P. and Fornabaio, L. (2017) Participation as the essence of good governance: some general reflections and a case study on the Arctic Council. *Arctic Review on Law and Politics* 8, 139–159.

Preuss, L. (1932) The dispute between Denmark and Norway over the sovereignty of East Greenland. *American Journal of International Law* 26(3), 469–487.

Prideaux, B., Thompson, M. and Pabel, A. (2020) Lessons from COVID-19 can prepare global tourism for the economic transformation needed to combat climate change. *Tourism Geographies* 22(3), 667–678.

Prior, T.L., Duyck, S., Heinämäki, L., Koivurova, T. and Stepien, A. (2013) *Addressing Climate Vulnerability: Promoting the Participatory Rights of Indigenous Peoples and Women through Finnish Foreign Policy*. Rovaniemi, Finland: University of Lapland Press.

Qaqatorq Museum (2019) Greenland during the second European period of colonization. Qaqatorq, Greenland: Museum presentation (viewed 28 July).

Rahbek-Clemmensen, J. (2016) 'An Arctic great power'? Recent developments in Danish Arctic policy. In: Heininen, L., Exner-Pirot, H. and Plouffe, J. (eds) *Arctic Yearbook 2016*. Akureyri, Iceland: Northern Research Forum, pp. 383–396.

Rahbek-Clemmensen, J. and Thomasen, G. (2018) *Learning from the Ilulissat Initiative: Power, Institutional Legitimacy, and Governance in the Arctic Ocean 2007–18*. Copenhagen: University of Copenhagen, Centre for Military Studies.

Rambøll Group (2014) *Hvor Skal Udviklingen Komme Fra? Potentialer og Faldgruber i den Grøndlandske Erhvervssektor frem Mod 2015*. Copenhagen: Rambøll Group for Greenpeace.

Rantala, O., de la Barre, S., Granås, B., Jóhannesson, G.T., Müller, D. *et al.* (2019) *Arctic Tourism in Times of Change: Seasonality*. Copenhagen: Norden/TemaNord/Nordic Council of Ministers.

Rantala, O., Hallikainen, V., Ilola, H. and Tuulentie, S. (2018) The softening of adventure tourism. *Scandinavian Journal of Hospitality and Tourism* 18(4), 343–361.

Rasmussen, K. (1925) *Fra Grønland til Stillehavet: Rejser og Mennesker. Fra 5.Thule Ekspedition 1921-24.* Copenhagen: Gyldendalske Boghandel, Nordisk Forlag.

Rasmussen, K.R. and Merkelsen, H. (2017) Post-colonial governance through securitization? A narratological analysis of a securitization controversy in contemporary Danish and Greenlandic uranium policy. *Politik* 20(3), doi: 10.7146/politik.v20i3.97183.

Rebhan, C. (2016) *North Atlantic Euroscepticism: The Rejection of EU Membership in the Faroe Islands and Greenland*. Tórshavn: Frođskapur Faroe University Press.

Reffstrup, T. and Christiansen, S.K. (2017) *Nordic Entrepreneurship Islands: Status and Potential Mapping and Forecasting Entrepreneurship Education on Seven Selected Nordic Islands*. Copenhagen: Nordic Council of Ministers.

Ren, C. (2010) Beyond hosts and guests: translating the concept of cultural misconception. *International Journal of Culture, Tourism and Hospitality Research* 4(4), 287–298.

Ren, C. (2012) Envisioning Greenland: contested naturecultures in the making. *Conditions* 11/12, 154–155.

Ren, C. (2016) Cool or hot Greenland? Exhibiting and enacting sustainable Arctic futures. *Journal of Cleaner Production* 111, 442–450.

Ren, C. and Bjørst, L.R. (2018) Cold facts, hot topics, and uncertain futures: political and industry responses to climate changes in Greenland. In: Dellasala, D.A. and Goldstein, M.I. (eds) *Encyclopedia of the Anthropocene Volume 2.* Amsterdam: Elsevier, pp. 501–503.

Ren, C., Bjørst, L. and Dredge, D. (2016) Composing Greenlandic tourism futures: an integrated political ecology and actor-network theory approach. In: Mostafanezhad, M., Shelton, E.J., Thompson-Carr, A. and Norum, R. (eds) *Political Ecology of Tourism: Community, Power and the Environment*. Abingdon, UK: Routledge, pp. 284–301.

Ren, C.B. and Blichfeldt, B.S. (2011) One clear image? Challenging simplicity in place branding. *Scandinavian Journal of Hospitality and Tourism* 11(4), 416–434.

Ren, C.B. and Chimirri, D. (2017) *Turismeudvikling i Grønland: Afdækning og Inspiration*. Aalborg, Denmark: Institut for Kultur og Globale Studier, Aalborg Universitet.

Ren, C. and Chimirri, D. (2018) Arctic tourism – more than an industry? *The Arctic Institute* (Washington, DC) 3 April. https://www.thearcticinstitute.org/arctic-tourism-industry/ (accessed 30 October 2019).

Ren, C., Chimirri, D. and Abildgaard, M.S. (2020) *Towards Culturally Sensitive Tourism: Report from Greenland*. Rovaniemi, Finland: Multidimensional Tourism Institute (MTI)/Lapland University Consortium (LUC).

Ren, C., Gad, U.P. and Bjørst, L.R. (2019) Branding on the Nordic margins: Greenland brand configurations. In: Cassinger, C., Lucarelli, A. and Gyimóthy, S. (eds) *The Nordic Wave in Place Branding: Poetics, Practices, Politics*. Cheltenham, UK: Edward Elgar, pp. 160–174.

Ren, C. and Gyimóthy, S. (2013) Transforming and contesting nation branding strategies: Denmark at the Expo 2010. *Place Branding and Public Diplomacy* 9(1), 17–29.

Ren, C., Jóhannesson, G.T., Kramvig, B., Pashkevich, A. and Höckert, E. (2021) Arctic and Indigenous cultures in Nordic tourism. *Scandinavian Journal of Hospitality and Tourism* 21(1), 111–121.

Ren, C. and Ooi, C-S. (2013) Auto-communicating micro-orientalism: articulating 'Denmark' in China at the Shanghai expo. *Asia Europe Journal* 11(2), 129–145.

Ren, C.B. and Rasmussen, R.K. (2017) 'Future games': enacting innovation in Greenland. *Arctic Yearbook* 6, 247–258.

Ren, C.B. and Thomsen, R.C. (2016) The 2016 Arctic Winter Games: 'Now We Do What We Do Best'. *The Arctic Institute* (Washington, DC) 30 March. https://www.thearcticinstitute.org/2016-arctic-winter-games (accessed 13 September 2020).

Renaud, L. (2020) Reconsidering global mobility – distancing from mass cruise tourism in the aftermath of COVID-19. *Tourism Geographies* 22(3), 679–689.

Resource Development Council (2020) *Alaska's Tourism Industry*. Anchorage: Resource Development Council for Alaska. https://www.akrdc.org/tourism (accessed 7 December 2020).

Richter-Menge, J., Jeffries, M.O. and Osborne, E. (eds) (2018) The Arctic. *Bulletin of the American Meteorological Society* 99(8), S143–S173.

Riddel-Dixon, E. (2017) *Breaking the Ice: Canada, Sovereignty, and the Arctic Extended Continental Shelf*. Toronto, ON: Dundurn.

Rignot, E. and Kanagaratnam, P. (2006) Changes in the velocity structure of the Greenland ice sheet. *Science* 311, 986–990.

Rink, H. ([1875] 1997) *Tales and Traditions of the Eskimo: With a Sketch of Their Habits, Religion, Language and Other Peculiarities*. New York, NY: Dover.

Ritter, C. (1938) *Eine Frau erlebt die Polarnacht* . Berlin: Propyläen Verlag. (1954/2019) *A Woman in the Polar Night*. [Trans. J. Degras] London: George Allen and Unwin/Pushkin Press.

Ritzau (2020) Danmarks største isproducent holder fast i Kæmpe Eskimo. *Ritzau* (Copenhagen) 15 July. https://www.foodsupply.dk/article/view/727739/danmarks_storste_isproducent_holder_fast_i_kaempe_eskimo (accessed 27 September 2020).

Robinson, M. and Silverman, H. (2015) Mass, modern, and mine. Heritage and popular culture. In: Robinson, M. and Silverman, H. (eds) *Encounters with Popular Pasts*. New York, NY: Springer, pp. 1–30.

Rodrigue, J-P. and Notteboom, T. (2013) The geography of cruises: itineraries, not destinations. *Applied Geography* 38, 31–42.

Roeben, V. and Azubuike, S.I. (2020) Climate change and responsibility: Arctic states' cooperation through the Arctic Council in climate change mitigation and adaptation effects. In: Heininen, L., Exner-Pirot, H. and Barnes, J. (eds) *Climate Change and the Arctic: Global Origins, Regional Responsibilities? Arctic Yearbook 2020.* Akureyri, Iceland: Arctic Portal, pp. 443–463.

Rogne, O., Rachold, V., Hacqebord, L. and Corell, R. (2015) *IASC After 25 Years – A Quarter of a Century of International Arctic Research Cooperation*. Potsdam, Germany: International Arctic Science Committee.

Romero-Nieva Santos, P., Holm, N., Olsen, J. and Miettinen, S. (2020) Gender and climate change research: moving beyond transformative adaptation. In: Heininen, L., Exner-Pirot, H. and Barnes, J. (eds) *Climate Change and the Arctic: Global Origins, Regional Responsibilities? Arctic Yearbook 2020.* Akureyri, Iceland: Arctic Portal, pp. 189–218.

Rosen, J. (2016) Mysterious, ice-buried Cold War military base may be unearthed by climate change. *Science* 4 August. https://www.sciencemag.org/news/2016/08/mysterious-ice-buried-cold-war-military-base-may-be-unearthed-climate-change# (accessed 1 September 2020).

Rosenthal, E. (2012) Race is on as ice melt reveals Arctic treasures. *New York Times* 18 September. www.nytimes.com/2012/09/19/science/earth/arctic-resources-exposed-by-warming-set-off-competition.html (accessed 2 October 2020).

Ross, R.S. (2009) China's naval nationalism: sources, prospects, and the US response. *International Security* 34(2), 46–81.

Rud, S. (2017) *Colonialism in Greenland: Tradition, Governance and Legacy*. Cham, Switzerland: Palgrave Macmillan.

Saarinen, J. and Rogerson, C.M. (2014) Tourism and the Millennium Development Goals: perspectives beyond 2015. *Tourism Geographies* 16(1), 23–30.

Saarinen, J., Rogerson, C.M. and Hall, C.M. (2017) Geographies of tourism development and planning. *Tourism Geographies* 19(3), 307–317.

Sam-Aggrey, H. and Lanteigne, M. (2020) Environental security in the Arctic: shades of grey? In: In: Hoogensen Gjørv, G., Lanteigne, M. and Sam-Aggrey, H. (eds) *Routledge Handbook of Arctic Security*. Abingdon, UK: Routledge, pp. 102–113.

Sampath, H.M. (1990) The changing pattern of Inuit suicide and attempted suicide. In: *Looking to the Future: Papers from the Seventh Inuit Studies Conference*. Quebec City, QC: Association Inuksiutit Katimajiit, pp. 141–148.

Sanz-Blas, S., Buzova, D. and Schlesinger, W. (2019) The sustainability of cruise tourism onshore: the impact of crowding on visitors' satisfaction. *Sustainability* 11(6), 1510, doi: 10.3390/su11061510.

Sanz-Ibáñez, C., Wilson, J. and Clavé, S.A. (2016) 'Moments' as catalysts for change in tourism evolutionary paths In: Brouder, P. Clavé, A.S., Gill, A. and Ioannides, D. (eds) *Dynamic Destinations. Tourism and Evolutionary Economic Geography.* Abingdon, UK: Routledge, pp. 81–102.

Sasgen, I., Wouters, B., Gardner, A.S., King, M.D., Tedesco, M. *et al.* (2020) Return to rapid ice loss in Greenland and record loss in 2019 detected by the GRACE-FO satellites. *Communications Earth & Environment* 1: 8, doi: 10.1038/s43247-020-0010-1.

Scambos, T., Stroeve, J. and Koenig, L. (eds) (2020) Greenland's 2020 melt season in review. *Greenland Ice Sheet Today* 10 November. http://nsidc.org/greenland-today/2020/11/greenlands-2020-melt-season-in-review/ (accessed 7 December 2020).

Schach, M. and Madlener, R. (2018) Impacts of an ice-free Northeast Passage on LNG markets and geopolitics. *Energy Policy* 122, 438–448.

Schaller, B. (2019) The forgotten spirit of Gorbachev. In: Heininen, L., Exner-Pirot, H. and Barnes, J. (eds) *Redefining Arctic Security: Arctic Yearbook 2019*. Akureyri, Iceland: Arctic Portal, pp. 87–88.

Schneider, P.P. and Vogt, C.A. (2012) Applying the 3M model of personality and motivation to adventure travelers. *Journal of Travel Research* 51(6), 704–716.

Schøler, M. (2020) Greenland decides to keep Egede statue in place. *Over The Circle* (Yellowknife, NWT) 22 July. https://overthecircle.com/2020/07/22/greenland-decides-to-keep-egede-statue-in-place/ (accessed 24 July 2020).

Schrot, O.G., Christensen, J.H. and Formayer, H. (2019) Greenland winter tourism in a changing climate. *Journal of Outdoor Recreation and Tourism* 27, doi: 10.1016/j.jort.2019.100224.

Schweinsberg, S., Wearing, S. and Lai, P-H. (2020) Host communities and last chance tourism. *Tourism Geographies* doi: 10.1080/14616688.2019.1708446.

Sciascia, R., Straneo, F., Cenedese, C. and Heimbach, P. (2013) Seasonal variability of submarine melt rate and circulation in an East Greenland fjord. *Journal of Geophysical Research: Oceans* 118, 2492–2506.

Scidmore, E.R. (1885) *Alaska: Its Southern Coast and the Sitkan Archipelago*. Boston, MA: D. Lothrop and Co.

Scott, D., Amelung, B., Becken, S., Ceron, J-P., Dubois, G. *et al.* (2008) *Climate Change and Tourism: Responding to Global Challenges*. Madrid: World Tourism Organization.

Scott, D., Dawson, J. and Jones, B. (2008) Climate change vulnerability of the US Northeast winter recreation - tourism sector. *Mitigation and Adaptation Strategies for Global Change* 13(5-6), 577–596.

Scott, D., Gössling, S., Hall, C.M. and Peeters, P. (2016) Can tourism be part of the decarbonized global economy? The costs and risks of alternate carbon reduction policy pathways. *Journal of Sustainable Tourism* 24(1), 52–72.

Scott, D. and McBoyle, G. (2007) Climate change adaptation options in the ski industry. *Mitigation and Adaptation Strategies for Global Change* 12, 1411.

Scott, D., McBoyle, G. and Mills, B. (2003) Climate change and the skiing industry in southern Ontario (Canada): exploring the importance of snowmaking as a technical adaptation. *Climate Research* 23, 171–181.

Scott, D.C., Hall, C.M. and Gössling, S. (2012) *Tourism and Climate Change: Impacts, Adaptation and Mitigation*. Abingdon, UK: Routledge.

Seaver, K.A. (1996) *The Frozen Echo: Greenland and the Exploration of North America, c.A.D. 1000-1500*. Stanford, CA: Stanford University Press.

Sejersen, F. (2015) *Rethinking Greenland and the Arctic in an Era of Climate Change: New Northern Horizons*. Abingdon, UK: Routledge.

Sejersen, F. (2019) Scaling sustainability in the Arctic. In: Gad, U.P. and Strandsbjerg, J. (eds) *The Politics of Sustainability in the Arctic: Reconfiguring Identity, Space, and Time*. Abingdon, UK: Routledge, pp. 94–107.

Sermitsiaq (2020) Tourist in your own magical Greenland. *Sermitsiaq* (Nuuk), May (translated from the original Danish).

Sevunts, L. (2019) Arctic Council: officials struggle to find climate language that U.S. can support. *The Barents Observer* 15 April. https://thebarentsobserver.com/en/arctic/2019/04/arctic-council-officials-strugglefind-climate-language-us-can-support (accessed 5 February 2020).

Shamonova, D. (2020) Majority in Nuuk votes to keep controversial statue. *CPH Post Online* (Copenhagen) 22 July. https://cphpost.dk/?p=116715 (accessed 28 July 2020).

Shapovalova, D. (2020) Oil spill in Siberia: are we prepared for permafrost thaw? *High North News* (Tromsø, Norway), 15 June. https://www.highnorthnews.com/en/oil-spill-siberia-are-we-prepared-permafrost-thaw (accessed 19 June 2020).

Shepherd, A., Ivins, E.R., Geruo, A., Barletta, V.R., Bentley, M.J. *et al.* (2012) A reconciled estimate of ice-sheet mass balance. *Science* 338, 1183–1189.

Shepherd, A., Ivins, E., Rignot, E., Smith, B., Broeke, M. van den *et al.* (2020) Mass balance of the Greenland Ice Sheet from 1992 to 2018. *Nature* 579, 233–239.

Simpson, J. (2018) How Greenland could become China's Arctic base. *BBC News* 18 December. https://www.bbc.co.uk/news/world-europe-46386867 (accessed 1 December 2020).

Sjørslev, I. (1998). Women, gender studies and the international indigenous movement. In: Vinding, D. (ed.) *Indigenous Women: The Right to a Voice*. Copenhagen: IWGIA (International Working Group for Indigenous Affairs), pp. 296–314.

Skydsbjerg, H. and Turnowsky, W. (2016) Namminiilivinnassamut angertut amerlanerussuteqarluartut. *Sermitsiaq* (Nuuk) 1 December. http://semitsiaq.ag/kl/ namminiilivinnassamut-angertut-amerlanerussuteqarluartut

Smed, K.M. (2014) The culture of nature: destination visitability in Ilulissat, Greenland. *The Northern Review* (Burnaby, BC) 38, 269–293.

Smed, K.M. (2017) Culture in nature: exploring the role of 'culture' in the destination of Ilulissat, Greenland. In: Viken, A. and Müller, D.K. (eds) *Tourism and Indigeneity in the Arctic*. Bristol, UK: Channel View, pp. 137–153.

Smith, L.C. and Stephenson, S.R. (2013) New Trans-Arctic shipping routes navigable by mid-century. *Proceedings of the National Academy of Sciences of the United* States 110(13), 4871–4872.

Smith, M.K. (2003) *Issues in Cultural Tourism Studies*. London: Routledge.

Smith, V.L. (1982) Tourism to Greenland: renewed ethnicity? *Cultural Survival Quarterly* 6(3), 26–27. https://www.culturalsurvival.org/publications/cultural-survival-quarterly/tourism-greenland-renewed-ethnicity (accessed 21 February 2020).

Snyder, J.M. (2007) The economic role of Arctic tourism. In: Snyder, J.M. and Stonehouse, B. (eds) *Prospects for Polar Tourism*. Wallingford, UK: CABI, pp. 102–122.

SOA (State Oceanic Administration) (1996) *China Ocean Agenda 21*. Beijing: Ocean Press.

Solana, J. and Ferrero-Waldner, B. (2008) Climate change and international security. Brussels: Paper from the High Representative and the European Commission to the European Council S113/08, 14 March.

Sommerkorn, M. and Hassol, S.J. (eds) (2009) *Arctic Climate Feedbacks: Global Implications*. Oslo: World Wildlife Fund International Arctic Programme.

Sonne, B. (2017) *Worldviews of the Greenlanders: An Inuit Arctic Perspective*. Fairbanks, AL: University of Alaska Press.

Sorensen, A. (2015) Taking path dependency seriously: an historical institutionalist research agenda in planning history. *Planning Perspectives* 30(1), 17–38.

Sørensen, A.K. (1995) Greenland: from colony to home-rule. In: Tägil, S. (ed.) *Ethnicity and Nation Building in the Nordic World.* London: C. Hurst, pp. 85–105.

Sørensen, A.K. (2007) *Denmark-Greenland in the Twentieth Century*. Copenhagen: Museum Tusculanum Press.

Sørensen, B.W. (1998) Explanations for wife beating in Greenland. In: Klein, R.C.A. (ed.) *Multidisciplinary Perspectives on Family Violence*. London: Routledge, pp. 153–175.

Sørensen, B.W. (2001) 'Men in transition': the representation of men's violence against women in Greenland. *Violence Against Women* 7(7), 826–847.

Sørensen, C.T.N. (2018a) China is in the Arctic to stay as a great power: how China's increasingly confident, proactive and sophisticated Arctic diplomacy plays into Kingdom of Denmark tensions. In: Bislev, A., Gad, U.P. and Zeuthen, J.W. (eds) (2018) *Arctic Yearbook 2018. Special Section: China & the Arctic*. Akureyri, Iceland: Northern Research Forum, pp. 43–58.

Sørensen, C.T.N. (2018b) Chinese investments in Greenland: promises and risk as seen from Nuuk, Copenhagen and Beijing. In: Kristensen, K.S. and Rahbek-Clemmensen, J. (eds) *Greenland and the International Politics of a Changing Arctic: Postcolonial Paradiplomacy Between High and Low Politics*. Abingdon, UK: Routledge, pp. 83–97.

Splettstoesser, J. (2000) IAATO's stewardship of the Antacrtic environment: a history of tour operators' concern for a vulnerable part of the world. *International Journal of Tourism Research* 2, 47–55.

Staalesen, A. (2018) The Chinese oilmen are back in Murmansk. *The Barents Observer* 27 June. https://thebarentsobserver.com/en/2018/06/chinese-oilmen-are-back-murmansk (accessed 2 October 2020).

State Council (of the People's Republic of China) (2018) *China's Arctic Policy.* Beijing: State Council Information Office of the PRC, 26 January. http://english.gov.cn/archive/white_paper/2018/01/26/content_281476026660336.htm (accessed 1 December 2020).

Statistics Greenland (2018) *Greenland in Figures 2018*. Nuuk: Statistics Greenland. http://www.stat.gl/publ/kl/GF/2018/pdf/Greenland%20in%20Figures%202018.pdf (accessed 27 December 2019).

Statistics Greenland (2020) *Greenland in Figures 2020*. Nuuk: Statistics Greenland. http://www.stat.gl/publ/kl/GF/2020/pdf/Greenland%20in%20Figures%202020.pdf (accessed 5 January 2021).

Steenholdt, N. C. and Chimirri, D. (2018) Tourism and quality of life in Greenland: exploration through farm stays in south Greenlandic settlements. In: Heininen, L. and Exner–Pirot, H. (eds) *Arctic Yearbook 2018: Arctic Development in Theory and Practice.* Akureyri, Iceland: Northern Research Forum, pp. 123–140.

Steenholdt, S. (2015) *Zombieland/Zombiet Nunaat*. Nuuk: Milik.

Stefansson, V. ([1921] 2015) *The Friendly Arctic*. New York, NY: Macmillan.

Stefansson, V. ([1922] 2017) *The Northward Course of Empire*. New York, NY: Harcourt, Brace and Company/Macmillan.

Steinberg, P.E., Bruun, J.M. and Medby, I.A. (2014) Covering Kiruna: a natural experiment in Arctic awareness. *Polar Geography* 37(4), 273–297.

Steinberg, P.E., Tasch, J. and Gerhardt, H. (2015) *Contesting the Arctic: Politics and Imaginaries in the Circumpolar North*. London: I.B. Tauris.

Stendel, M., Christensen, J.H. and Petersen, D. (2008) Arctic climate and climate change with a focus on Greenland. *Advances in Ecological Research* 40, 13–43.

Stepien, A., Kauppila, L., Kopra, S., Käpylä, J., Lanteigne, M. *et al.* (2020) China's economic presence in the Arctic: realities, expectations and concerns. In: Koivurova, T. and Kopra, S. (eds) *Chinese Policy and Presence in the Arctic*. Leiden, The Netherlands: Brill, pp. 90–136.

Stevenson, T.C., Davies, J., Huntington, H.P. and Sheard, W. (2019) An examination of trans-Arctic vessel routing in the central Arctic Ocean. *Marine Policy* 100, 83–89.

Stewart, E.J., Draper, D. and Dawson, J. (2010) Monitoring patterns of cruise tourism across Arctic Canada. In: Lück, M., Maher, P.T. and Stewart, E.J. (eds) *Cruise Tourism in the Polar Regions: Promoting Environmental and Social Sustainability?* London: Earthscan, pp. 133–145.

Stewart, E.J., Draper, D. and Dawson, J. (2011) Coping with change and vulnerability: a case study of resident attitudes toward tourism in Cambridge Bay and Pond Inlet, Nunavut, Canada. In: Maher, P. T. Stewart, E.J. and Lück, M. (eds) *Polar Tourism: Human, Environmental and Governance Dimensions*. Wallingford, UK: CABI, pp. 33–53.

Stewart, E.J., Draper, D. and Johnston, M.E. (2005) A review of tourism research in the polar regions. *Arctic* 58(4), 383–394.

Stewart, E.J., Howell, S.E.L., Draper, D., Yackel, J. and Tivy, A. (2007) Sea ice in Canada's Arctic: implications for cruise tourism. *Arctic* 60(4), 370–380.

Stewart, E.J. and Liggett, D. (2018) Polar tourism: status, trends, futures. In: Nuttall, M., Christensen, T.R. and Siegert, M. (eds) *The Routledge Handbook of the Polar Regions*. Abingdon, UK: Routledge, pp. 357–370.

Stewart, E.J., Liggett, D. and Dawson, J. (2017) The evolution of polar tourism scholarship: research themes, networks and agendas. *Polar Geography* 40(1), 59–84.

Stockinger, G. (2013) Archaeologists uncover clues to why Vikings abandoned Greenland. *Spiegel International* 10 January. https://www.spiegel.de/international/zeitgeist/archaeologists-uncover-clues-to-why-vikings-abandoned-greenland-a-876626.html (accessed 10 May 2020).

Stonehouse, B. and Snyder, J.M. (2007) Polar tourism in changing environments. In: Snyder, J.M. and Stonehouse, B. (eds) (2007) *Prospects for Polar Tourism*. Wallingford, UK: CABI, pp. 32–48.

Stonehouse, B. and Snyder, J.M. (2010) *Polar Tourism: An Environmental Perspective*. Clevedon, UK: Channel View.

Strambach, S. and Halkier, H. (2013) Reconceptualizing change - path dependency, path plasticity and knowledge combination. *Zeitschrift für Wirtschaftsgeographie* 57(1–2), 1–14.

Strandsbjerg, J. (2014) Making sense of contemporary Greenland: indigeneity, resources and sovereignty. In: Powell, R. and Dodds, K. (eds) *Polar Geopolitics? Knowledges, Resources and Legal Regimes.* Cheltenham, UK: Edward Elgar, pp. 259–276.

Straneo, F. and Heimbach, P. (2013) North Atlantic warming and the retreat of Greenland's outlet glaciers. *Nature* 504, 36–43.

Stroeve, J. and Notz, D. (2018) Changing state of Arctic Sea ice across all seasons. *Environmental Research Letters* 13(10): 103001.

Struzik, E. (2018) As ice recedes, the Arctic isn't prepared for more shipping traffic. *The Conversation* 4 September. http://theconversation.com/as-ice-recedes-the-arctic-isntprepared-for-more-shipping-traffic-102312 (accessed 10 January 2020).

Sung, H.H. (2004) Classification of adventure travelers: behaviour, decision making, and target markets. *Journal of Travel Research* 42(4), 343–356.

Supasudaca [*sic*], Bruno, S.S., Delgado, M. and Madrazo, F. (2020) From HERITAGE to Feritage: how economic path dependencies in the Caribbean cruise destinations are distorting the uses of heritage architecture and urban form. In: Hein, C. (ed.) *Adaptive Strategies for Water Heritage: Past, Present and Future*. Cham, Switzerland: Springer, pp. 363–381.

Sweet, V. (2014) Rising waters, rising threats: the human trafficking of Indigenous women in the Circumpolar region of the United States and Canada. *Yearbook of Polar Law* 6, 162–188.

Sýkora, L. and Bouzarovski, S. (2012) Multiple transformations: conceptualising the post-communist urban transition. *Urban Studies* 49 (1), 43–60.

Taagholt, J. and Brooks, K. (2016) Mineral riches: a route to Greenland's independence. *Polar Record* 52(264), 360–371.

Takahashi, M., Kawana, S., Saitou, K., Koizumi, Y., Hateruma, S. *et al.* (2019) Autonomy and military bases: USAF Thule Base in Greenland as the study case. In: Heininen, L., Exner–Pirot, H. and Barnes, J. (eds) *Redefining Arctic Security: Arctic Yearbook 2019.* Akureyri, Iceland: Arctic Portal, pp. 43–57.

Tedesco, M. and Fettweis, X. (2020) Unprecedented atmospheric conditions (1948–2019) drive the 2019 exceptional melting season over the Greenland ice sheet. *The Cryosphere* 14, 1209–1223.

Tedstone, A.J., Bamber, J.L., Cook, J.M., Williamson, C.J., Fettweis, X. *et al.* (2017) Dark ice dynamics of the south-west Greenland Ice Sheet. *The Cryosphere* 11, 2491–2506.

Têtu, P-L. (2016) *Géopolitique de l'Approvisionnement de la Chine en Matières Premières: L'Arctique, une Région Prioritaire?* Louvain-la-Neuve, Belgium: Université Catholique de Louvain, Centre d'Etudes des Crises et Conflits Internationaux.

Têtu P-L. and Lasserre, F. (2017) Chinese investment in Greenland's mining industry: toward a new framework for foreign direct investment. *The Extractive Industries and Society* 4(3), 661–671.

Theroux, M. (2009) *Far North*. London: Faber and Faber.

Thisted, K. (1992) Greenlandic literature: suicide as an interpretation of modern life. In: Blom, G., Graves, P., Kruse, A. and Thorup Thomsen, B. (eds) *Minority Languages: The Scandinavian Experience*. Oslo: Nordic Language Secretariat, pp. 141–153.

Thisted, K. (2015) Pioneering nation: new narratives about Greenland and Greenlanders launched through arts and branding. In: Evengård, B., Nymand Larsen, J. and Paasche, Ø. (eds) *The New Arctic*. Cham, Switzerland: Springer, pp. 23–38.

Thisted, K. and Gremaud, A–S. (eds) (2020) *Denmark and the New North Atlantic: Narratives and Memories in a Former Empire*. Århus, Denmark: Aarhus University Press.

Thomson, C. and Thomson, J.S. (2006) Arctic cruise ship island tourism. In: Baldacchino, G. (ed.) *Extreme Tourism: Lessons from the World's Cold Water Islands*. Oxford, UK: Elsevier, pp. 169–178.

Thuesen, S.T. (2016) The formation of Danish Eskimology: from William Thalbitzer to the Greenland Home Rule era. In: Krupnik, I. (ed.) *Early Inuit Studies: Themes and Transitions, 1850s – 1980s.* Washington, DC: Smithsonian Institution Press, pp. 245–264.

Thurnell-Read, T. (2017) 'What's on your Bucket List?' Tourism, identity and imperative experiential discourse. *Annals of Tourism Research* 67, 58–66.

Timothy, D.J. and Olsen, D.H. (2001) Challenges and opportunities of marginality in the Arctic: a case of tourism in Greenland. *Tourism* (Zagreb) 49(4), 299–308.

Toka, L. (ed.) (2002) *Taking Wing Conference Report: Conference on Gender Equality and Women in the Arctic*. Helsinki: Ministry of Social Affairs and Health.

Tomaselli, A. and Koch, A. (2014) Implementation of Indigenous rights in Russia: shortcomings and recent developments. *International Indigenous Policy Journal* 5, 1–23.

Tribe, J. (2008) Tourism: a critical business. *Journal of Travel Research* 46(3), 245–255.

Trusel, L.D., Das, S.B., Osman, M.B., Evans, M.J., Smith, B.E. *et al.* (2018) Nonlinear rise in Greenland runoff in response to post-industrial Arctic warming. *Nature* 564(7734), 104–108.

Tse, T.S. (2013) Chinese outbound tourism as a form of diplomacy. *Tourism Planning and Development* 10(2), 149–158.

Tse, T.S. (2015) A review of Chinese outbound tourism research and the way forward. *Journal of China Tourism Research* 11(1), 1–18.

Tsing, A.L. (2005) *Friction: An Ethnography of Global Connection*. Princeton, NJ: Princeton University Press.

Tsui, E. (2019) The Agreement on Enhancing International Arctic Scientific Co-operation: legal and practical consequences. In: Heininen, L., Exner-Pirot, H. and Barnes, J. (eds) *Redefining Arctic Security: Arctic Yearbook 2019.* Akureyri, Iceland: Arctic Portal, pp. 332–339.

Turnowsky, W. (2017) Ajornerulertoqassanngippat aatsaat namminersulivikkusupput. *Sermitsiaq* (Nuuk) 1 April.

Turnowsky, W. (2018) Naalakkersuisoqatigiit isasoorput: Partii Naleraq Naalakkersuisunit tunuarpoq. *Sermitsiaq* (Nuuk) 9 September.

Tuulentie, S. (2006) The dialectic of identities in the field of tourism. The discourses of the indigenous Sami in defining their own and the tourists' identities. *Scandinavian Journal of Hospitality and Tourism* 6(1), 25–36.

Tveter, E. (2017) The effect of airports on regional development: evidence from the construction of regional airports in Norway. *Research in Transportation Economics* 63, 50–58.

Tysiachniouk, M.S. and Petrov, A.N. (2018) Benefit sharing in the Arctic energy sector: perspectives on corporate policies and practices in northern Russia and Alaska. *Energy Research and Social Science* 39, 29–34.

UArctic (2020) *How we serve the North*. Rovaniemi, Finland: University of the Arctic. https://www.uarctic.org/about-uarctic/how-we-serve-the-north/ (accessed 6 October 2020).

Udenrigsministeriet/Udenrigsdirektoratet (2011) *Danmark, Grønland & Færøerne: Kongeriget Danmarks Strategi for Arktis* 2011–2020. Copenhagen: Udenrigsministeriet; Nuuk: Udenrigsdirektoratet, Grønlands Selvstyre; Tórshavn: Udenrigsministeriet, Færøernes Landsstyre.

UNBC (University of Northern British Columbia) (2020) What is Indigenous tourism? Prince George, BC: UNBC Outdoor Recreation and Tourism Management. https://www.unbc.ca/outdoor-recreation-tourism-management/indigenous-tourism-about (accessed 14 October 2020).

UNDP (United Nations Development Programme) (1995) *Human Development Report 1994*. Oxford, UK: Oxford University Press.

UNDRIP (2007) *United Nations Declaration on the Rights of Indigenous People*. New York, NY: United Nations. http://www.un.org/esa/socdev/unpfii/documents/DRIPS_en.pdf (accessed 17 February 2019).

UNEP (United Nations Environment Programme) and International Ecotourism Society (2007) *Tourism in the Polar Regions: The Sustainability Challenge*. Paris: UNEP.

UNESCO (2004/2019) *World Heritage List: Ilulissat Icefjord*. Paris: UNESCO World Heritage Centre. https://whc.unesco.org/en/list/1149 (accessed 29 September 2019).

UNESCO (2018) *Aasivissuit – Nipisat. Inuit Hunting Ground between Ice and Sea.* Paris: UNESCO. https://whc.unesco.org/en/list/1557/ (accessed 7 February 2020).

United Nations (UN) (1994) *United Nations Convention on the Law of the Sea*. New York, NY: United Nations.

United Nations (UN) (2007) *United Nations Declaration of Rights of Indigenous Peoples*. New York, NY: United Nations. https://www.un.org/development/desa/indigenouspeoples/wp-content/uploads/sites/19/2018/11/UNDRIP_E_web.pdf (accessed 11 July 2020).

United Nations (UN) (2015) *2030 Agenda for Sustainable Development and its 17 Sustainable Development Goals*. New York, NY: United Nations. www.un.org/sustainabledevelopment/development-agenda (accessed 25 January 2016).

UNWTO (United Nations World Tourism Organization) (2016) *Sustainable Cruise Tourism Development Strategies*. Madrid: UNWTO.

UNWTO (United Nations World Tourism Organization) (2018) *FAQ – Climate Change and Tourism.* Madrid: UNWTO.

US Arctic Research Commission (1990) *Logistic Support of United States Research in Greenland: Current Situation and Prospects*. Washington, DC: US Arctic Research Commission.

US Coast Guard (2020) *Coast Guard Oversees Disembarkation of 250,000 from Cruise Ships to Reduce Risks Under COVID–19 Emergency, Medevac'd 31*. Washington DC: United States Coast Guard 4 April. https://content.govdelivery.com/accounts/USDHSCG/bulletins/284ce7b (accessed 6 June 2020).

USGS (United States Geological Survey) (2007) *Assessment of Undiscovered Oil and Gas Resources of the East Greenland Rift basins Province*. Washington DC: USGS, Fact Sheet 2007–3077.

USGS (United States Geological Survey) (2008) *Assessment of Undiscovered Oil and Gas Resources of the West Greenland – East Canada Province*. Washington, DC: USGS, Fact Sheet 2008–3014.

Vanderklippe, N. (2016) Minister 'surprised' at backlash over tweets stating climate change hurts women most. *The Globe and Mail* (Toronto, ON) 16 May. https://www.theglobeandmail.com/news/politics/minister-surprised-at-backlash-over-tweets-stating-climate-change-hurts-women-most/article30029885/ (accessed 2 September 2020).

Viggo Jakobsen, P. (2019) *A New Ilulissat Declaration on Security and Defence is Needed in the Arctic*. Odense, Denmark: University of Southern Denmark, Department of Political Science and Public Management. https://www.sdu.dk/en/cws/cws_activities/2018/new+ilulissat+declaration (accessed 28 June 2020).

Viken, A. (2006) Sami tourism and Sami identity: an analysis of the tourism-identity nexus in a Sami community. *Scandinavian Journal of Hospitality and Tourism* 6(1), 7–24.

Viken, A. (2013) What is Arctic tourism, and who should define it? Implication of an Anglo-American hegemony in academic writing. In: Müller, D.K., Lundmark, L. and Lemelin, R.H. (eds) *New Issues in Polar Tourism: Communities, Environments, Politics*. Dordrecht, Netherlands: Springer, pp. 37–50

Viken, A. and Jørgensen, F. (1998) Tourism on Svalbard. *Polar Record* 34(189), 123–128.

Viken, A. and Müller, D.K. (2017) Indigeneity and indigenous tourism. In: Viken, A. and Müller, D.K. (eds) *Tourism and Indigeneity in the Arctic*. Bristol, UK: Channel View, pp. 16–32.

Vincent, W.F. (2020) Arctic climate change: local impacts, global consequences, and policy implications. In: Coates, K.S. and Holroyd, C. (eds) *The Palgrave Handbook of Arctic Policy and Politics*. Cham, Switzerland: Palgrave Macmillan, pp. 507–526.

Vincent, W.F., Lemay, M. and Allard, M. (2017) Arctic permafrost landscapes in transition: towards and integrated Earth system approach. *Arctic Science* 3(2), 39–64.

VisitBritain (2019) *Market and Trade Profile China*. London: VisitBritain. https://www.visitbritain.org/sites/default/files/vb-corporate/markets/visitbritain_marketprofile_china_2019_0.pdf (accessed 2 December 2020).

Visit Denmark (2020) *About Chinavia*. Copenhagen: Visit Denmark. https://www.wonderfulcopenhagen.com/chinavia/about-chinavia/about-chinavia (accessed 2 August 2020).

Visit Finland (2017) *Chinese overnights in Finland*. Helsinki: Visit Finland. http://www.visitfinland.fi/wp-content/uploads/2018/01/Wu-and-Ahola-Whats-up-in-China.pdf?dl. (accessed 2 August 2020).

Visit Greenland (2013) *Spring 2013 Visitor Survey Report: Greenland*. Nuuk: Visit Greenland.

Visit Greenland (2016a) Greenland: be a pioneer. Nuuk: Visit Greenland. www.greenland.com/en (accessed 28 August 2019).

Visit Greenland (2016b) *Tourism Strategy 2016–2019*. Nuuk: Visit Greenland. http://corporate.greenland.com/media/8043/turismestrategi_eng_light_senest.pdf (accessed 28 August 2019).

Visit Greenland (2017a) Ski touring. https://visitgreenland.com/things-to-do/skitouring/# (accessed 26 January 2020).

Visit Greenland (2017b) Heliskiing. https://visitgreenland.com/things-to-do/heliskiing/ (accessed 26 January 2020).

Visit Greenland (2017c) Cross-country skiing. https://visitgreenland.com/things-to-do/cross-country-skiing/ (accessed 26 January 2020).

Visit Greenland (2018) Trophy hunting. https://visitgreenland.com/things-to-do/trophy-hunting/ (accessed 26 January 2020).

Visit Greenland (2020a) *Nunarput Nuan: Udforsk Nunarput Nuan Vores Vidunderlige Grønland*. Nuuk: Visit Greenland. https://traveltrade.visitgreenland.com/da/seneste-nyt/udforsk-nunarput-nuan-vores-vidunderlige-groenland/ (accessed 8 June 2020).

Visit Greenland (2020b) *Nunarput Nuan: Vores Vidunderlige Grønland*. Nuuk: Visit Greenland. https://nunarputnuan.gl/da/ (accessed 8 June 2020).

Visit Greenland (2020c) *Take Advantage of the Wait in the Kangerlussuaq Airport: Waiting in Kangerlussuaq is Never a Waste of Time*. Nuuk: Visit Greenland. https://visitgreenland.com/articles/take-advantage-kangerlussuaq-airport/ (accessed 27 March 2020).

Visit Greenland (2020d) *The Big Arctic Five – Five Core Things to Do in Greenland*. Nuuk: Visit Greenland. https://visitgreenland.com/articles/big-arctic-five/ (accessed 27 March 2020).

Visit Greenland (2020e) *Towards More Tourism – A Joint Effort for Growth, Innovation and Responsible Tourism*. Nuuk: Visit Greenland.

Visit Sweden (2019) *Kina Marknadsprofil 2019*. Stockholm: Visit Sweden.

Voigt, C., Lamprecht, R.E., Marushchak, M.E., Lind, S.E., Novakovskiy, A. *et al.* (2017) Warming of subarctic tundra increases emissions of all three important greenhouse gases – carbon dioxide, methane and nitrous oxide. *Global Change Biology* 23(8), 3121–3138.

Waits, A., Emelyanova, A., Oksanen, A., Abass, K. and Rautio, A. (2018) Human infectious diseases and the changing climate in the Arctic. *Environment International* 121(1), 703–713.

Walker, K. and Moscardo, G. (2002) The impact of interpretation on passengers of expedition cruises. In: Dowling, R.K. (ed.) *Cruise Ship Tourism*. Wallingford, UK: CABI, pp. 105–114.

Walker, K., and Moscardo, G. (2016) Moving beyond sense of place to care of place: the role of Indigenous values and interpretation in promoting transformative change in tourists' place images and personal values. *Journal of Sustainable Tourism* 24(8/9), 1243–1261.

Walsh, J. (2020) Protests highlight Europe's own problems. *New European* 18 June, pp. 20–21.

Wang, L. and Alasuutari, P. (2017) Co-construction of the tourist experience in social networking sites: two forms of authenticity intertwined. *Tourist Studies* 17(4), 388–405.

Wang, S., Mu, Y., Zhang, X. and Xie, J. (2020) Polar tourism and environment change: opportunity, impact and adaptation. *Polar Science* doi: 10.1016/j.polar.2020.100544.

Wang, W. and Nan, Y. (2012) Polar travel quietly arrived. *China Nature* 2, 12–16.

Watt-Cloutier, S. (2015) *The Right to be Cold: One Woman's Story of Protecting Her Culture, The Arctic and the Whole Planet*. Toronto, ON: Penguin Canada.

Watts, J. (2020) Arctic find 'a new page' in climate emergency. *The Guardian* 28 October, pp. 1–2.

WCED (World Commission on Environment and Development) (1987) *Report of the World Commission on Environment and Development: Our Common Future*. (The Brundtland Report). New York: United Nations.

WCN Editorial Team (2020) NIB to provide £57m loan to redevelop two airports in Greenland. *World Construction Network* 22 July, https://www.worldconstructionnetwork.com/news/nib-to-provide-57m-loan-to-redevelop-two-airports-in-greenland (accessed 1 December 2020).

WCRP (World Climate Research Programme, Global Sea Level Budget Group) (2018) Global sea-level budget 1993–present. *Earth System Science Data* 10, 1551–1590.

Weaver, A. (2004) The McDonaldization thesis and cruise tourism. *Annals of Tourism Research* 32(2), 346–366.

Weaver, A. (2006) *The Disneyization of Cruise Travel*. Wallingford, UK: CABI.

Weaver, D. (2016) Beyond convention: reimaging indigenous tourism. *Tourism Geographies* 18, 28–37.

Weaver, D.B. and Lawton, L.J. (2017) Degree of peripherality in the production and consumption of leisure tourism in Greenland. In: Lee, Y-S., Weaver, D.B. and Prebensen, N.K. (eds) *Arctic Tourism Experiences: Production, Consumption and Sustainability*. Wallingford, UK: CABI, pp. 56–66.

Weir, B. (2017) Climate change and tourism – are we forgetting lessons from the past? *Journal of Hospitality and Tourism Management* 32, 108–114.

Welch, D.A. (2020) The Arctic and geopolitics. In: Coates, K.S. and Holroyd, C. (eds) *The Palgrave Handbook of Arctic Policy and Politics*. Cham, Switzerland: Palgrave Macmillan, pp. 473–486.

Welling, J.T., Árnason, P. and Ólafsdóttir, R. (2015) Glacier tourism: a scoping review. *Tourism Geographies* 17(5), 635–662.

Wenneke, C.W., Jacobsen, R.B. and Ren, C. (2019) Motivations for Indigenous island entrepreneurship: entrepreneurs and behavioural economics in Greenland. *Island Studies Journal* 14(2), 43–60.

Wesche, S. and Chan, H. (2010) Adapting to the impacts of climate change on food security among Inuit in the Western Canadian Arctic. *EcoHealth* 7(3), 361–373.

White, P., Morgan, N., Pritchard, A. and Heimtun, B. (2019) Framing the land of the Northern Lights. *Annals of Tourism Research* 78, 102739.

Williams, W.M. (1859) Through Norway with a Knapsack. London: Smith, Elder and Co.

Wilson, E. (2016) What is the social license to operate? Local perceptions of oil and gas projects in Russia's Komi Republic and Sakhalin Island. *The Extractive Industries and Society* 3(1), 73–81.

Wilson, E., Hansen, A.M. and Rowe, E.W. (2017) Imagining the future; local perceptions of Arctic extractive projects that didn't happen. In: Kelman, I. (ed.) *Arcticness: Power and Voice for the North*. London: UCL Press, pp. 130–149.

Wiseman, M.S. (2020) The future of the Arctic Council. In: Coates, K.S. and Holroyd, C. (eds) *The Palgrave Handbook of Arctic Policy and Politics*. Cham, Switzerland: Palgrave Macmillan, pp. 439–452.

Wood, R.E. (2000) Caribbean cruise tourism: globalization at sea. *Annals of Tourism Research* 27(2), 345–370.

Woodall, S. (2020) 5 Types of Nature Travellers. Nuuk: Visit Greenland. https://visitgreenland.com/articles/5-types-of-nature-travelers/ (accessed 31 December 2020).

Woon, C.Y. (2020) Framing the 'Polar Silk Road' (冰上丝绸之路): critical geopolitics, Chinese scholars and the (re)positionings of China's Arctic interests. *Political Geography* 78, 102141.

Worldometers (2019) China population (LIVE). *Worldometers*. https://www.worldometers.info/world-population/china-population/ (accessed 15 June 2019).

WTTC (World Travel and Tourism Council) (2004) *The Caribbean. The Impact of Travel and Tourism on Jobs and the Economy.* London: WTTC.

Wüst, C. (2019) Cruise ships look to clean up. *Spiegelonline* 6 June. https://www.spiegel.de/international/zeitgeist/cruise-ship-lines-look-to-clean-up-a-1270968.html (accessed 22 August 2019).

WWF (2018) *Last Ice Area – Similijuaq*. Toronto, ON: World Wildlife Fund. https://arctic-wwf.org/site/asset/files/2021/lia-brochure.pdf (accessed 29 October 2020).

Xanthaki, A. (2007) *Indigenous rights and United Nations standards: self-determination, culture, land.* New York, NY: Cambridge University Press.

Xinhua (2018) *China's Arctic Policy*. Beijing: The State Council Information Office of the People's Republic of China, 26 January. http://english.gov.cn/archive/white_paper/2018/01/26/content_281476026660336.htm (accessed 26 September 2019).

Xu, H., Wang, K. and Song, Y.M. (2020) Chinese outbound tourism and soft power. *Journal of Policy Research in Tourism, Leisure and Events* 12(1), 34–49.

Yallop, M., Anesio, A.M., Perkins, R.G., Cook, J., Telling, J. *et al.* (2012) Photophysiology and albedo-changing potential of the ice algal community on the surface of the Greenland ice sheet. *The ISME Journal* 6, 2302–2313.

Yarnal, C.M. and Kerstetter, D. (2005) Casting off: an exploration of cruise ship space, group tour behavior, and social interaction. *Journal of Travel Research* 43, 368–379.

Yttergren, Å. (2012) Swedish Gender Equality for Trafficked Women? : Radical Official Remedies and Ethnic Otherness. Umeå, Sweden: Umeå Universitet, Juridiskt Forum.

Yumashev, D., Hopes, C., Schaefer, K., Riemann-Campe, K., Iglesias-Suarez, F. *et al.* (2019) Climate policy implications of the nonlinear decline of Arctic land permafrost and other cryosphere elements. *Nature Communications* 10(1), 1–11.

Zeuthen, J. (2017) Part of the master plan? Chinese investment in rare earth mining in Greenland. In: Heininen, L., Exner-Pirot, H. and Plouffe, J. (eds) *Arctic Yearbook 2017: Change and Innovation in the Arctic: Policy, Society and Environment.* Akureyri, Iceland: Northern Research Forum, pp. 302–325.

Zou, K. (2012) China's ocean policymaking: practice and lessons. *Coastal Management* 40(2), 145–160.

Zurick, D.N. (1992) Adventure travel and sustainable tourism in the peripheral economy of Nepal. *Annals of the Association of American Geographers* 82, 608–628.

Index

Note: Page numbers in **bold** type refer to **figures;** Page numbers in *italic* type refer to *tables*